Notes on Geoplasticity

Notes on Geoplasticity

William G. Pariseau
Malcolm McKinnon Endowed Chair – Emeritus
Department of Mining Engineering
University of Utah, Salt Lake City, UT, USA

CRC Press
Taylor & Francis Group
Boca Raton London New York Leiden

CRC Press is an imprint of the
Taylor & Francis Group, an **informa** business
A BALKEMA BOOK

Cover photo: a portion of a 45 deg slope in an open pit copper mine in the western USA where mining depth is approaching 1000 m. The benign appearance of the slope masks complex geology replete with joints, faults, dikes, underground workings and groundwater. Poro-elastic/plastic analysis details are given by Schmelter and Pariseau, 1997 (see reference in Chapter 8).

CRC Press/Balkema is an imprint of the Taylor & Francis Group, an informa business

Typeset by Apex CoVantage, LLC

Library of Congress Cataloging-in-Publication Data

Applied for

Published by: CRC Press/Balkema
Schipholweg 107c, 2316 XC Leiden, The Netherlands
e-mail: Pub.NL@taylorandfrancis.com
www.crcpress.com – www.taylorandfrancis.com

ISBN: 978-1-138-37005-0 (Hbk)
ISBN: 978-0-429-42827-2 (eBook)

DOI: https://doi.org/10.1201/9780429428272

Contents

Prologue

My interest in rock mechanics in general and soil plasticity in particular – geoplasticity – began some time ago in graduate school at the University of Minnesota. My research topic focused on ore passes. Ore passes are long, steeply inclined "tunnels" in underground mines. They are conduits used to transport broken ore and waste rock ("muck" in miner's language) under the action of gravity from one mine level to a level below. Backfill may also be transported underground in this manner. Ore passes malfunction on occasion because of "hang-ups," stoppages caused by interlocking blocks and cohesive action of "fines." Piping is also a malfunction when flow only occurs in the central portion of an ore pass. Flow segregation can be problematic. Soil plasticity appeared to be a logical choice, if not the only choice, for a rational analysis of the "muck" mechanics involved, although my interest was in flow rather than stability.

The timing was fortunate because of the resurgent interest and development of theoretical soil plasticity, especially at Brown University in the Department of Applied Mathematics and Mechanics, where oft-cited publications were produced by such luminaries as W. Prager, P. G. Hodge, D. C. Drucker, A. H. Greenberg, R.T. Shield, and others.

Design of bins, hoppers, chutes, and silos for handling bulk materials in numerous industrial operations pose problems closely related to ore pass design. Although less well known, the development and application of practical approaches to design of bins and hoppers were advanced at the University of Utah Engineering Experiment Station. Not glamorous but of considerable importance in many industries, the technology was intensively studied under the guidance of Professor A. Jenike and his student J. R. Johanson, subsequent principals in the highly respected firm of Jenike and Johanson with specialization in bulk materials handling solutions for industrial processes.

An initial interest and exploration of soil and metal plasticity, where extrusion and finite strain have some similarity to flow in ore passes, quite naturally led to possible applications to rock. Although, at the time, common knowledge considered rock a brittle material that failed catastrophically by fast fracture with any attempt to continue loading beyond the elastic limit. The elastic limit was thus the "end of the story." Much laboratory test experience confirmed this view. However, casual observations of rock at the engineering scale – for example, in road cuts – revealed numerous fractures, cracks, faults, and joints. Certainly, fracture was not the end of the story at the engineering scale. Noticeable "squeeze" in deep hardrock mines also indicated inelasticity. Plasticity theory based on first principles offered a way forward.

An important objective of this book is to present a clear but compact account of "sliplines." The technical literature is replete with numerous diagrams of sliplines, especially in discussions of foundations on soils, but the relevant mathematics is often lacking and with it a genuine understanding. A short introduction to this book is presented in Chapter 1. But there is no point to proceeding to theoretical analyses without physical justification. Hence, the physical foundations for application of plasticity theory to rock are examined in Chapter 2. No justification for application to soils is needed. A brief review of continuum mechanics principles is given in Chapter 3. Chapter 4 focuses on plane plastic strain and goes to the origins of this book. Examples illustrate application of theory to traditional geomechanics problems such as computation of retaining wall forces in soils, foundation-bearing capacity of soil and rock, wedge penetration of rock under confining pressure, and others. Axial symmetry is relegated to an appendix despite superficial similarities to plane strain analysis. Chapter 4 also contains a brief exposition of boundary value problem types and numerical solutions. Limit theorems that were of great interest before the digital era but are now largely of academic interest are discussed in Chapter 5. Discussions of anisotropy, viscoplasticity, and poroplasticity are presented in Chapters 6, 7, and 8. Anisotropy is as much the rule as the exception in geomechanics. Consideration of a viscous component of deformation introduces a rate and therefore a time dependency. Slipline analyses do not do justice to the importance of fluid flow in porous rock and soil. The chapter on poroplasticity is intended to open the door to this important phenomenon in geomechanics.

The advent of the computer and concurrent development of numerical methods such as the discrete element method (DEM), discontinuous deformation analysis (DDA), and variants has since revolutionized design analysis involving bulk materials handling. Indeed, the popular finite element method (FEM) has made routine complex elastic-plastic analyses of soil and rock mechanics problems. Mechanics of ice and snow could be added to the mix of geo-materials. Indeed, geoplasticity is now well-known and widely accepted technology in the realm of geotechnical engineering.

These "notes" were begun shortly after graduation and formed a basis for graduate courses in geoplasticity and post-graduate research as well. They have evolved in time. The digital age prompted development of a second course, one in numerical modeling, with focus on the finite element method. A brief appendix outlines the fundamentals of the method. Example problems are found in Chapters 7 and 8. Much has been accomplished by the technical community since these notes were begun, especially in the realm of numerical methods that allow for site-specific design analyses of a great variety of geotechnical problems, which potentially involve deformation beyond the purely elastic range, that is, geoplasticity.

William G. Pariseau
Salt Lake City, Utah, USA
October 2018

Acknowledgements

Grateful acknowledgement is made to the Department of Mining Engineering at the University of Utah and to Professor M. Nelson, Department Chair, for financial support when most needed. Highly skillful drafting of figures to publication standards by Stephen Adams is much appreciated. Expert assistance by K. Ehrmann with the demanding task of obtaining permission to use figures and tables found in the technical literature is also appreciated, as is the assistance of G. Guilinger with the important task of indexing.

Geoplasticity is necessarily based on experimental evidence that is found throughout an extensive technical literature. Sampling of this literature and illustrating effects of confining pressure, temperature, strain rate, and time required figures from a variety of sources, including:

- American Institute of Mining, Metallurgical and Petroleum Engineers
- American Rock Mechanics Association
- American Society of Mechanical Engineers
- American Society of Petroleum Geologists (*Bulletin of the American Association of Petroleum Geologists*)
- Elsevier Science Ltd (*International Journal of Rock Mechanics and Mining Sciences*)
- Geological Society of America
- John Wiley & Sons
- Society of Mining Metallurgy and Exploration
- Springer-Verlag (*Rock Mechanics and Rock Engineering*)
- Taylor & Francis, CRC Press
- University of Chicago Press (*Journal of Geology*)

Permission to use figures from these publications is most gratefully acknowledged; each individual source is recognized in the figure caption.

Appreciation is also extended to authors of figures obtained from government publications and others that are not copyrighted. Sources of such figures are also recognized and cited in figure captions.

Finally, the dedication and inspiration provided by graduate students at the Pennsylvania State University in the beginning, later at the Montana College of Mineral Science and Technology, and subsequently at the University of Utah is gratefully acknowledged. Their diligence and perseverance in the tedious mapping of sand model flows from photographs and film, testing rock and joints in the laboratory, and doing individual class notebooks on geoplasticity without benefit of a textbook was hard work.

Chapter 1

Introduction

Theories in the mechanics of solids, as a practical matter, are simplified descriptions of actual material behavior. They are continuum models. Plasticity theories describe the response of model solids strained beyond an elastic limit. Prior to reaching this limit, the response is purely elastic in accordance with Hooke's law. As in any continuum theory, motion of the material must be consistent with physical laws and kinematics, often expressed in terms of stress equations of motion and strain-displacement relationships. A plasticity theory is completed when the elastic limit and the stress-strain relationship beyond the elastic limit are specified. Stress at the elastic limit may be defined as material strength. At the elastic limit, the material is at yield. The constitutive equations (material laws) are then Hooke's law, a strength criterion, and an elastic-plastic stress-strain relationship. Generally, the resulting system of nonlinear partial differential equations leads to boundary value problems that are too complicated for much analytical progress and for solution to problems of engineering interest. An exception occurs in plane analyses and to a much lesser extent in axial symmetry. In this regard, plane strain analyses are of particular interest and form the focus of much that follows in the form of *limiting equilibrium analyses*.

Plasticity is sometimes thought of rather casually as a material that flows slowly or squeezes like toothpaste. However, plastic and viscous flows are different in mechanics. An elastic-plastic material is a solid with a well-defined yield point; shear loading below the yield point results in a limited and purely elastic strain. A viscous material is a fluid and cannot sustain a shear stress; flow continues as long as the stress is applied. More complicated material models containing various elastic, viscous, and plastic elements can be hypothesized. In one dimension, such models are often referred to as rheological models. Rheological models help to guide conceptual thinking about stress-strain relationships, but are difficult to generalize to three dimensions.

Strength criteria are also referred to as a yield conditions or failure criteria and are based on a combination of experimental observations and physical constraints. Not any mathematical function is a candidate for the description of material strength, even though a good fit to test data is obtained. The term *loading function* also appears in this context. Most strength or failure criteria in rock and soil mechanics are simple polynomials in stress and are often associated with names, for example, Mohr-Coulomb (MC), Drucker-Prager (DP), and Hoek-Brown (HB).

Elastic-plastic stress-strain laws may be *total* or *differential*. Total laws relate stress to strain in much the same way that Hooke's law relates (total) stress to (total) strain. Differentiation of Hooke's law results in a differential law. The differential form may be integrated back to a total law without further consideration. Differential elastic-plastic stress-strain laws are nonlinear and often preclude integration to obtain a total law.

Elastic-plastic stress-strain laws may be further distinguished with respect to time. Time-independent plasticity requires an attempt at additional loading for additional straining. Time-dependent plasticity implies possible creep, increasing strain at constant stress, and possible relaxation, decreasing stress at constant strain. Time-dependent plasticity may also involve a viscosity of sorts and give rise to viscoplasticity, perhaps better classified within a context of rheology. Poroelastic-plastic behavior also gives rise to a time dependency related to fluid flow.

Differential elastic-plastic stress-strain relationships are generally obtained from a *plastic potential*. A plastic potential is a function that when differentiated with respect to stress leads to a corresponding plastic strain increment or rate, as the case may be. The plastic potential (yield condition) and failure criterion need not be the same. When the plastic potential and failure criterion (strength criterion) are the same, the plastic stress-strain relationships are known as *associated rules of flow*; when they are not, as is often the case in geomechanics, the rules of flow are *nonassociated.*

Consideration of plasticity models for soil and rock requires some justification, especially in view of the common assumption that plasticity is synonymous with ductile deformation of metals such as mild steel, a most important material in mechanics of structures. Of course, concrete, a brittle material, is also of importance to structures. Both are reasonably considered linearly elastic solids with a range of purely elastic deformation that is limited by material strength. In the case of mild steel, the range of elastic-plastic behavior is limited by rupture strength, reached only after considerable plastic strain. Collapse of the structure occurs at rupture. In the case of concrete, the range of elastic-plastic behavior is much less. Indeed, rupture strength of concrete may also be the strength at first yield (ideally brittle). However, applicability of plasticity theory is not necessarily dependent upon the amount of *plastic* strain that may occur before collapse.

The micro-mechanisms of plastic strain in ductile steel and brittle concrete are quite different. Dislocations are largely responsible for ductility of mild steel; while microcracking is the dominant mechanism of inelasticity in nominally brittle materials such as concrete and rock. In soils, microcracking of grains is possible at high stress, but generally, soil plasticity is associated with interparticle translations and rotations. In this regard, one may consider soil not at all an elastic material. However, the elastic model proves quite useful in soil mechanics, despite a limited range of purely *elastic* strain.

Soil and rock are geo-materials that have much in common. Indeed, gradation from intact rock to weathered rock to soil makes the distinction difficult under some circumstances. However, there are significant differences between soil and rock at the laboratory scale and even more so at the field scale. The field scale is the scale of engineering applications, that is, the scale of excavation, which ranges from small drill holes of centimeter scale to the scale of hundreds of meters in large, open pit mines. Laboratory-scale soil samples may be representative of the associated field-scale soil mass or of a soil construct, for example, a built-up embankment. However, it is unlikely for a laboratory-scale rock sample to be representative of a field-scale rock mass. The reason is the presence of geologic features in the rock mass that are absent in laboratory test specimens, especially joints.

Natural rock masses generally contain many structural discontinuities such as cracks, fractures, joints, and bedding planes that profoundly affect rock mass behavior relative to the behavior of intact, laboratory test specimens taken between the discontinuities ("joints").

Conventional wisdom indicates that elastic moduli and strengths of rock masses are lower than values obtained from laboratory test measurements. The results are "scale effects" in rock mechanics. Engineering-scale rock masses may not only have different property values, but may also respond differently to load compared with laboratory-scale test pieces. While intact rock tested in the laboratory may fail violently at the limit to elasticity, field-scale rock masses – already fractured and jointed – may experience slip and separation along joints and appear ductile up to collapse.

Motion of hard, well-jointed rock masses under light loads may appear to defy a continuum description and thus require mechanics of a different kind, certainly not a plasticity theory. However, there is no genuine choice to make here, nor is a real choice possible. What is required is a more complicated continuum description that allows for slip and separation of joints as well as for the motion of individual rock blocks defined by multiple intersecting joints. Simple rock block material models that do so are implemented in a variety of ways in computer programs. These rock block models are sometimes referred to as brick-wall models, for obvious reasons. In this regard, a long-standing issue in rock mechanics is the question of just what to do about the joints.

The rock mass combination of intact rock and joints suggests separate material models. Both can be sampled and tested in the laboratory for elastic and strength properties. Multiple joint sets having different properties may be present, so the rock mass is a heterogeneous composite of joints and intact rock. Averaging the heterogeneity in a way that relates average stress to average strain in a sample volume leads to a composite with *effective* or *equivalent* properties. Equivalent properties thus relate averages of stress and strain at a macroscopic level in a sample volume despite the heterogeneity of joints and intact rock within the sample. Although progress has been made in recent years with the aid of computer models, there is no consensus in the technical community as to how equivalent properties should be computed. In this regard, estimation of macroscopic material behavior from microstructural observations is also helpful in determining features of material behavior that appear at various scales of aggregation. Whether by microstructural modeling or an equivalent properties approach, the practical result is an averaging up of heterogeneity to a level of aggregation amenable to practical computation using numerical methods.

An important complication that arises in rock mechanics is the generation of new fractures and cracks and the extension of existing joints. The same complication occurs in progressive failure of concrete structures. Such events are irreversible and therefore plastic in some sense, but the mixture of brittle fracture and plastic flow is not easily reconciled. A similar situation occurs during laboratory testing as new microcracks are formed and existing ones extended in an otherwise sample of intact rock. One approach to describing the collective behavior of many fractures is through the concept of *damage*. Damage affects elastic moduli. As damage increases, the elastic moduli decrease. Collapse of the test specimen may follow as *instability* in consequence of unbounded growth of the fracture network rather than as strength failure. Instability is associated with an *energy* maximum; strength failure with a *stress* maximum. Collapse implies loss of load-bearing capacity. Collapse may also follow as a consequence of *localization* of deformation into a thin zone of intense shear that transects the test specimen. To be sure, localization is not fracture; localization involves a shear zone of finite thickness, although there are some superficial similarities. As in rheological models, one may hypothesize combinations of elasticity, damage, and plasticity.

In structural mechanics involving manufactured materials such as steel and concrete, what material model is apt for the problem at hand is usually settled by testing *representative* material samples under conditions that span the range of stress, strain rate, temperature, moisture, and so forth that are anticipated in practice. In geomechanics, the same guidance would seem applicable. However, the nature of the testing program is not clear because of the need for representative samples. A sample of intact rock, say, granite taken during an exploratory diamond drilling operation, is several centimeters in diameter with a length that is twice the diameter when tested. Grains are of millimeter size, so the test cylinder contains many grains and may be considered representative, but only of intact rock between joints. Testing according to standards such as those of the American Society for Testing and Materials (ASTM) or the International Society for Rock Mechanics (ISRM) then allows for direct determination of elastic moduli and strengths. Joints may also be sampled in the field and tested in the laboratory, although operational difficulties make the practice much less common than testing of diamond drill core for intact rock properties. Somewhat larger volumes of rock may be tested with borehole devices. However, when joints are spaced at the meter scale, the size of a representative sample of jointed rock is much too large for laboratory-like determination of rock mass properties. An alternative to direct testing is to rely on a technically sound theory for the computation of equivalent properties. Another alternative is to postulate the constitutive equations outright and use field observations to estimate material properties, a process known as back analysis, which may be done empirically or within a more rigorous methodology.

Many materials fail under uniaxial compression at about 0.1% to 0.2% strain. For example, a granite with a Young's modulus of 103 GPa or $15(10^6)$ psi and an unconfined compressive strength of 2.07 GPa or $30(10^4)$ psi would fail at a strain ε_f of 0.002. This strain ε_f is small compared with unity in the sense that the square $(\varepsilon_f)^2 = 0.000\,004 << 1.0$. If ε_f were an order of magnitude larger, say, 2%, the square $(\varepsilon_f)^2 = 0.000\,400$ would still be small relative to unity. Strains that are small compared with unity are *infinitesimal* strains or simply "small" strains.

Strains that do not meet the smallness requirement are *finite* or large strains. The terminology can be ambiguous because finite strains may be small, that is, have a low value in the context of finite strain, but still not be small relative to unity. The same discussion applies to strain increments. Finite strain leads to an enormous increase in complexity of analysis, mainly in the kinematical features of theory. One potential complication is a multiplicative decomposition of elastic and plastic strains rather than an additive decomposition available when strain increments are small. Fortunately, there are many problems of interest in geomechanics that are amenable to small strain analysis.

References are given at the end of each chapter. Only sources actually consulted are cited. The literature is enormous, and undoubtedly there are many worthwhile sources not cited. However, those that are form a reasonable background for developing an initial interest in geoplasticity. The early and authoritative book by Hill (1950) was inspirational in forming the first draft of these notes. A more mathematical treatment by Thomas (1961) was most helpful in clarifying a proper invariance requirement for strain rate and later discussion of "localization." The book by Prager and Hodge (1951) was helpful in understanding error control during numerical analyses. The prodigious two-volume work by Nadai (1950, 1963) is notable especially in discussion of envelopes of sliplines. Journal articles by Prager (1948), Drucker (1951, 1959), Drucker *et al.* (1951), Drucker and Prager (1952), Shield (1953), and Koiter (1953) were essential to development of theory. Continuum mechanics books by Prager (1961), Fung (1965), and Jaunzemis (1967) were most helpful with understanding general concepts of solid mechanics in the broader context of continua. Malvern (1969) is also worthy of mention. A most noteworthy book is one by Sokolovski (1960 translation) that is replete with numerical results for a variety of problems in soil mechanics that were obtained before the advent of the computer. A later book on plasticity by Kachanov (1971) proved helpful in understanding various types of boundary value problems. The modern treatment of plasticity theory by Kahn and Huang (1995) was especially helpful in providing guidance to finite strain and to understanding the interesting strain cycle formulation of plasticity originally advanced by Ilyushin (1961). Recent books by Chen and Han (1988), Davis and Selvadurai (2002), Chen (2008), and Pietruszczak (2010) should also be helpful in furthering one's interest in plasticity theory for metals, soils, concrete, and rock.

References

Chen, W.F. (2008) *Limit Analysis and Soil Plasticity*. Wai-Fah Chen, J. Ross Publishing, Fort Lauderdale, Florida.

Chen, W.F. & Han, D.J. (1988) *Plasticity for Structural Engineers*. Springer-Verlag, New York.

Davis, R.O. & Selvadurai, A.D.S. (2002) *Plasticity and Geomechanics*. Cambridge University Press, Cambridge.

Drucker, D.C. (1951) A more fundamental approach to plastic stress-strain relations. *Proceedings of the First U.S. National Congress of Applied Mechanics, ASME*, pp. 487–491.

Drucker, D.C. (1959) A definition of stable inelastic material. *Journal of Applied Mechanics*, 26, 101–106.

Drucker, D.C. & Prager, W. (1952) Soil mechanics and plastic analysis or limit design. *Quarterly of Applied Mathematics*, 10, 157–164.

Drucker, D.C., Prager, W. & Greenberg, H.J. (1951) Extended limit design theorems for continuous media. *Quarterly of Applied Mathematics*, 9, 381–389.

Fung, Y.C. (1965) *Foundations of Solid Mechanics*. Prentice-Hall, Englewood Cliffs, NJ.

Hill, R. (1950) *The Mathematical Theory of Plasticity*. Clarendon Press, Oxford.

Ilyushin, A.A. (1961) On the postulate of plasticity. *Journal of Applied Mechanics*, 25(3), 746–752.

Jaunzemis, W. (1967) *Continuum Mechanics*. Macmillan, New York.

Kachanov, L.M. (1971) *Foundations of the Theory of Plasticity*. North-Holland, Amsterdam.

Kahn, A.S. & S. Huang (1995) *Continuum Theory of Plasticity*. Wiley, New York.

Koiter, W.T. (1953) Stress-strain relations, uniqueness and variational theorems for elastic-plastic materials with a singular yield surface. *Quarterly Applied Mathematics*, 11, 350–355.

Malvern, L.E. (1969) *Introduction to the Mechanics of a Continuous Medium*. Prentice-Hall, Englewood Cliffs, NJ.

Nadai, A. (1950) *Theory of Flow and Fracture of Solids*, Volume 1. McGraw-Hill, New York.

Nadai, A. (1963) *Theory of Flow and Fracture of Solids*, Volume 2, McGraw-Hill, New York.

Pietruszczak, S. (2010) *Fundamentals of Plasticity in Geomechanics*. CRC Press, New York.

Prager, W. (1961) *Introduction to Continuum Mechanics*. Dover, New York.

Prager, W. (1948) The stress-strain laws of the mathematical theory of plasticity – a survey of recent progress. *Journal of Applied Mechanics*, 15, 226–233.

Prager, W. & Hodge, P.G. (1951) *Theory of Perfectly Plastic Solids*. Wiley, New York.

Shield, R.T. (1953) Mixed boundary value problems in soil mechanics. *Quarterly of Applied Mathematics*, 11, 61–75.

Sokolovski, V.V. (1960) *Statics of Soil Media* (Translated by D.H. Jones & A.N. Schofield). Butterworths, London.

Thomas, T.Y. (1961) *Plastic Flow and Fracture in Solids*. Academic Press. New York.

Chapter 2

Physical foundations of theory

The physical foundations and justification of plasticity theory for application to rock and soil, geo-materials, is necessarily based on experimental evidence. For many years conventional wisdom indicated rock was a brittle material that failed violently by fast fracturing at the elastic limit. Plasticity theory was for ductile materials that showed large post-yield strain. This view was strongly conditioned by laboratory testing of intact rock cylinders that, indeed, often shattered at failure with an audible bang. Of course, no further plotting of a force-displacement (stress-strain) graph was possible; the "story" of the sample response to load was ended. However, the amount of plastic strain that occurs after reaching the elastic limit is not a requirement of plasticity theory as such. Moreover, rock masses in nature are almost always fractured, "jointed," and certainly the story of the rock mass response to load is not ended. Additional elastic deformation may be possible in response to further loading, perhaps induced by excavation. Deformation beyond the range of a purely elastic response is also possible and indeed may even be likely. A plasticity theory is needed to describe deformation beyond the elastic limit and thus to continue the "story."

Two important questions need to be answered in formulating a plasticity theory. The first question is to describe the limit to elasticity considering three-dimensional stress states; the second is to describe the plastic portion of a now elastic-plastic stress-strain law. Guidance toward answering the first question is obtained mainly from experimental results obtained in the laboratory under controlled conditions. The second question is addressed mainly through consideration of physical constraints. Both are eventually disciplined by results obtained in application of theory to practice.

Experimental evidence concerning the response of rock to load is derived almost entirely from direct static compression of solid, cylindrical specimens of intact rock under controlled laboratory test conditions that often include application of a confining pressure. Thus, the bulk of experimental observations refer to axially symmetric states of stress and strain. In the traditional triaxial test, a specimen is compressed between two opposing platens in the presence of a laterally applied confining pressure.[1] Usually, the test cylinder is jacketed to prevent penetration of the specimen by the chamber fluid, which is under a confining pressure. Jacketed, porous rock specimens (and soils) admit of a pore fluid pressure that may differ from the chamber or confining pressure. In unjacketed tests on saturated, permeable geo-materials, the confining and pore fluid pressure are the same during "slow" testing, that is, under quasi-static application of load that allows the pore fluid to come to equilibrium.

In addition to the effects of confining and pore fluid pressure on behavior of rock, effects of temperature and strain rate have also been observed. Compression tests have been made

1 Only two loads are controlled in the conventional triaxial test, the axial load and the confining pressure; thus, the test is not a true triaxial test involving control of three applied loads, say, the three principal stresses.

under more than 60,000 psi confining pressure and 29,000 psi pore fluid pressure. Temperatures as high as 800 °C and as low as −320 °F have been used. Strain rates used range from 10^{-7} per second to 10^{-1} per second and higher. All major rock types – igneous, sedimentary, and metamorphic – have been tested, as have all major soil types –sand, silt, and clay. Jointed rock and rock joints have also been tested; the latter mainly in direct shear.

The effect of an increase in confining pressure is almost always an increase in compressive strength, whereas an increase in pore pressure usually decreases strength. An increase in strain rate generally increases strength. Increasing temperature generally lowers strength. The ability of geo-materials to deform without fracture is usually enhanced by an increase in confining pressure and temperature and by a decrease in strain rate and a decrease in pore fluid pressure. Although an increase in strain rate increases strength, the deformation tends towards more brittle behavior.

A sampling of the literature illustrates experimental behavior of rock, rock joints, jointed rock, and soil as observed in the laboratory. Experimental work was at the cutting edge of the rapidly developing subdiscipline of solid mechanics in the 1950s – rock mechanics. An enormous accumulation of literature has developed in specialty journals and symposia since then. However, any attempt to summarize such would be foolish indeed.

The sampling format here is first a citation of the source work, then a summary of the experimental variables, followed by a view of associated stress-strain plots and in some case plots of shear stress – normal stress at failure (yield functions, Mohr envelopes). Unless otherwise stated, the test specimens are cylinders of intact rock with a height-to-diameter ratio of two. All tests were performed without the use of friction reducers between testing machine platens and test specimens ends.

The terminology is as follows:

Confining pressure: chamber pressure, pressure applied to the lateral surface of the test specimen.

Pore fluid pressure: pressure of the fluid contained in the connected pore space in the test specimen.

Effective pressure: the difference between confining and pore fluid pressure; applies to porous, permeable specimens only.

Stress difference: the difference between the axial stress and confining pressure.

Effective stress difference: same as the stress difference.

Effective stress: the (total) stress less the pore fluid pressure, also the Terzaghi effective stress.

Yield strength: the stress obtained at the limit of elastic deformation.

Ultimate strength: the stress obtained at the highest point of the stress-strain plot.

Rupture strength: the stress obtained at final fracture.

Ductility: the permanent strain obtained at the rupture point, sometimes measured by total strain when the yield strength is difficult to define or determine.

Brittle behavior: fractures at the limit of elastic deformation without intervening permanent or plastic strain.

Hydrostatic pressure: the average of the three normal stresses, a mean normal stress.

Dilatancy: refers to relative expansion or relative increase in volume under generally compressive stress.

Plastic strain: refers to strain following yield that is not recovered upon unloading and is thus "permanent" (time-independent) strain.

Some of these terms are illustrated in Figures 2.1, 2.2, and 2.3. Figure 2.1 illustrates the concepts of strain hardening, ideal or perfect plasticity, and strain softening. Figure 2.2 illustrates the main features of the "triaxial" test. Figure 2.3 illustrates some additional concepts and associated vocabulary, including elastic unloading from a plastic state and reloading to the same. An additive decomposition of strain into a sum of elastic and plastic parts is also illustrated in the figure.

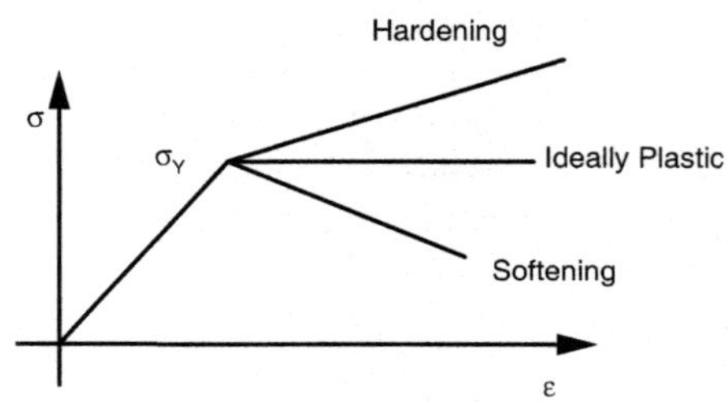

Figure 2.1 Plasticity concepts of strain hardening, ideal plasticity, and strain softening.

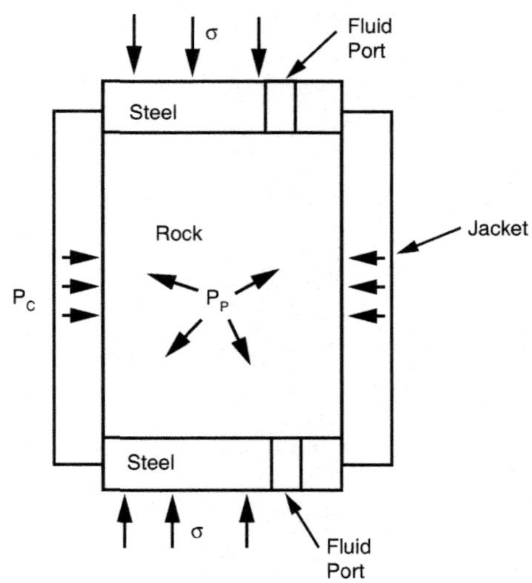

Figure 2.2 Triaxial test schematic showing axial stress (σ), confining pressure (P_C), and pore pressure (P_P).

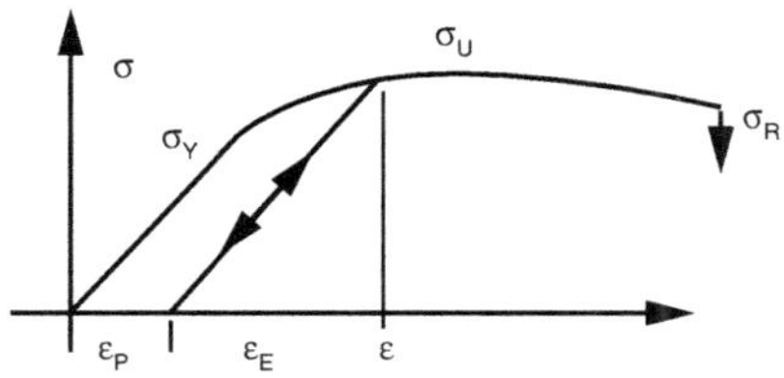

Figure 2.3 More plasticity concepts showing initial yield strength (σ_Y), ultimate strength (σ_U), and rupture strength (σ_R). Also shown are unloading and reloading with elastic (ε_E), and plastic strains (ε_P) that constitute total strain (ε).

Figure 2.4 shows that rock is certainly capable of plastic flow. The figure shows generation of boudinage (sausage) structure in a laboratory experiment.

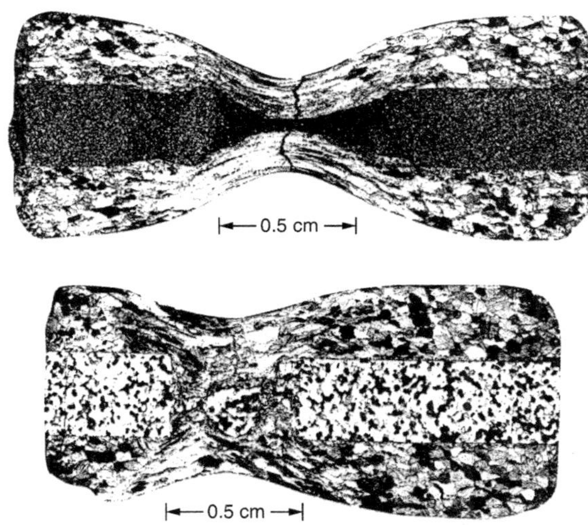

Figure 2.4 Laboratory experiment demonstrating generation of boudinage structure. Top: Lining dolomite encased in Yule marble. Bottom: Inner core is Eureka quartzite; outer core is Yule marble extended 47% (Griggs and Handin, 1960).

2.1 Intact rock response to load

The response of intact rock to load includes effects of confining pressure, pore fluid pressure, temperature, and strain rate. Differences between (1) axial stress and pore fluid pressure and (2) confining pressure and pore fluid pressure influence strength of saturated, porous rock (and soil) through a concept of *effective stress*. Effective stress is the force per unit area transmitted through the solid skeleton of a porous material. The solid provides material strength; the pore fluid does not. Consider the forces acting normal to a plane in a saturated porous material as shown in Figure 2.5.

The total force is simply the sum $F = F_s + F_f$. The total stress is the force per unit area:

$$\begin{aligned}\sigma &= F / A \\ &= (F_s + F_f) / A \qquad (2.1)\\ \sigma &= \sigma_s + \sigma_f\end{aligned}$$

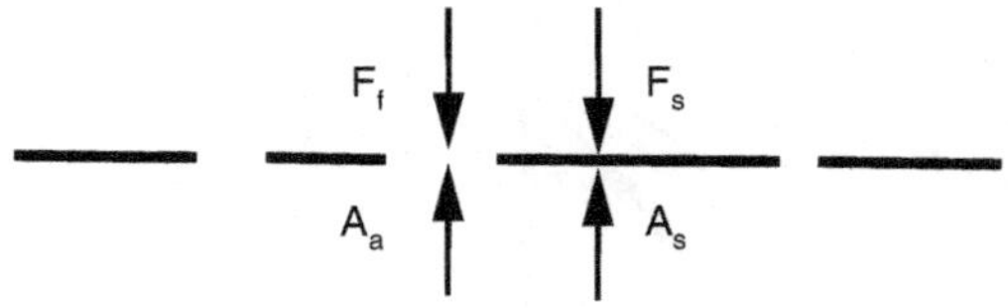

Figure 2.5 Forces acting across a section of a porous solid.

The effective stress is thus:

$$\sigma_s = \sigma - \sigma_f \text{ or} \qquad \text{(2.2a,b)}$$
$$\sigma' = \sigma - \alpha p_f$$

where the prime is commonly used to denote effective stress, α is a coefficient and p_f is the pore fluid pressure that would be indicated by a manometer connected to the pore space of the material. The three stresses in Equation (2.2a) are forces per unit of *total* area as Equation (2.1) shows. In Equation (2.2b), the effective stress and total stress are reckoned per unit of total area, but the fluid pressure is per unit of fluid area.

In soil mechanics, where contact between grains is small and idealized as point contacts, the fluid area becomes a total area. This view gives rise to the Terzaghi concept of effective stress:

$$\sigma' = \sigma - p_f \qquad (2.3)$$

In the triaxial test, the effective axial stress is given by Equation (2.3). There is also an effective confining pressure that is given by the difference:

$$p'_c = p_c - p_f \qquad (2.4)$$

With compression positive, the major principal stress and minor principal stresses are given by Equations (2.3) and (2.4), respectively. A consequence of the cylindrical geometry, the intermediate principal stress is equal to Equation (2.4) also. If the confining pressure exceeds the axial load, then the test is an *extension* test and the roles of Equations (2.3) and (2.4) are reversed with Equation (2.4) being the major principal stress.

A comment about strain is in order as one views experimental data. Many materials reach an elastic limit at 0.1% strain ($0.001 = 10^{-3}$) or so. This is a "small" or "infinitesimal" strain in the sense that a square of this strain is 10^{-6} and would be negligible relative to the failure strain. At 1% strain (10^{-2}), the square is 10^{-4} and is marginally negligible. At 10% strain, the square of such a strain would not be negligible.

Confining pressure

Gnirk, P.F. & Cheatham, J.B., Jr. (1965) An experimental study of single bit-tooth penetration into dry rock at confining pressures to 5000 psi. *Transactions of the American Institute of Mining Metallurgical and Petroleum Engineers*, 234, 117–129.

Direct compression of solid cylinders
Confining pressure to 15,000 psi (103 MPa)

Room temperature tests
Constant strain rate
Dry testing, zero pore pressure
One sandstone, one limestone, two marbles, one dolomite, one schist

Comments: (a) brittle-ductile transition observed with increase in confining pressure, (b) effect of confining pressure on ultimate strength is greater than on initial yield strength, and (c) pronounced curvature of yield envelopes near the origin in the normal stress–shear stress plane.

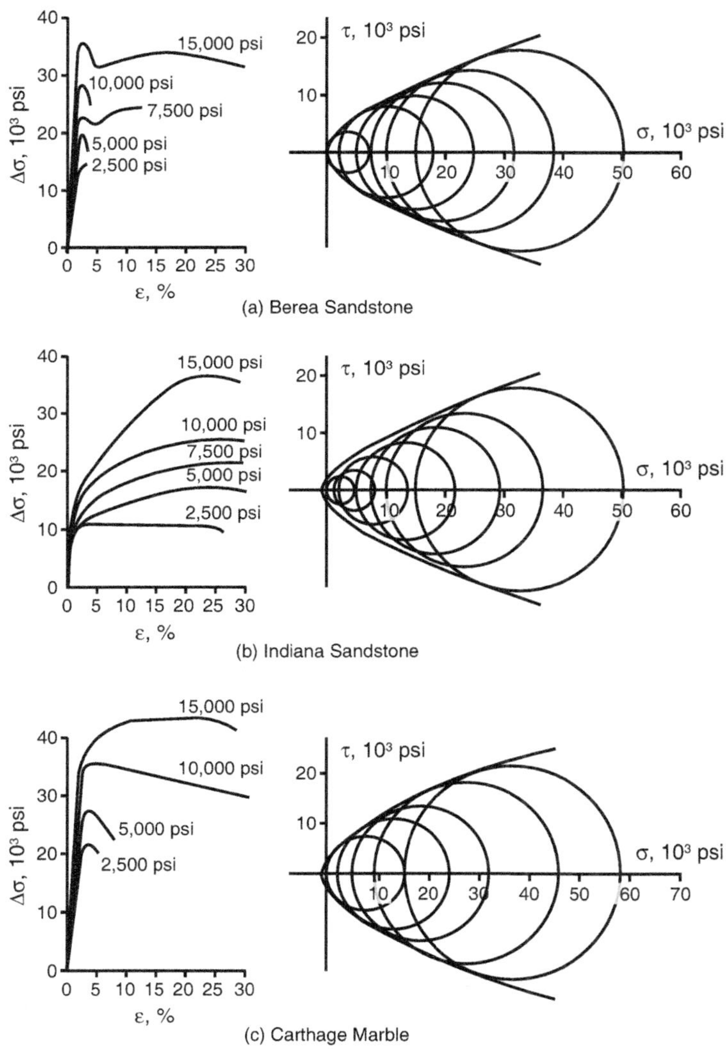

Figure 2.6 Stress-strain plots and Mohr envelope for several rock types (Gnirk and Cheatham, 1965). Stress = axial stress − confining pressure.

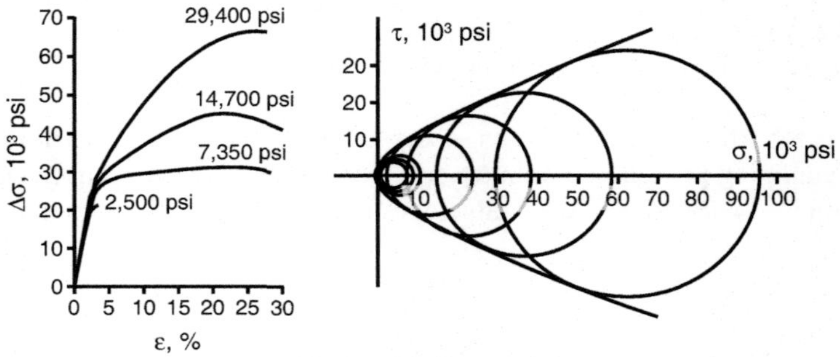

(d) Danby Marble

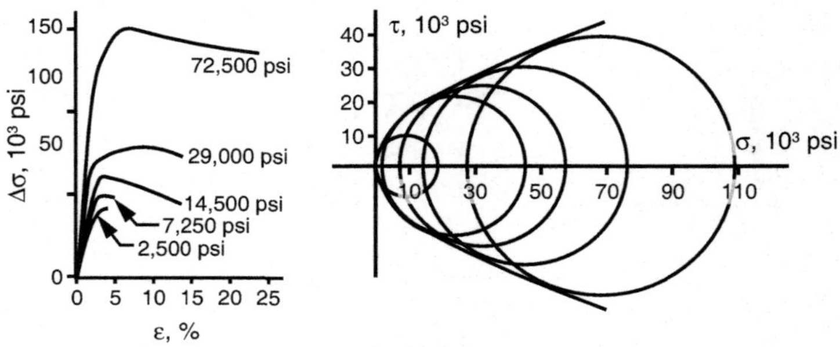

(e) Hasmark Dolomite

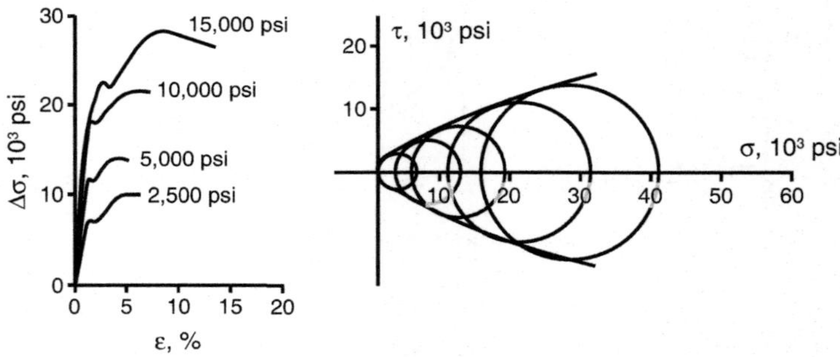

(f) Virginia Greenstone

Figure 2.6 continued.

Source: **Schwartz, A.E.** (1964) Failure of rock in triaxial shear. *Proceedings Sixth Symposium on Rock Mechanics*. University of Missouri, Rolla, MO, pp.109–151.
Direct compression of solid cylinders
Confining pressure to 10,000 psi (69 MPa)
Pore fluid pressure to 5000 psi (34 MPa)
Strain rate approximately 0.5% per minute
Room temperature tests
One limestone, one sandstone, one marble, one granite
Comments: (a) elastic response initially, (b) brittle-ductile transition occurs as confining pressure (effective) increases, (c) dilatancy at lower effective pressure. Note: 1 ksi = 6.9 MPa.

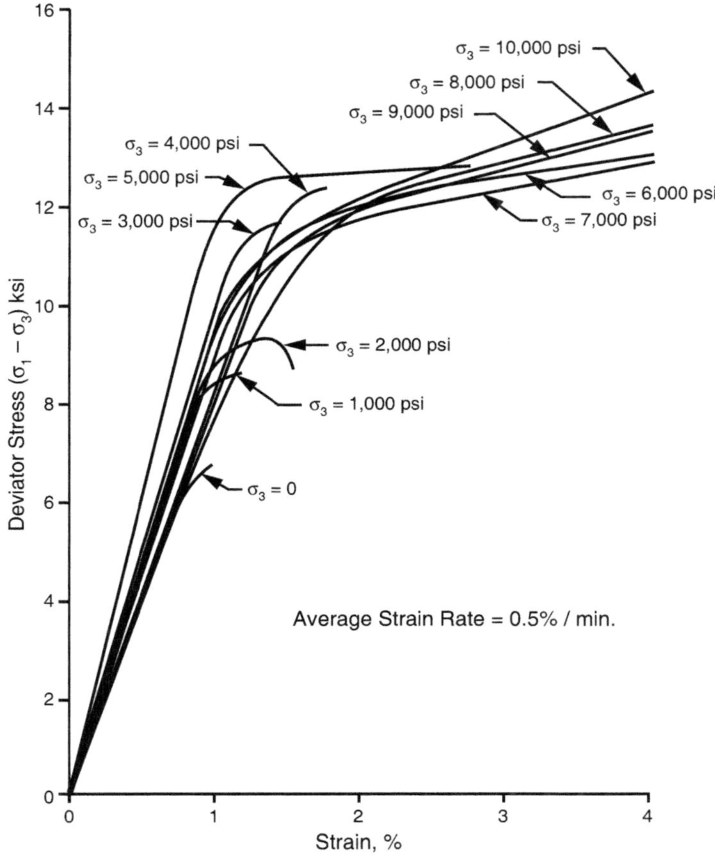

Figure 2.7a Indiana limestone.

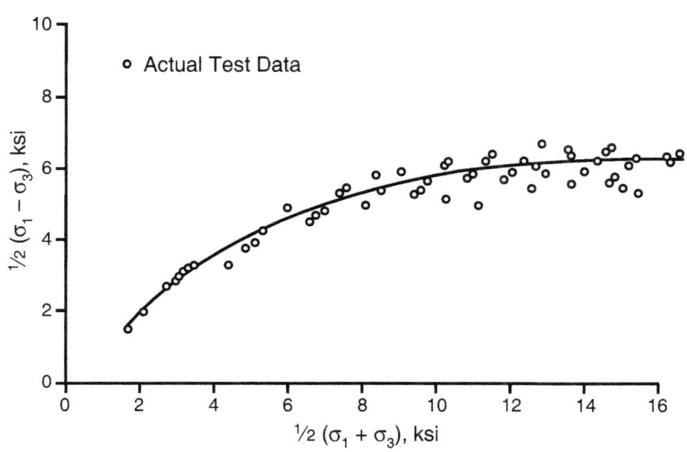

Figure 2.7b Indiana limestone.

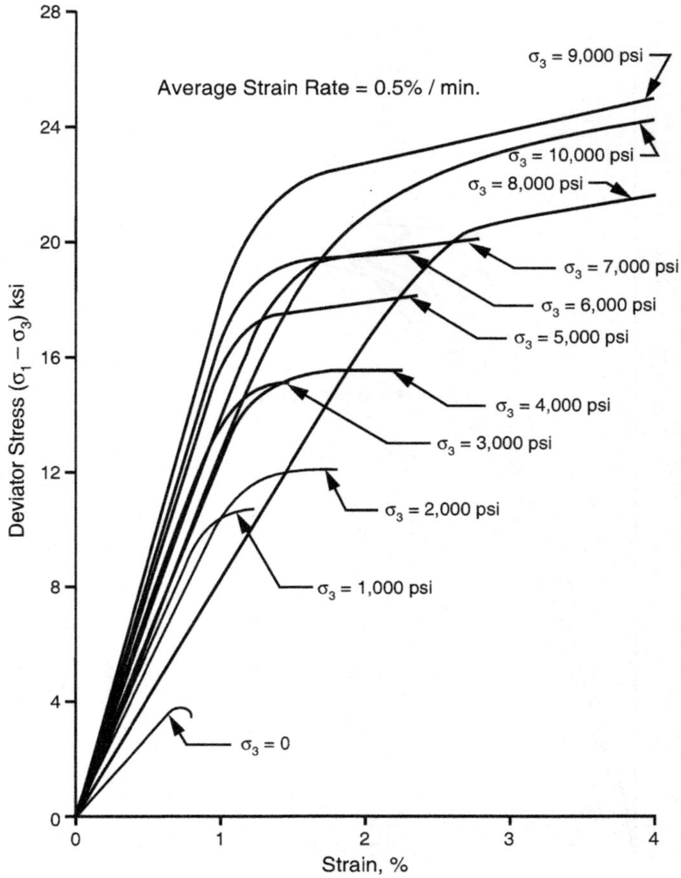

Figure 2.8a Pottsville sandstone.

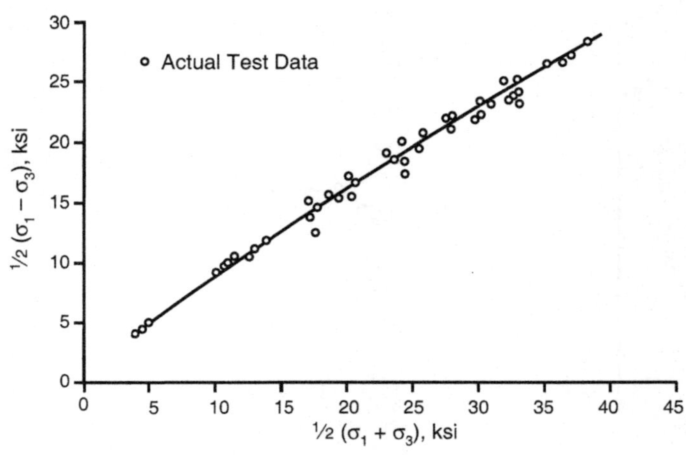

Figure 2.8b Pottsville sandstone.

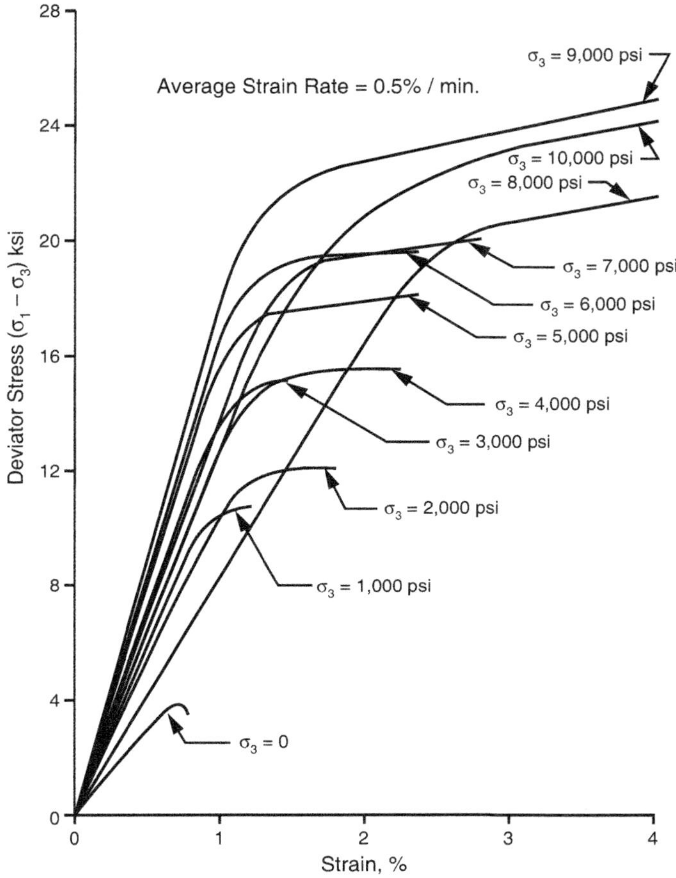

Figure 2.9a Georgia marble.

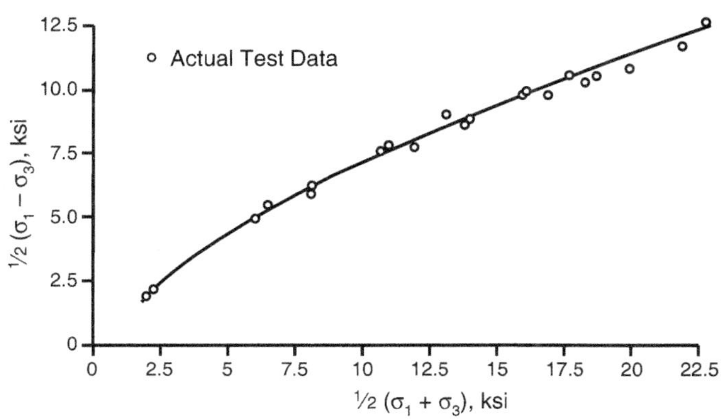

Figure 2.9b Georgia marble.

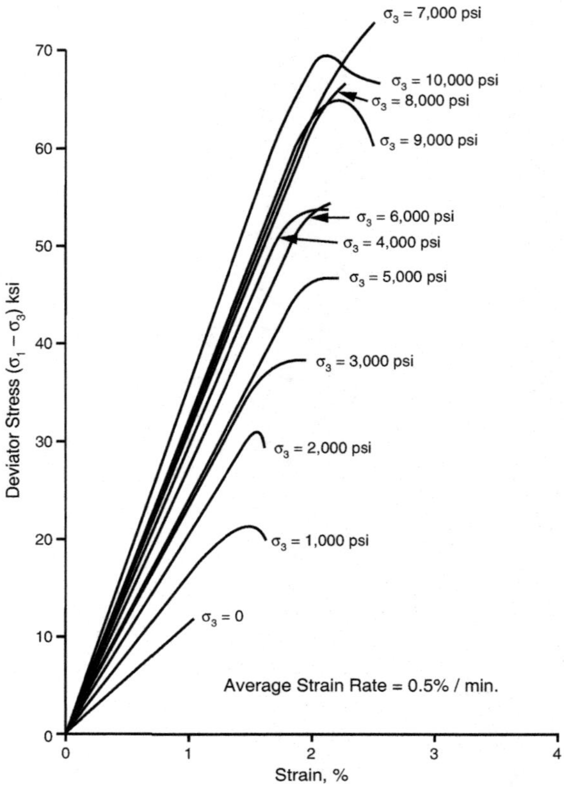

Figure 2.10a Stone Mountain granite.

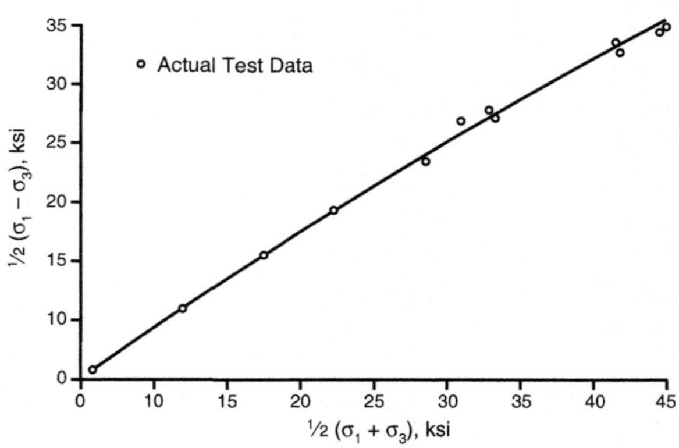

Figure 2.10b Stone Mountain granite.

Source: **Handin, J. & Hager, R.V., Jr. (1957)** Experimental deformation of sedimentary rocks under confining pressure: Tests at room temperature on dry samples. *Bulletin America Associate Petroleum Geology*, 41(1), 1–50.

Confining pressure to 29,000 psi (200 MPa)

Jacketed specimens under direct compression

Twenty-three rock types, dry samples

One siltstone, three shale, four sandstone, six limestone, six dolomite, one quartzite, one slate, one anhydrite

Constant strain rate at 1% per minute

Comments: *Confining pressure in atmospheres.* 1 atm=0.1MPa=1 kg/cm^2. An increase in confining pressure induces a transition from brittle to ideally plastic to work-hardening response.

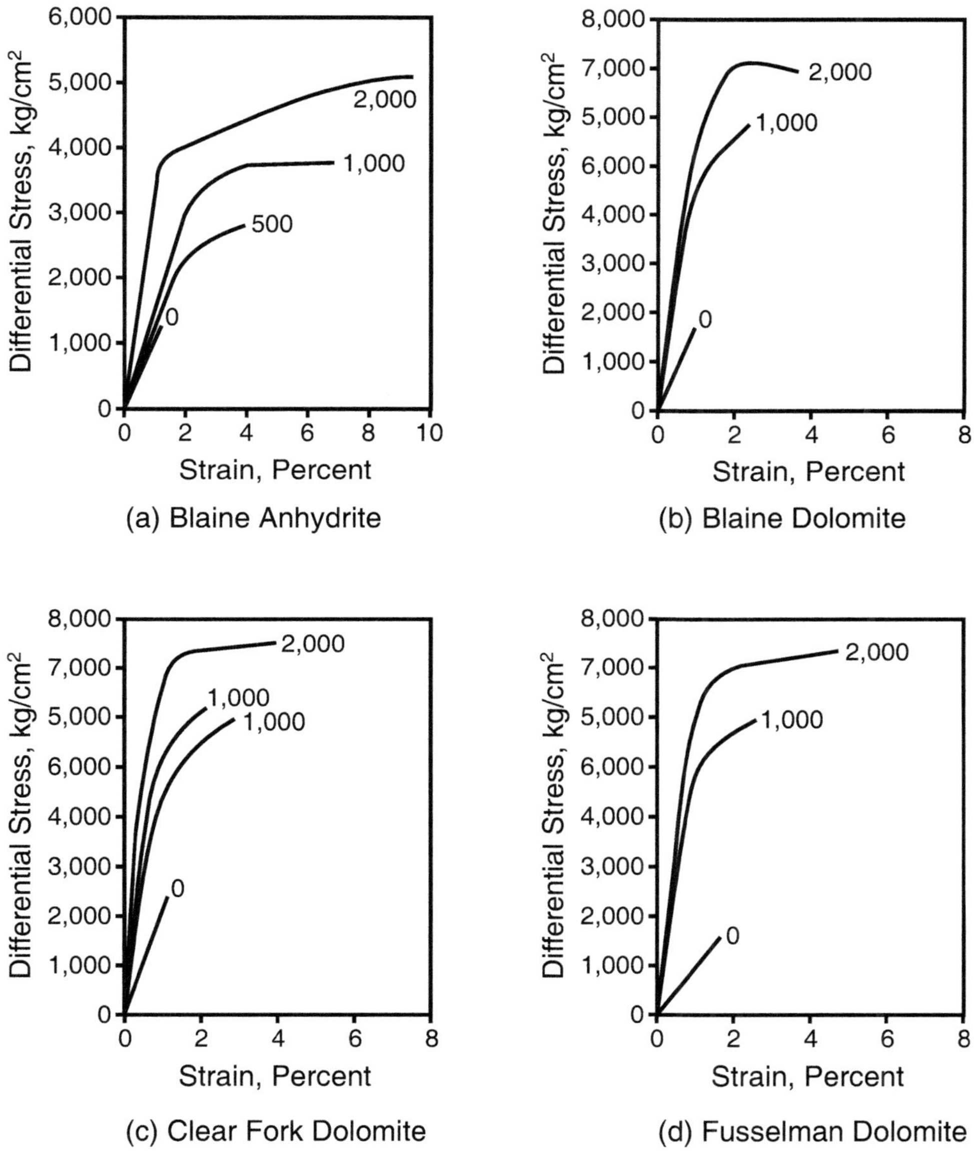

Figure 2.11 (a) Blaine anhydrite, (b) Blair dolomite, (c) Clear Fork dolomite, (d) Fusselman dolomite

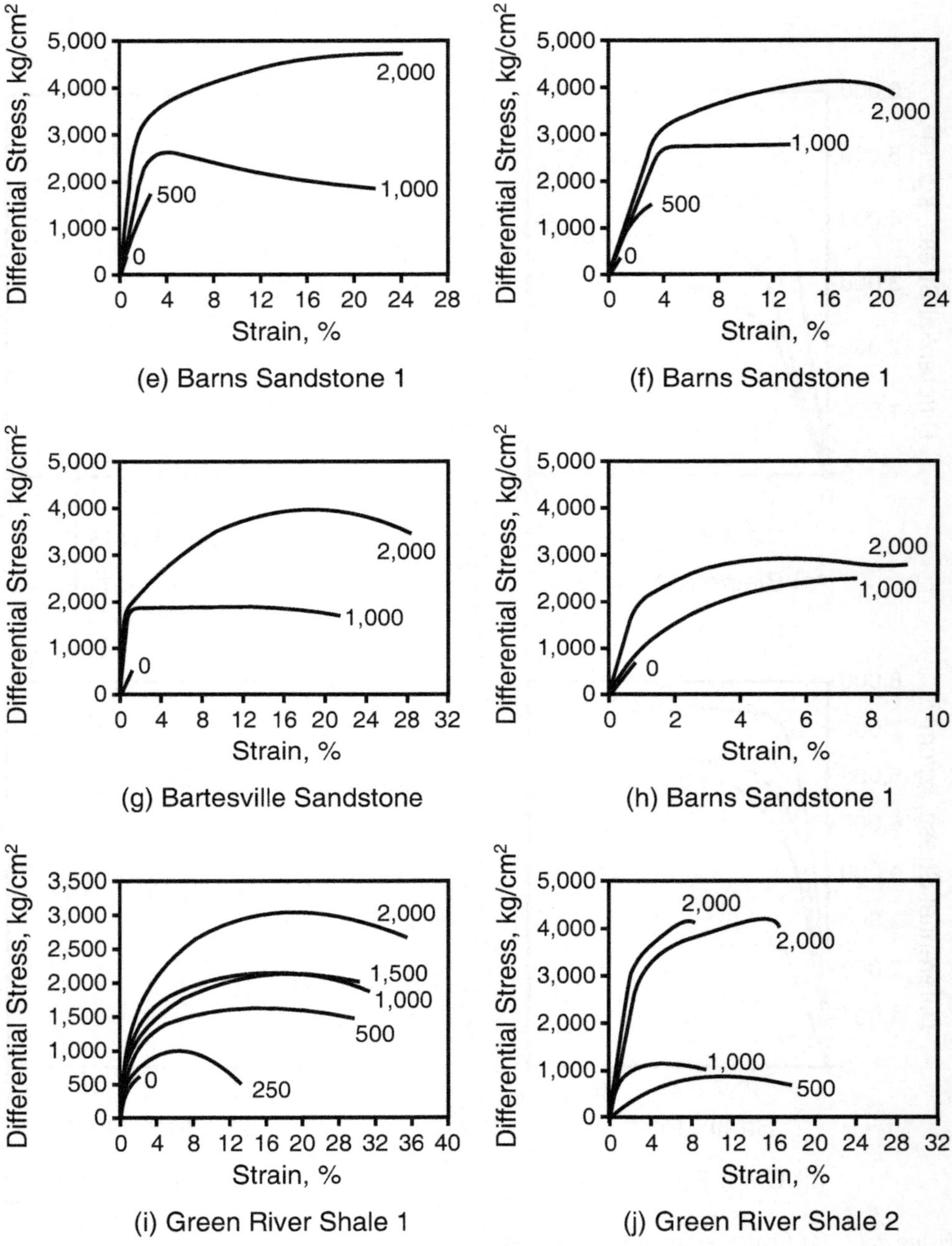

Figure 2.11 (e) Barns sandstone 1, (f) Barns sandstone 2, (g) Bartlesville sandstone, (h) shale, (i) Green River shale 1, (j) Green River shale 2

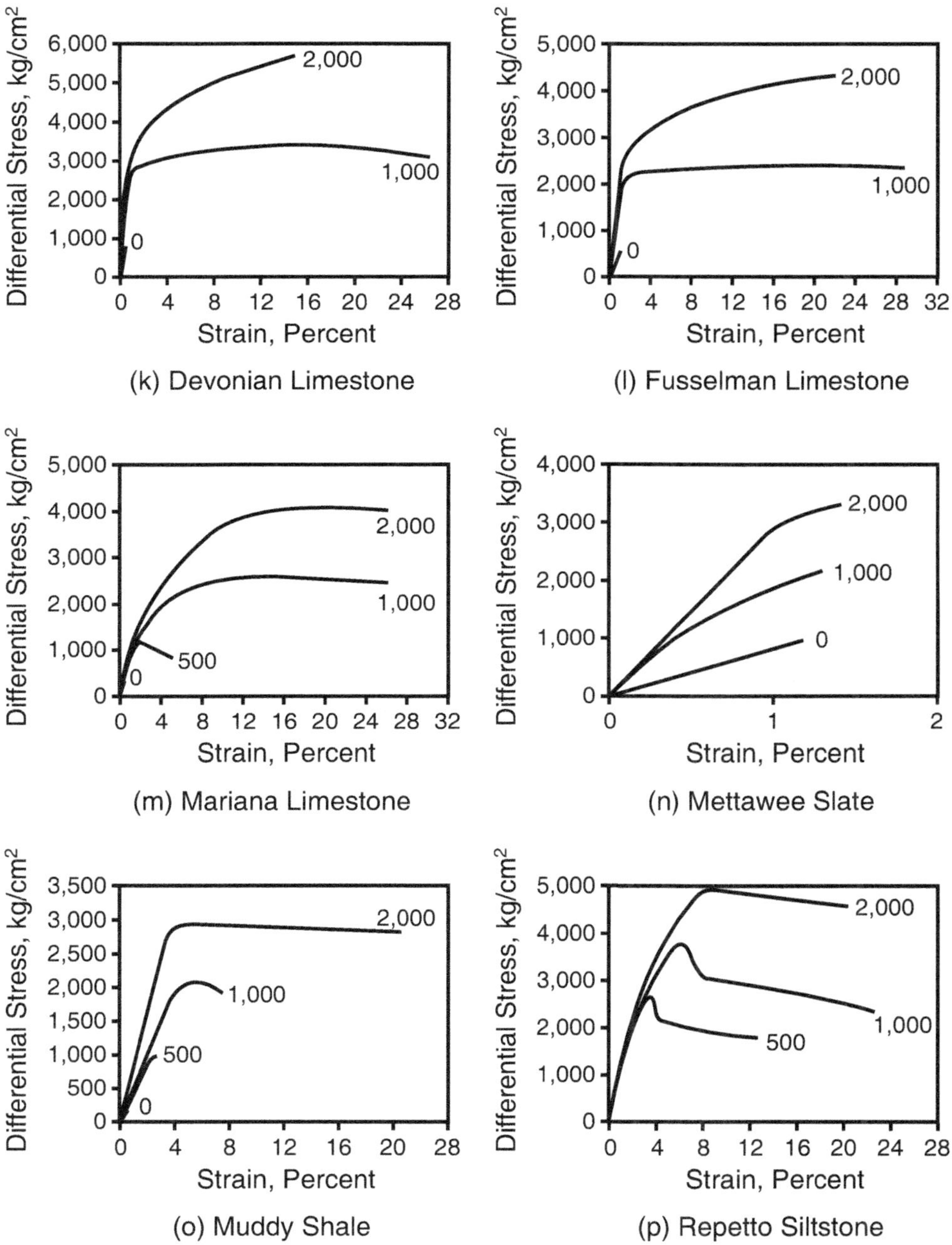

Figure 2.11 (k) Devonian limestone, (l) Fusselman limestone, (m) Mariana limestone, (n) Mettawee slate, (o) muddy shale, (p) Repetto siltstone.

Figure 2.11 (q) Repetto siltstone after testing. Faulting is associated with peak -residual response.

Source: **Robertson, E.C.** (1955) Experimental study of the strength of rocks. *Bulletin Geology Society America*, 66, 1275–1314.
Direct compression of solid cylinders
Confining pressure to 60,000 psi (414 MPa)
Jacketed specimens
Room temperature tests
Limestone, marble, granite, slate, pyrite, dolomite stress-strain plots
Comments: (a) all rocks exhibited elastic linearity, (b) above 15,000 psi (103 MPa) failed by plastic deformation.

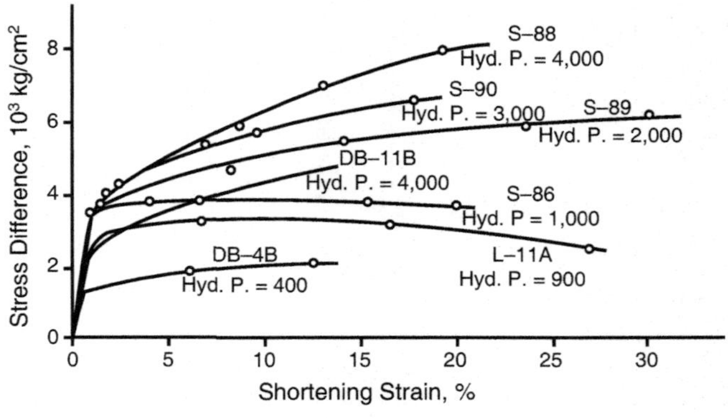

Figure 2.12 Hyd. P.=confining pressure.

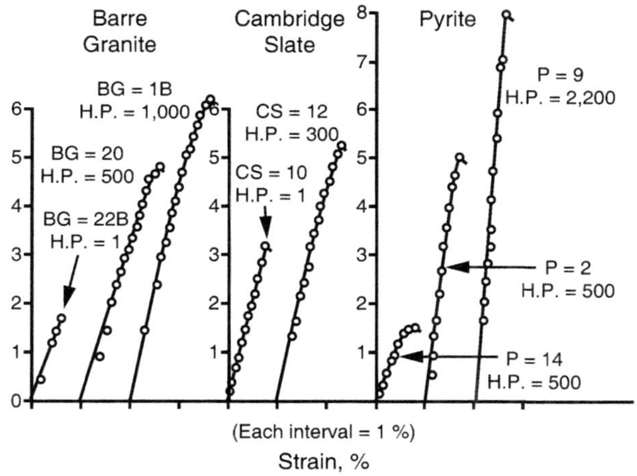

Figure 2.13 Hyd. P.=confining pressure

Source: **Paterson, M.S.** (1958) Experimental deformation and faulting in Wombeyan marble. *Bulletin Geology Society America*, 69, 465–476.
Confining pressures to 15,000 psi (103 MPa)
Direct compression of marble cylinders
Coarse-grained marble.

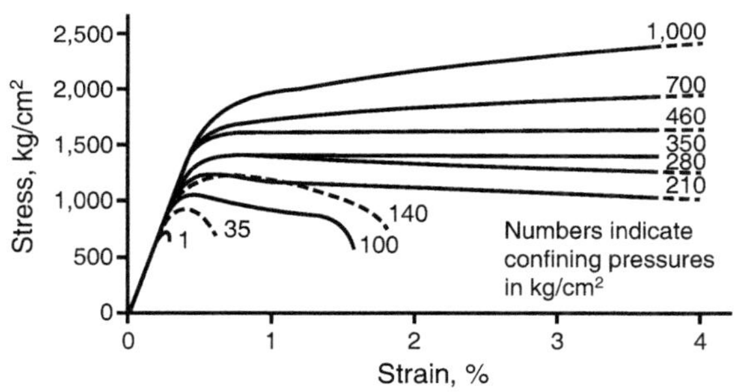

Figure 2.14 Compressive stress-strain plots under confining pressure.

Pore pressure

Handin, J., Hager, R.V. Jr., Friedman, M. & Feather, J. (1963) Experimental deformation of sedimentary rocks under confining pressure: Pore pressure tests. *Bulletin America Associate Petroleum Geology*, 47(5), 717–755.

Direct compression tests
Confining pressure to 29,000 psi (200 MPa)
Sandstone, limestone, dolomite, siltstone, shale
Room temperature
Constant strain rate at 1% per minute

Comments: (a) effective pressure concept verified when applicable (interconnected void space, slow test), (b) at high effective pressure (15,000 psi, 103 MPa), porosity decreases with strain, whereas at low effective pressure (3000 psi, 21 MPa), porosity increases, that is, specimen becomes dilatant.

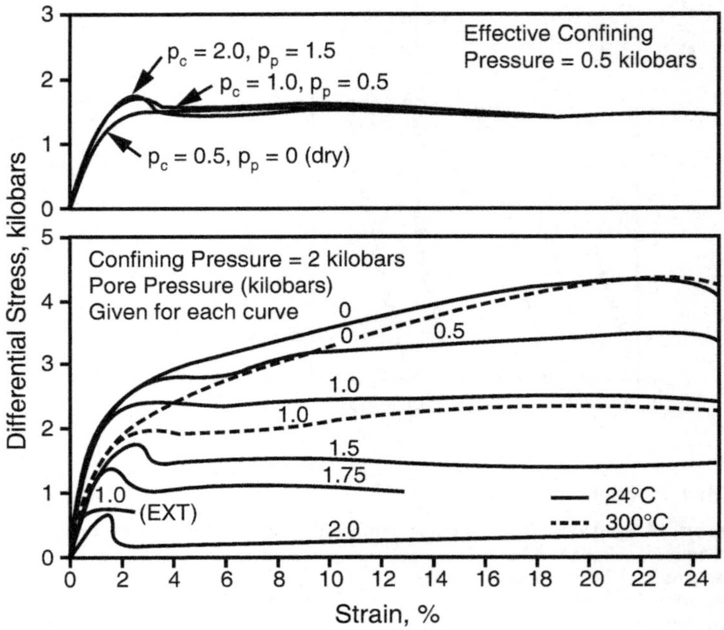

Figure 2.15 Berea sandstone.

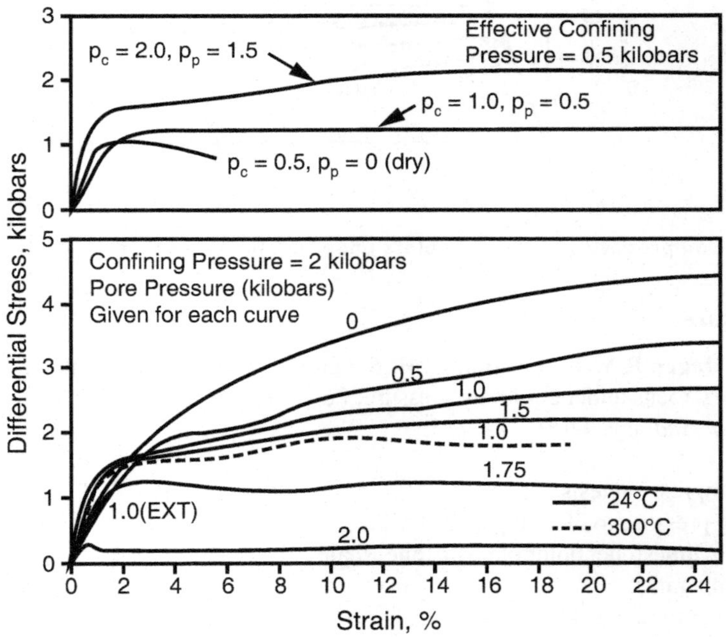

Figure 2.16 Mariana limestone.

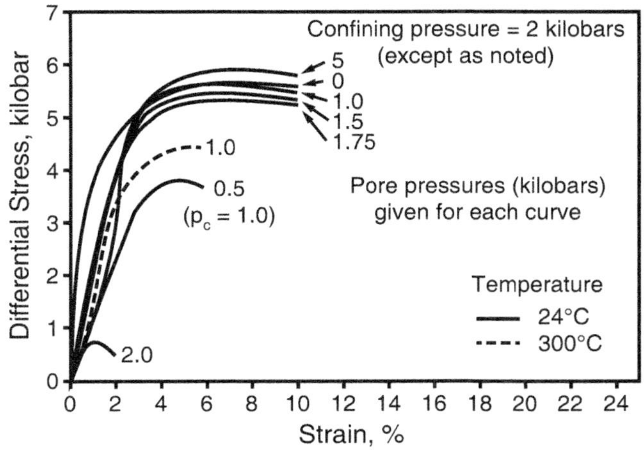

Figure 2.17 Hasmark dolomite.

Source: **Robinson, L.H., Jr.** (1959) The effect of pore and confining pressure on the failure process in sedimentary rock. *Proceedings Third Symposium on Rock Mech.* Colorado School of Mines, Golden. Also: *Colorado School of Mines Quarterly*, 54(3), 177–198.
Direct compression of solid cylinders
Confining pressure to 10,000 psi (69 MPa)
Room temperature tests, constant strain rate at 0.4% per minute
Two limestones, two sandstones, one shale
Comments: (a) effective pressure concept verified, shale impermeable, (b) transition from ductile to brittle behavior as pore fluid pressure approaches confining pressure, (c) dilatancy apparent at low effective pressures.

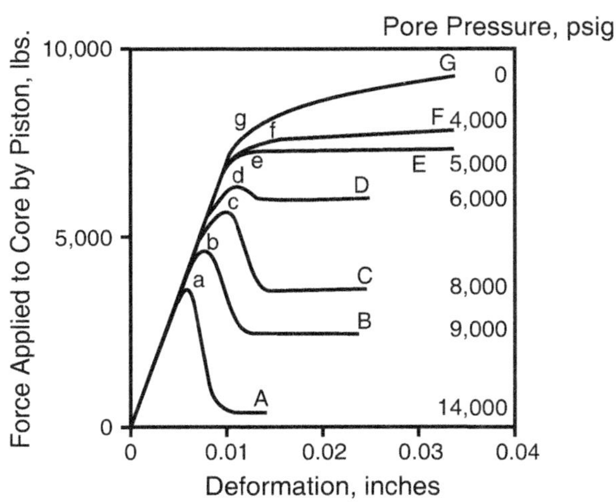

Figure 2.18 Indiana limestone.
Temperature and Strain Rate

Source: **Griggs, D.T., Turner, F.J. & Heard, H.C.** (1960) Deformation of rocks at 500° to 800°C. In: Griggs, D.T. & Handin J. (eds) *Geology Society America Memoir*, Volume79, Waverly Press, Baltimore, pp.59–104.
Direct compression and extension of solid cylinders
Confining pressure at 75,000 psi (517 MPa)
Strain rate 2% to 4% per minute
Temperatures 500 °C to 800 °C (932 °F to 1462 °F)
Rock types: dunite, pyroxenite, basalt, granite, dolomite, marble
Comments: (a) strength decreases with temperature increase as does work-hardening effect, (b) no sudden fracture observed in any rock above 500 °C, (c) cataclasis is important in hard rocks (dunite, pyroxenite, granite), (d) large strain is reflected in fabric geometry.

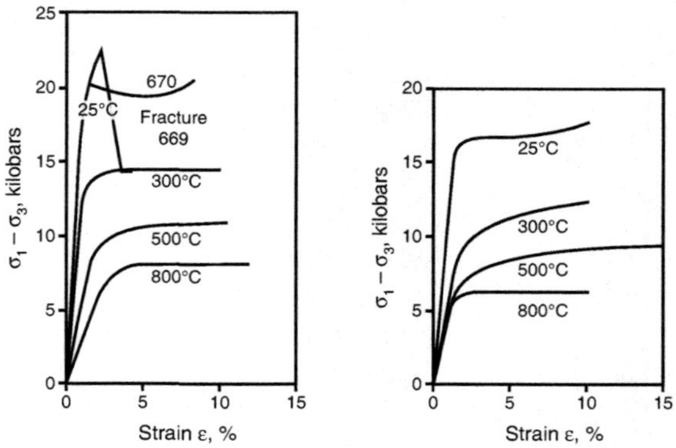

Figure 2.19 Dun Mountain dunite and pyroxenite.

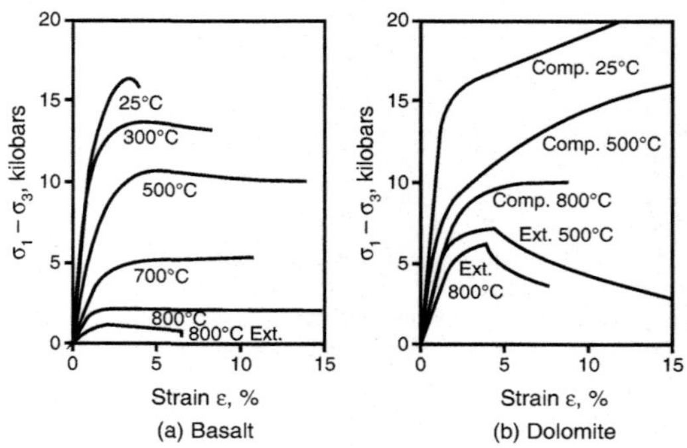

Figure 2.20 (a) Basalt and (b) dolomite.

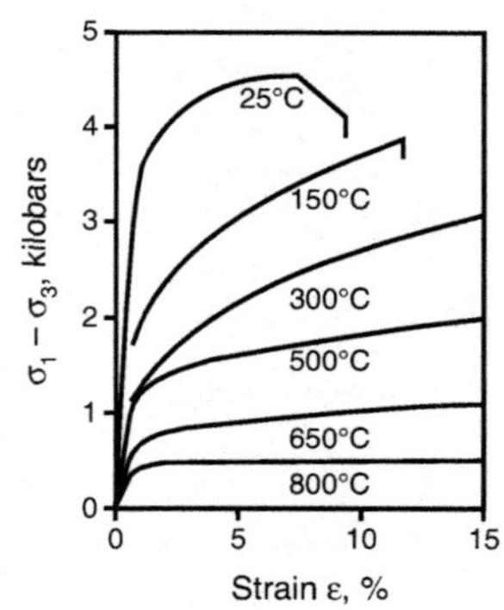

Figure 2.21 Yule marble.

Source: **Serdengecti, S. & Boozer, G.D.** (1961) The effects of strain rate and temperature on the behavior of rocks subjected to triaxial compression. *Proceedings Fourth Symposium on Rock Mechanics.* The Pennsylvania State University, University Park, PA, pp.83–97.
Direct compression of solid cylinders
Confining pressure to 20,000 psi (138 MPa)
Pore pressure to 20,000 psi (138 MPa)
Temperature to 300 °F (149 °C)
Strain rate from 10^{-3} to 100% per second
One sandstone, one limestone, one gabbro tested
Comments: (a) 0.2% offset yield strength used or ultimate strength, (b) concept of effective pressure demonstrate at all strain rates and temperatures for porous rocks, (c) sharp drop in stress-strain curve typical of failure by fracture along a single shear plane, (d) brittle-ductile transition occurs at various combinations of state variables.

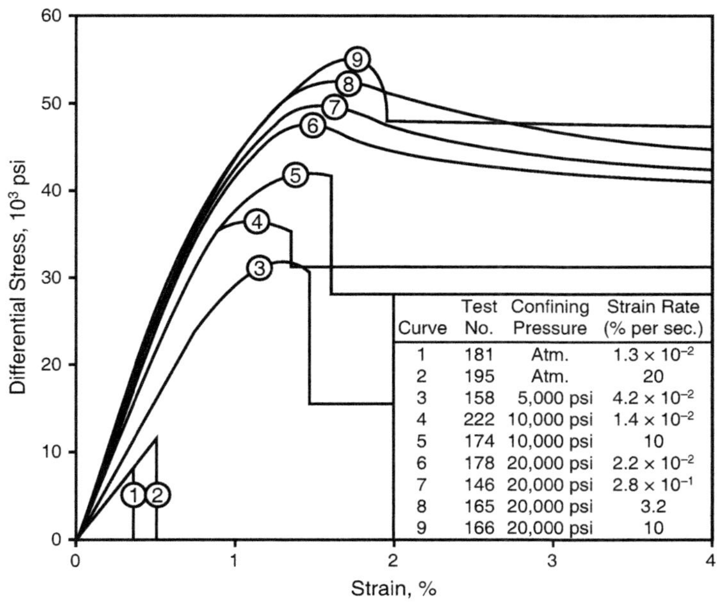

Figure 2.22 Berea sandstone.

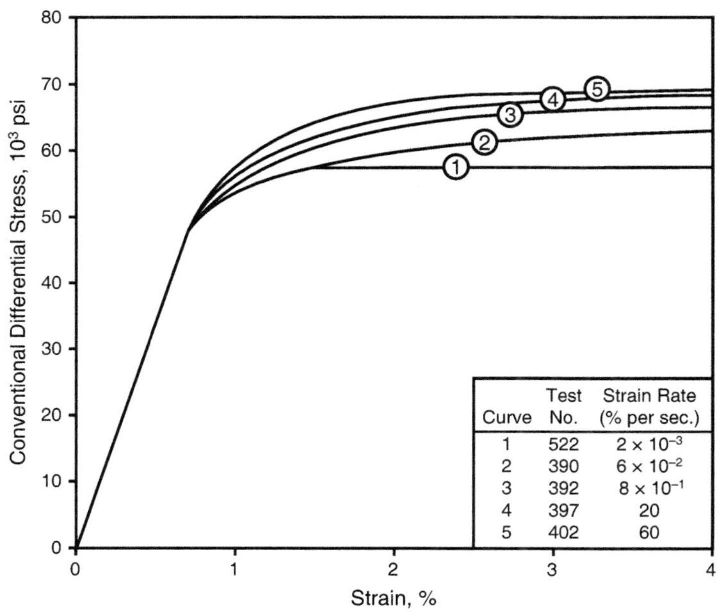

Figure 2.23 Solenhofen limestone.

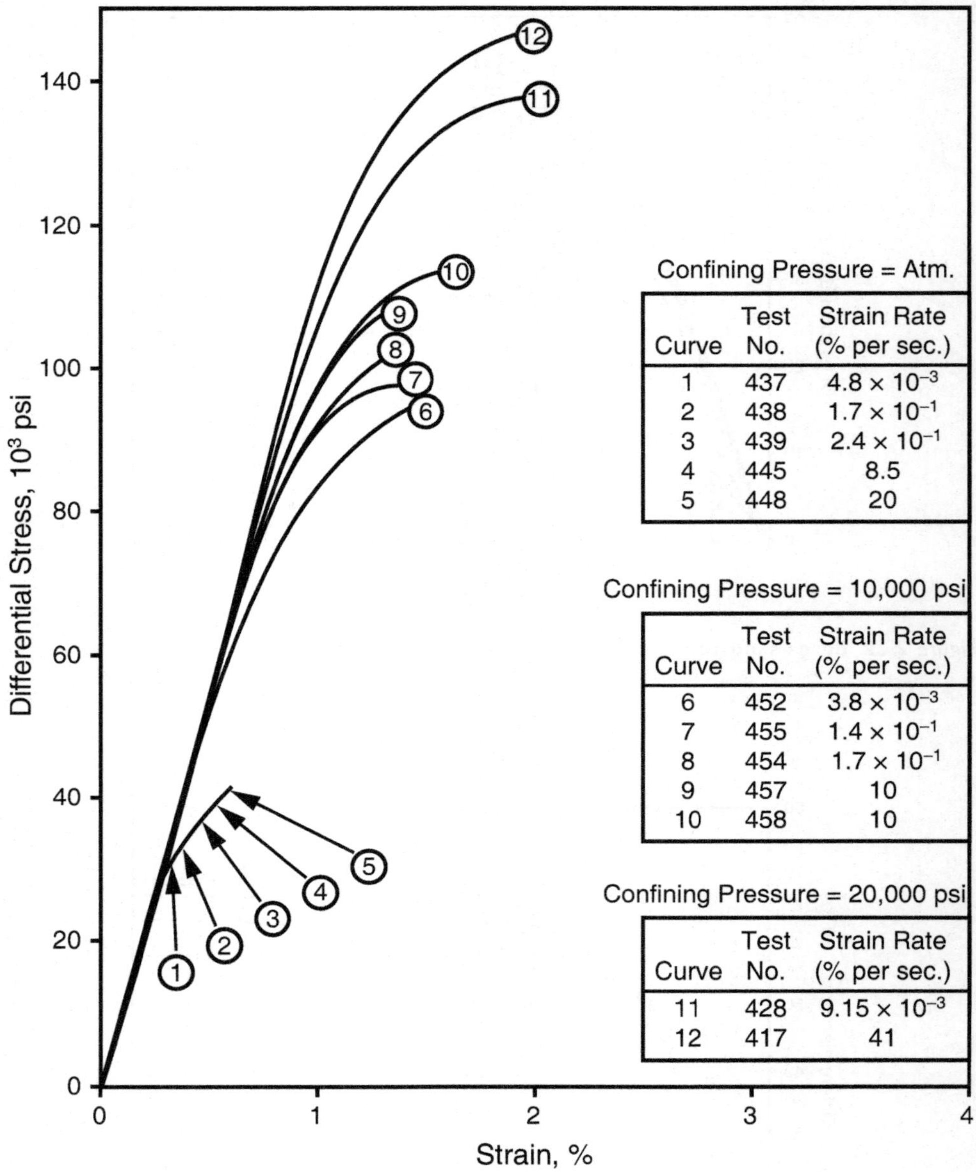

Figure 2.24 Pola gabbro.

Source: **Heard, H.C.** (1962) Effect of large changes in strain rate in experimental deformation of yule marble. *Journal of Geology*, 71(2), 162–196.
Confining pressure to 75,000 psi (517 MPa)
Temperature to 500 °C (932 °F)
Strain rates for 0.4 to 3 × 10^{-8} per second
64 tests on jacketed cylinders; parallel, at 45° and perpendicular to "bedding"
Comments: (a) samples deformed 10% with test durations ranging from 0.25 seconds to 35 days, (b) a decrease in strain rate tends to decrease hardening and is more pronounced at elevated temperature, (c) strength at 10% strain is lowered 25%, 50%, and 75% at 300 °C, 400 °C, and 500 °C when strain rates are lowered for 10^{-1} to 10^{-8} (over a million-fold decrease).

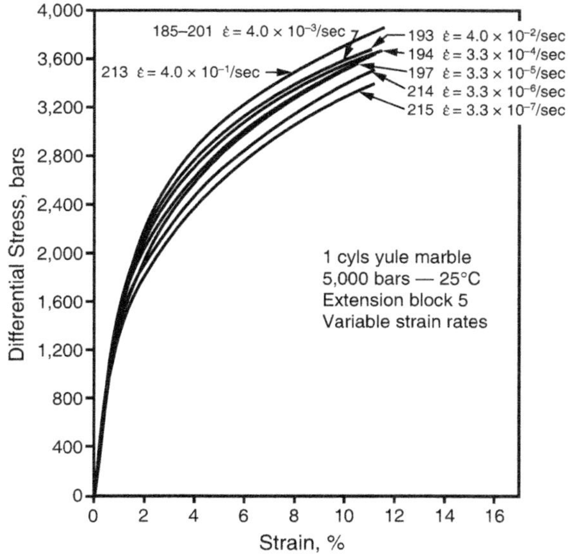

Figure 2.25 Yule marble at 25 °C.

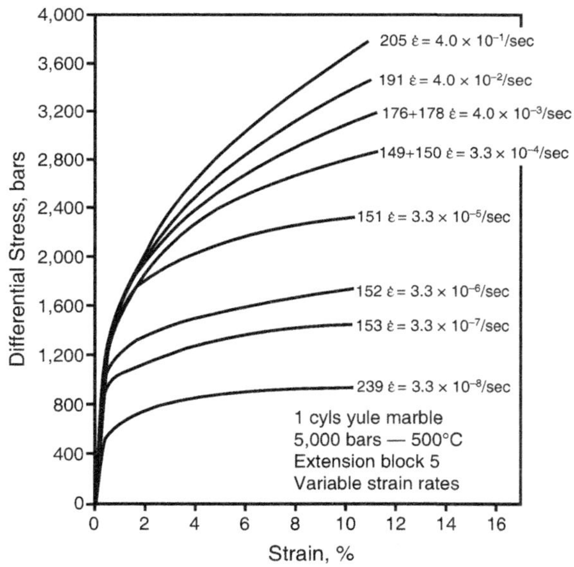

Figure 2.26 Yule marble at 500 °C.

Source: **Heard, H.C.** (1960) Transition from brittle fracture to ductile flow in Solenhofen limestone as a function of temperature, confining pressure and interstitial fluid pressure. In: Griggs, D.T. & Handin, J. (eds) *Geology Society America Memoir*, Volume 79,
Waverly Press, Baltimore, pp. 193–196.
Direct compression and extension of solid cylinders, 115 tests
Confining pressure to 75,000 psi (517 MPa)
Pore fluid pressure to 75,000 psi (517 MPa)
Strain rate constant at 10^{-4} per second
Temperature to 700 °C (1292 °F)
Comments: (a) transition defined as occurring between 3% and 5% strain, (b) transitional behavior typified by failure over a single shear surface.

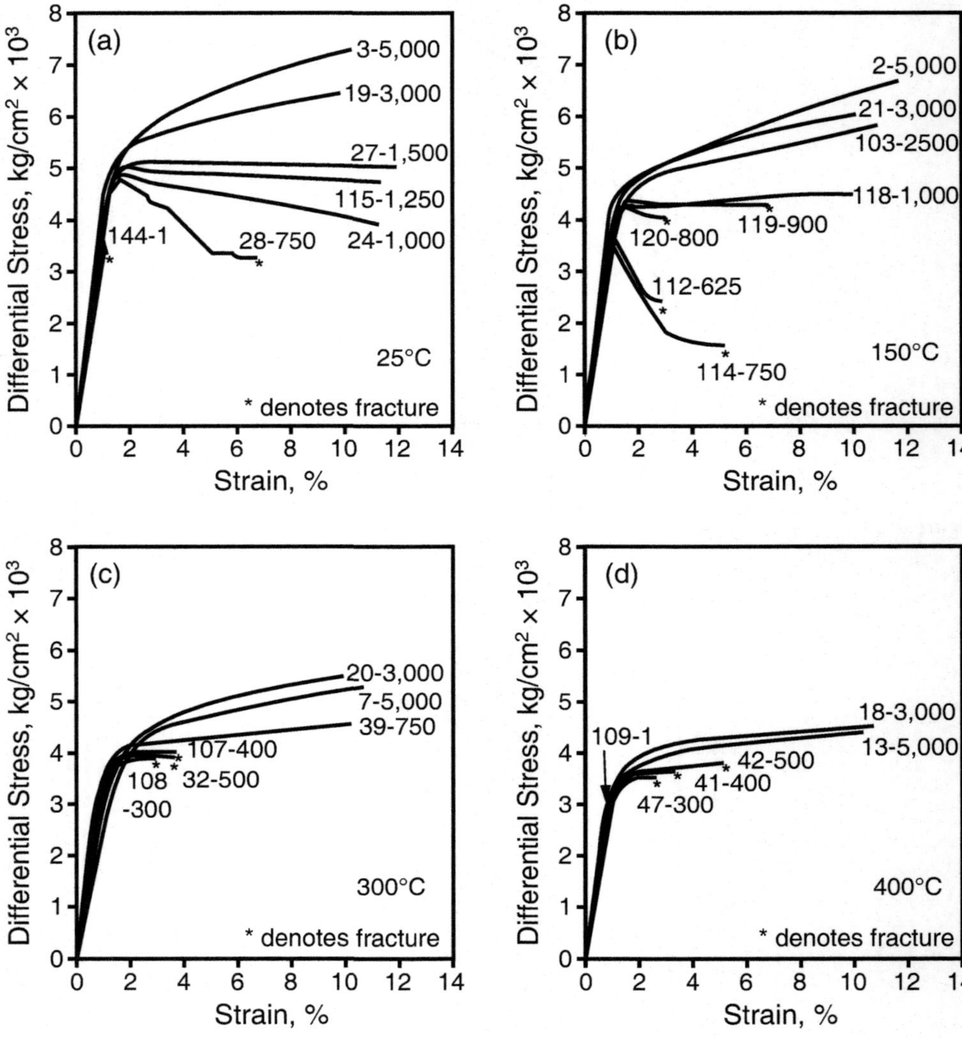

Figure 2.27 Solenhofen limestone: A=25 °C, B=150 °C, C=300 °C, D=400 °C.

Source: **Griggs, D.T. & Handin, J.** (1960) Observations on fracture and a hypothesis of earthquakes. In: Griggs, D.T. & Handin J. (eds) *Geological Society of America Memoir,* Volume 79, Waverly Press, Baltimore, pp.347–364.

Comment: Figure 2.28 is a compact summary of the mode of failure as a function of strain to failure. In this regard, failure by faulting in Cases 3 and 4 is not always accompanied by peak-residual behavior. The gray areas in the stress-strain curves indicate the possibility of faulting without reduction in stress. This event occurs when fault friction is equal to the angle of internal friction (inclination of the Mohr envelope to the normal stress axis) that characterizes strength before faulting. Splitting failures in Cases 1 and 2 are not likely to occur unless care is taken to reduce end friction in a test. Otherwise, end friction provides confinement with faulting likely to occur instead of splitting.

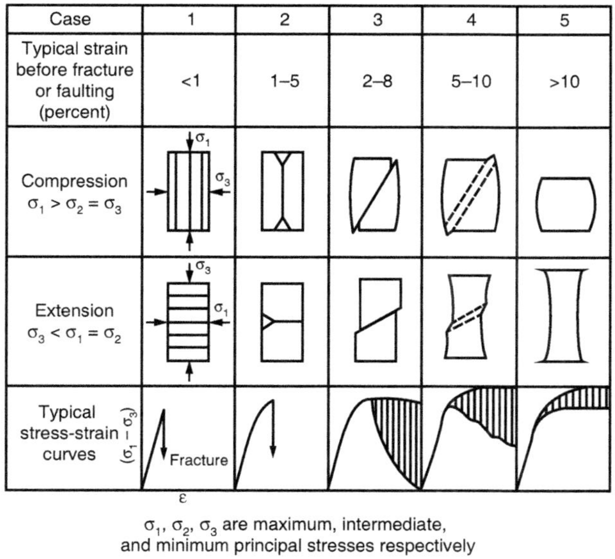

Figure 2.28 "Schematic representation of the spectrum from brittle fracture to ductile flow."

2.2 Rock joints response to load

Rock formations at excavation scales of meters and more contain numerous structural discontinuities such as fractures, cracks, joints, and faults ("joints" for brevity) that are absent at the laboratory scale of a few centimeters. Consequently, rock formations are often composite materials of joints and intact rock between joints. For this reason, joint properties are also essential to characterizing the response of rock formations to load.

Important properties of joints are (1) strengths that characterize the limit to elasticity when a small increase in load tends to produce a large increase in displacement and (2) joint stiffnesses that characterize reversible, elastic displacements perpendicular and parallel to a joint.

Joints are usually tested in direct shear, although triaxial testing is also done. Figure 2.29 illustrates a direct shear test of a rock joint. The shear box is usually circular with sample area A but may also be rectangular. A normal load N is applied perpendicular to the joint plane before application of a shear load T that acts parallel to the joint plane. The normal load is usually kept constant during a test. Displacements Δv and Δu are monitored during a test.

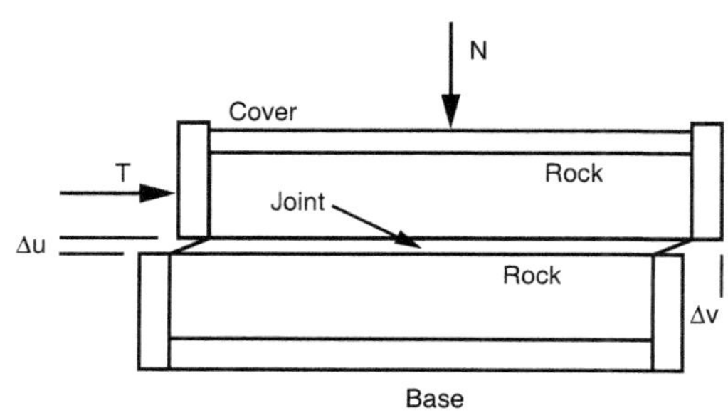

Figure 2.29 Schematic illustration of a direct shear box test of a rock joint.

Figure 2.30 illustrates a concept of a rough joint with asperities of varying height with a "thickness" h. Asperities are important features of joint roughness; in fact, they define joint roughness. Assigning a thickness to joints is well within physical reality, especially when considering filled joints. In case of fresh joints with no filling, asperity details lead to consideration of a thickness defined by the influence of asperities, say, equal to twice asperity height. Certainly, the material within the near vicinity of a joint surface is a different material than the more remote interior volume beyond the nominal mathematical plane that serves to define joint location. This detailed physical view of joints offers a distinct advantage in consideration of numerical models, because joints can then be treated as volumes of a material of a particular type distinct from adjacent intact rock volumes.

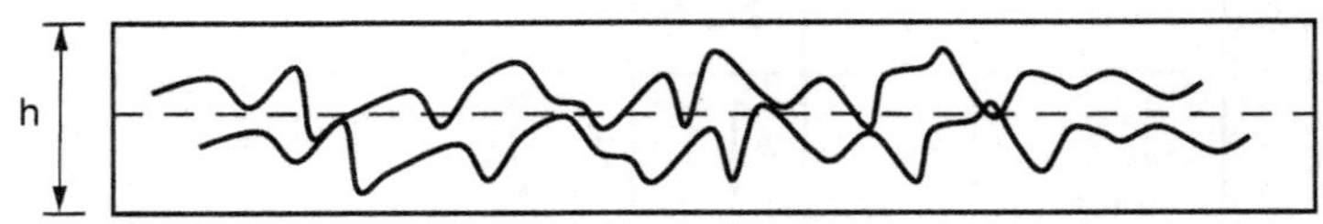

Figure 2.30 Detailed concept of a rough joint with "thickness" h.

Fresh tensile fractures or joints that have not been sheared generally show a fast-rising shear load with shear displacement until a peak shear load is reached. Joint asperities begin to fail following peak loading. Failed material tends to fill valleys between asperity peaks that in turn tends to reduce the shear load required for further shear displacement. The early fast-rising portion of the shear stress–shear displacement plot is often accompanied by joint dilatation as the asperities ride up and over one another. Continued shearing may result in joint closure to a limited extent and eventually to a constant normal displacement condition. Figure 2.31 illustrates a shear stress–shear displacement plot associated with testing of a joint formed by tensile fracture. Peak-residual behavior may also be observed during direct shear testing of a filled joint. However, peak-residual behavior is generally not observed during shear of a previously sheared joint.

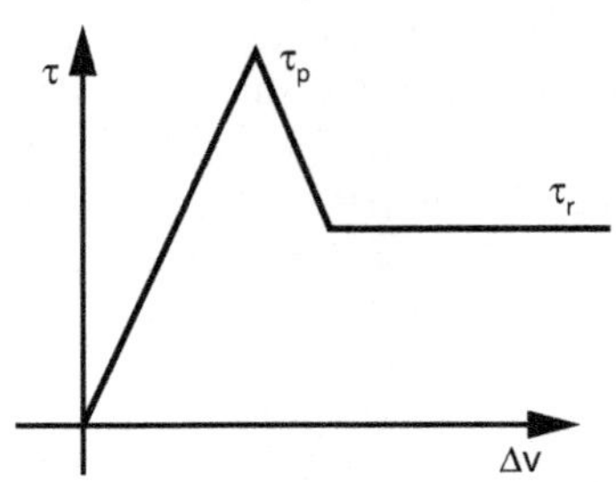

Figure 2.31 Schematic shear stress–shear displacement plot of a tensile fracture (fresh joint). Peak and residual strengths are indicated.

Actual joint test data obtained in triaxial apparatus are shown in Figure 2.32. Linear regression analyses show high correlation and give equations of the form:

$$\tau = \sigma \tan(\phi) + c \tag{2.5}$$

where ϕ and c are friction angle and cohesion, respectively. A coefficient of sliding friction μ is by definition the tangent of the friction angle: $\mu = \tan(\phi)$. The cohesion c is an "apparent" cohesion and is an artifact of a linear fit. Inspection of the figure indicates a curvature near the origin. The curvature in the data would likely be more pronounced at lower confining pressure. An alternative to the linear relationship Equation (2.5) for rock joints is a power law:

$$\tau = \mu_o \sigma^n \tag{2.6}$$

where μ_o and n are determined by experimental measurements.

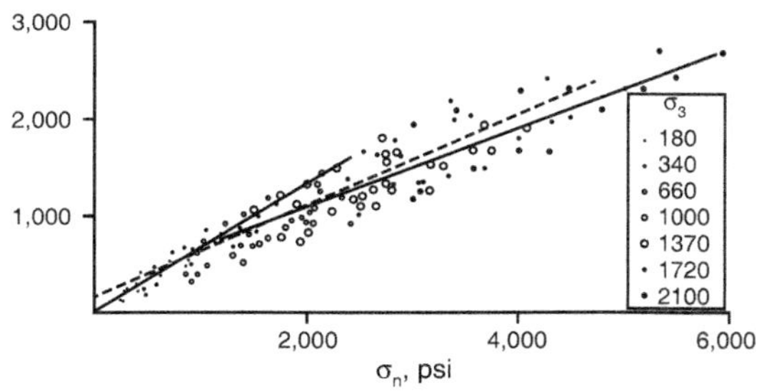

Figure 2.32 Example of joint shear stress as a function of normal stress obtained from triaxial testing (Jaeger, 1970).

Joint stiffnesses relate joint stresses to displacements. Thus:

$$\begin{aligned} \sigma &= k_n \Delta u \\ \tau &= k_s \Delta v \end{aligned} \tag{2.7a,b}$$

where σ, k_n, Δu, τ, k_s, and Δv are normal stress acting across a joint plane, joint normal stiffness, relative displacement normal to the joint plane, joint shear stress, joint shear stiffness and relative displacement parallel to the joint plane. All are determined from experimental test measurements. Convenient units of joint stiffnesses are 10^5 psi/inch (272 MPa/cm) and range over several orders of magnitude, perhaps from 0.01x to $10\text{x}10^5$ psi/inch and higher (2.7x to 2720 MPa/cm). Stiffnesses may interact; a more comprehensive joint stress-displacement statement is then:

$$\begin{Bmatrix} \sigma \\ \tau \end{Bmatrix} = \begin{bmatrix} k_{nn} & k_{ns} \\ k_{sn} & k_{ss} \end{bmatrix} \begin{Bmatrix} \Delta u \\ \Delta v \end{Bmatrix} \tag{2.8}$$

which is often nonlinear with stress or displacement-dependent stiffnesses.

Joint stiffnesses are related to joint elastic moduli. Consider the normal stress – displacement relationship $\sigma = k_n \Delta u$ with displacement expressed in terms of a normal strain ε and a joint thickness h, so that $\sigma = k_n \Delta u = k_n (\varepsilon h) = E\varepsilon$ in a one-dimensional elastic model (E = Young's modulus). In case of shear stiffness, $\tau = k_s \Delta v = k_s (\gamma h) = G\gamma$ where γ is engineering shear strain and G = shear modulus. Thus:

$$E = k_n h, \quad G = k_s h \tag{2.9}$$

which relates elastic moduli to joint stiffnesses.

2.3 Jointed rock response to load

Joints in rock formations are almost always spaced on a meter scale, and they have major effects on the response of rock to excavation-induced loads. Certainly, the response of intact rock between joints is readily determined in laboratory testing at a centimeter scale. But joint spacing precludes testing of composites of intact rock and joints under controlled laboratory conditions. The necessary apparatus would be too large as a practical matter. There are exceptions in case of very densely jointed rock and rock that has been fractured prior to laboratory experiments.

An oft-cited exception is the work of Jaeger (1970) involving 6-inch (15-cm) diameter drill core obtained by careful core drilling to avoid disturbing samples taken for laboratory experimental testing. Figure 2.33 shows several cylinders of Panguna andesite that were subsequently tested in axial compression. Stress-strain curves are shown in Figure 2.34 with confining pressure as a parameter. The stress-strain curves are indicative of elastic ideally plastic behavior at all confining pressures, 100 to 5000 psi, (0.7 to 28 MPa). Although the test results are at a centimeter scale, a reasonable supposition is that the same behavior would be observed at meter-scale joint spacing in meter-size volumes of jointed rock.

Figure 2.33 Photographs of 6-inch densely jointed rock cylinders (Jaeger, 1970).

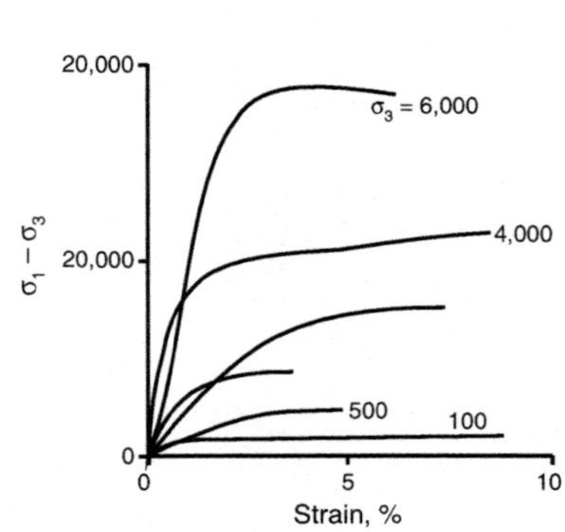

Figure 2.34 Stress difference as a function of strain with confining pressure as a parameter (Jaeger, 1970).

Pre-fractured rock cylinders are test pieces that are loaded axially under strong lateral restraint. The axial load is raised until fracturing of the test cylinder occurs at perhaps two to three times unconfined compressive strength. The fractured test specimen is then tested triaxially without removal of a jacket attached to the specimen before pre-fracturing. Pre-fractures occur parallel, perpendicular, and at angle to the axial load. Fragments are not allowed freedom of movement as individual blocks. Figures 2.35 and 2.36 show stress difference versus strain results from Swanson and Brown (1971) for Cedar City tonalite and Westerly granite. Stress difference is the difference between axial stress and confining pressure. With the exception of Westerly granite at the highest confining pressure of 100 ksi (MPa) where stick-slip oscillation is evident, the results indicate elastic-plastic behavior with negligible work hardening. Figure 2.37 shows two cycles of loading and unloading of Westerly granite that indicates reproducibility of the first cycle stress-strain response.

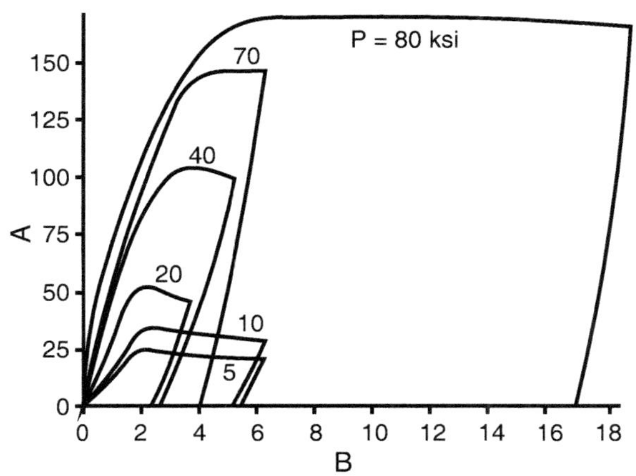

Figure 2.35 Axial stress difference as a function of strain for pre-fractured Cedar City tonalite with confining pressure as a parameter (Swanson and Brown, 1971).

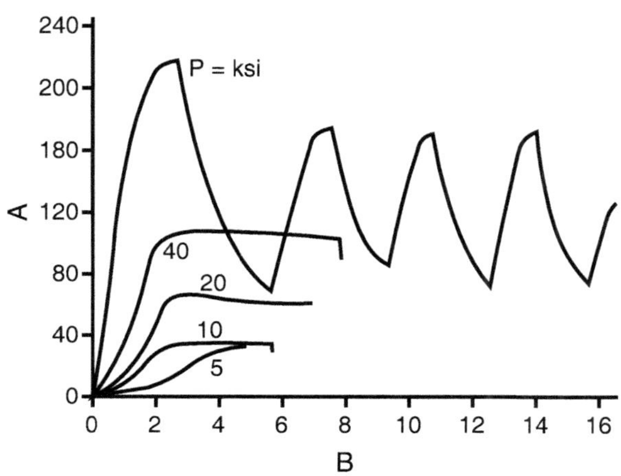

Figure 2.36 Axial stress difference as a function of strain for pre-fractured Westerly granite with confining pressure as a parameter (Swanson and Brown, 1971).

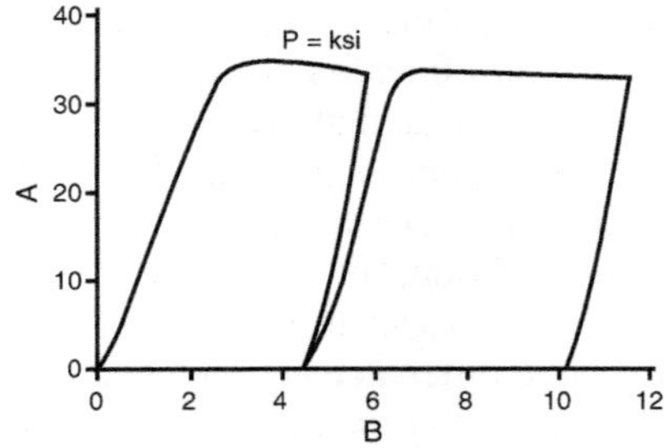

Figure 2.37 Two cycles of loading-unloading of pre-fractured Westerly granite (Swanson and Brown, 1971).

2.4 Equivalent jointed rock models

Although the few data available describing jointed rock response to load indicate an elastic-plastic formulation is reasonable, a reliable model based on first principles is needed in the general case. One well-known model proposed early in the development of the finite element method of analysis (Duncan and Goodman, 1968) is a model based on addition of displacements under equilibrium conditions. The concept is illustrated in Figure 2.38, which shows a combination of rock and a single joint under a load applied normal to the joint.

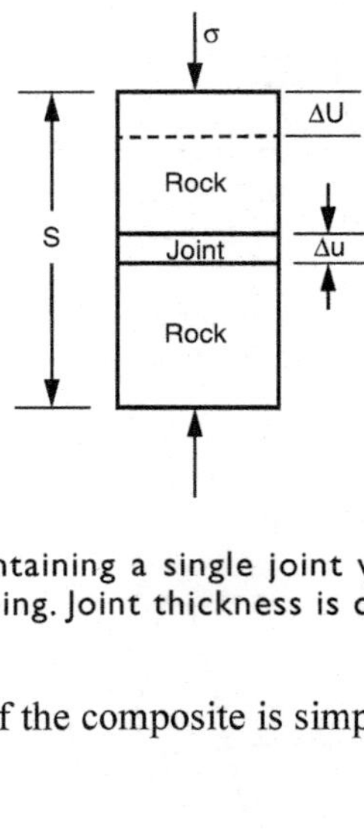

Figure 2.38 A rock test piece containing a single joint with load applied perpendicular to the joint. S=joint spacing. Joint thickness is considered negligible.

Total displacement of the top of the composite is simply the sum $U_t = \Delta U + \Delta u$. The corresponding strain:

$$\begin{aligned} \varepsilon_t &= U_t / S \\ &= (\Delta U / S) + \Delta u / S \\ \varepsilon_t &= \varepsilon_r + \Delta u / S \end{aligned} \tag{2.10}$$

Equilibrium requires the normal load acting on the joint to be equal to the normal load acting on the rock. In consideration of Hooke's law in one dimension:

$$\varepsilon_t = \sigma / E_r + \sigma / k_n S \tag{2.11}$$

where E_r = Young's modulus for intact rock. An equivalent modulus E_e follows:

$$\varepsilon_t / \sigma = 1/E_e = 1/E_r + 1/k_n S \tag{2.12}$$

Thus, Young's moduli for joints in series add as electrical resistances do in parallel.

An alternative formulation may be derived in consideration of loading applied parallel to a joint as illustrated in Figure 2.39. In this case, the displacement of the rock and joint are equal. However, the stresses are not. The total applied load is the sum of loads applied to the rock and joint. Thus:

$$\begin{aligned} F_t &= F_r + F_j \\ \sigma A_r &= \sigma_r A_r + \sigma_j A_j \\ E_e \varepsilon_t &= E_r \varepsilon_r A_r / A + E_j \varepsilon_j A_j / A \end{aligned} \tag{2.13a}$$

The strains are equal as the figure indicates. Thus, in this model:

$$E_e = E_r f_r + E_j f_j \tag{2.14}$$

where f_r and f_j are volume fractions of rock and joint, respectively. This model is *compatible* because of the continuity of displacements. In this model, Young's moduli for joints in series add as electrical resistances do in parallel.

Volume fractions also appear in the equilibrium model when recognition of joint thickness is made. In this case, Equation (2.13a) takes the form:

$$1/E_e = (1/E_r)f_r + (1/E_j)f_j \tag{2.13b}$$

In both Equation (2.13b) and Equation (2.14), the joint modulus is given in terms of joint stiffness, $E_j = k_k/h$ where the stiffness k_j is normal or shear stiffness as the case may be. A similar derivation gives similar results for equivalent shear moduli. Interestingly, in a rigorous analysis, the equilibrium and compatible models are lower and upper bounds to equivalent moduli.

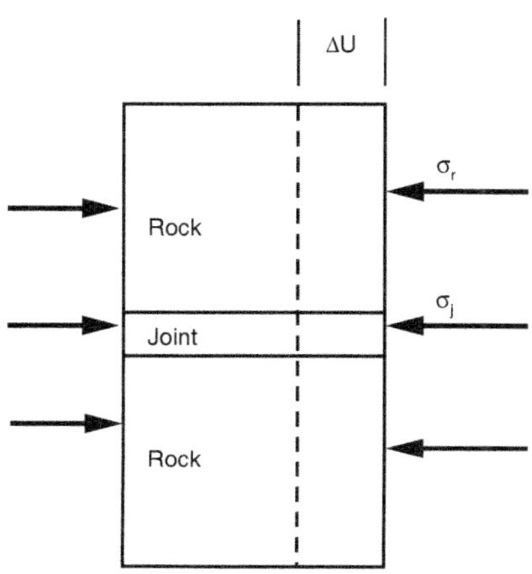

Figure 2.39 A rock test piece containing a single joint with load applied parallel to the joint. S=joint spacing. Joint thickness is considered negligible.

In case of an inclined joint, the requirements of equilibrium and compatibility, essentially requirements for continuity of tractions and displacements across a joint, are less easily imposed in a rigorous analysis. In case of several joints with different material properties and inclinations, the analysis becomes even more complex, but can still be done (Pariseau, 1999).

Jointed rock models have also been proposed in the context of *equivalent* properties. An extensive literature concerning composite manufactured material abounds in equivalent properties models that are based on a concept of a representative volume element (RVE) or a representative elementary volume (REV). However, an element of sufficient size to meet RVE (REV) requirements cannot always be obtained. In this case, equivalent properties may still be defined by homogenizing a given heterogeneous volume. The concept is illustrated in Figure 2.40. Equivalent properties are shown in Figure 2.41. Also shown in the figure are the responses of the joints, intact rock between joints, and equivalent properties shear stress-strain plots. Some joints are yielding in the figure as evidenced in nonlinearity. Agreement between detailed modeling and equivalent properties (NRVE and EQ) is excellent. Details are discussed in the chapter concerning finite element modeling.

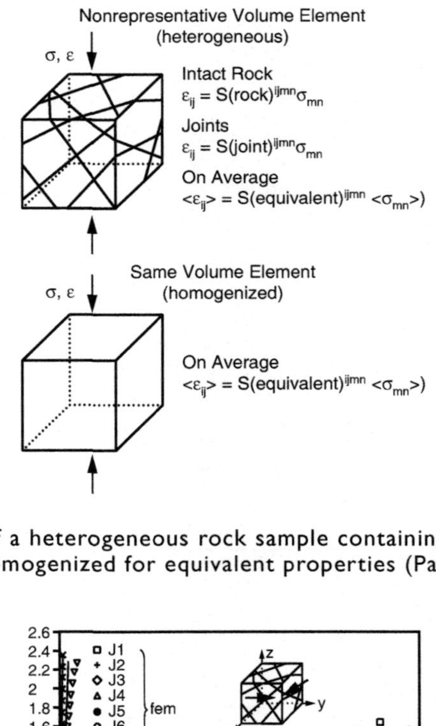

Figure 2.40 Schematic of a heterogeneous rock sample containing seven joints that is subsequently homogenized for equivalent properties (Pariseau, 1999).

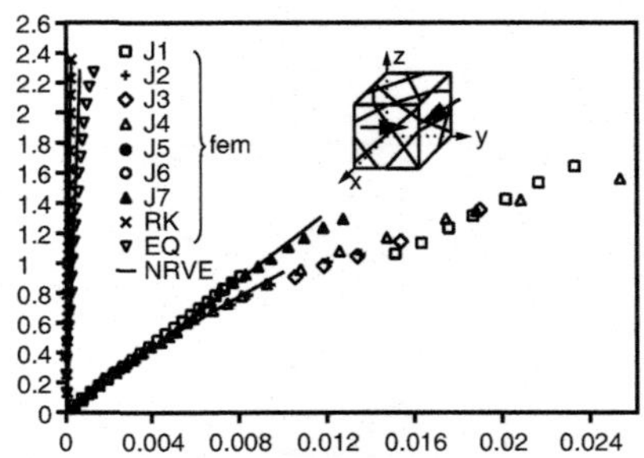

Figure 2.41 Stress-strain plots for each joint (J1–J7), intact rock (RK), equivalent shear properties (EQ), and model results (NRVE=nonrepresentative volume element) under shear loading (Pariseau, 1999).

2.5 Soil response to load

The response of soil to load is largely determined by two factors: (1) grain size and (2) moisture content. Grain size allows for classification of soil as sand, silt, or clay. Figure 2.42 is one such classification scheme. Intergranular forces in sand are predominantly gravitational and forces from external loads such as those from a bearing plate. Clay-size particle experience electrostatic surface forces as well as gravitational forces and forces from external loads. Interparticle silt forces are somewhat of a mix of forces. Moisture content brings pore water pressures into consideration. Pore water pressure may be static or associated with seepage. Seepage or flow causes a viscous drag on the solid particles and thus produces forces in addition to gravity. Surface tension of the pore fluid, mainly water, causes a capillary rise and a negative pore pressure (compression positive).

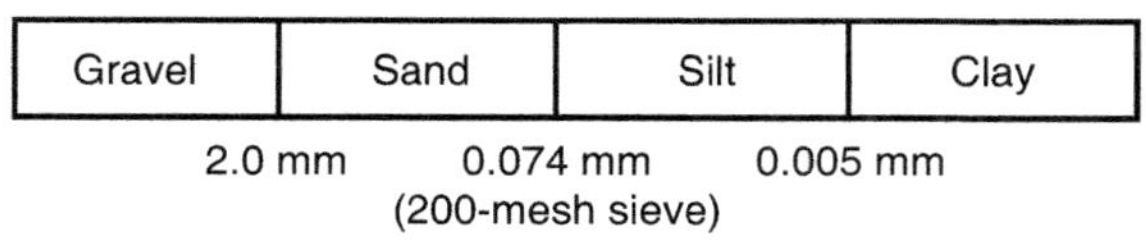

Figure 2.42 A particle size soil classification scheme. AASHO = American Association of State Highway Officials (Hough, 1957).

Laboratory testing of soil may be done in a conventional triaxial test or in direct shear. Both are complicated by consideration of moisture content. A partially saturated compression test may lead to saturation during the test, and testing a saturated soil may lead to out-of-equilibrium conditions in consideration of the distribution of load between pore fluid and solid grains. If free drainage of the sample is allowed, then a different stress-strain response occurs than if drainage is not allowed. Whether the sample acquired in the field is tested in an undisturbed state or is remolded during test preparation is also an important consideration. The reason relates to the stress state *in situ* where the sample was acquired. Density of a sample also affects soil test results. Figure 2.43 illustrates typical soil stress-strain data.

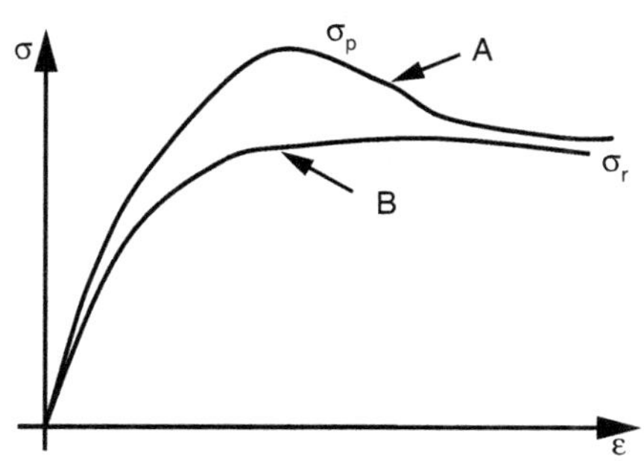

Figure 2.43 Typical stress-strain soil test results showing peak and residual strengths. A = initially dense, B = initially loose.

From the mechanical viewpoint, the difference between soil and intact rock is more one of degree than of kind. Both are aggregates of solid particles enclosing voids of varying shape and size. Both share physical laws and kinematics, of course. But also, both share material laws. Hooke's law finds applications in soil and rock mechanics, as does plasticity theory. The main difference is quantitative. Soil strengths are typically several orders of magnitude smaller than intact rock strength. Soil cohesion is usually measured in kPa (lbf/ft^2); rock cohesion is usually measured in MPa (lbf/in.^{2}). Elastic moduli of soils are also much smaller than moduli of intact rock. Elastic behavior of soil also tends to be nonlinear, so determination, say, of Young's modulus is more difficult and often requires distinguishing between secant and tangent moduli. Overall, the effects of confining pressure, pore pressure, and strain rate on soil are similar to those on intact rock, rock joints, and jointed rock formations.

References

Duncan, J.M. & Goodman, R.E. (1968) *Finite Element Analyses of Slopes in Jointed Rock.* U. S. Army Engineer Waterways Experiment Station Contract Report S-68–3, U. of California, Berkeley, Report No. TE-68–1.

Griggs, D.T. & Handin, J. (1960) Observations on fracture and a hypothesis of earthquakes. In: Griggs, D.T. & Handin J. (eds) *Geological Society of America Memoir,* Volume 79, Waverly Press, Baltimore, pp.347–364.

Griggs, D.T., Turner, F.J. & Heard, H.C. (1960) Deformation of rocks at 500° to 800 °C. In: Griggs, D.T. & Handin, J. (eds) *Geological Society of America Memoir*, Volume 79, Waverly Press, Baltimore, pp.59–104.

Handin, J. & Hager, R.V., Jr. (1957) Experimental deformation of sedimentary rocks under confining pressure: Tests at room temperature on dry samples. *Bulletin America Associate Petroleum Geology*, 41(1), 1–50.

Handin, J., Hager, R.V. Jr., Friedman, M. & Feather, J. (1963) Experimental deformation of sedimentary rocks under confining pressure: Pore pressure tests. *Bulletin America Associate Petroleum Geology*, 47(5), 717–755.

Heard, H.C. (1960) Transition from brittle fracture to ductile flow in Solenhofen limestone as a function of temperature, confining pressure and interstitial fluid pressure. In: Griggs, D.T. & Handin, J. (eds) *Geological Society of America Memoir*, Volume 79, Waverly Press, Baltimore, pp.193–1960.

Heard, H.C. (1962) Effect of large changes in strain rate in experimental deformation of yule marble. *Journal of Geology*, 71(2), 162–196

Hough, B.K. (1957) *Basic Soils Engineering*. Ronald Press, New York.

Jaeger, J.C. (1970). Behavior of closely jointed Rock. *Proceedings 11th Symposium on Rock Mechanics*. AIME/SME, New York, NY, USA, pp.57–68.

Paterson, M.S. (1958) Experimental deformation and faulting in Wombeyan marble. *Bulletin Geology Society America*, 69, 465–476.

Pariseau, W.G. (1999) An equivalent plasticity theory for jointed Rock masses. *International Journal Rock Mechanics & Mining Science*, 36(7), 907–918.

Robertson, E.C. (1955) Experimental study of the strength of rocks. *Bulletin Geology Society America*, 66, 1275–1314.

Robinson, L.H., Jr. (1959) The effect of pore and confining pressure on the failure process in sedimentary rock. *Proceedings Third Symposium on Rock Mechanics*. Colorado School of Mines, Golden. Also: *Colorado School of Mines Quarterly*, 54(3), 177–198.

Schwartz, A.E. (1964) Failure of rock in triaxial shear. *Proceedings Sixth Symposium on Rock Mechanics*. University of Missouri, Rolla, MO, pp109–151.

Swanson, S.R. & Brown, W.S. (1971) The mechanical response of pre-fractured Rock in compression. *Rock Mechanics*, 3(4), 208–216.

Chapter 3

Elements of three-dimensional theory

The general objective of an analysis in the mechanics of solids is computation of material motion in response to load that may be a combination of applied forces and specified displacements. A continuum theory is needed for this purpose. Elements of any continuum theory follow from considerations of physical laws, kinematics, and material laws.

Physical laws are usually conservation or balance laws such as conservation of mass and balances of linear and angular momentum. The momentum balances lead to an analysis of stress. Other conservation laws or balances are certainly possible, for example, a balance of energy. Another example is conservation of charge. Such laws are derived from even more fundamental requirements expressed as invariance principles.

Kinematics refers to the geometry of motion. Motion of a material element may be considered as rigid body translation and rotation accompanied by volume and shape changes. Deformation refers to volume and shape change only and leads to an analysis of strain. Physical laws and kinematics apply to all materials regardless of constitution.

However, equations that follow from physical laws and kinematics are less than the number of unknowns and therefore generally are not adequate for the computation of material motion. Additional equations need to be introduced without introduction of additional unknowns. In this regard, the independent variables describing motion are position and time (x, y, z, t); other variables such as stress and strain are dependent variables. Thus, the additional equations needed must be relations among the dependent variables; these are constitutive equations. Constitutive equations are also referred to as material laws or simply as stress-strain laws. Hooke's law in linear elasticity is a well-known example of a constitutive equation and is the simplest possible, nontrivial constitutive equation in mechanics of solids. Ohm's law in electricity and Darcy's law in groundwater analyses are other examples of constitutive equations.

3.1 Stress review

Consider the internal surface of a material body shown in Figure 3.1. The volume V of the body is bounded by a surface S. External loads are applied, and the body reacts to equilibrate the external loads by internal forces. A small force ΔF acts on a small internal surface element ΔA. The limit:

$$\lim_{\Delta A \to 0} \frac{\Delta F}{\Delta A} = T(x, y, z, n) \tag{3.1}$$

is a *stress vector* that depends not only on the position but also on the orientation of the area element at a given point. Here n represents an outward normal vector of unit length. When the considered point is at the surface of the body of interest, T is referred to as a *surface traction*. A state of stress is defined as the set of stress vectors acting at a point. There are an infinite number of orientations at a given point, so this definition is not particularly helpful.

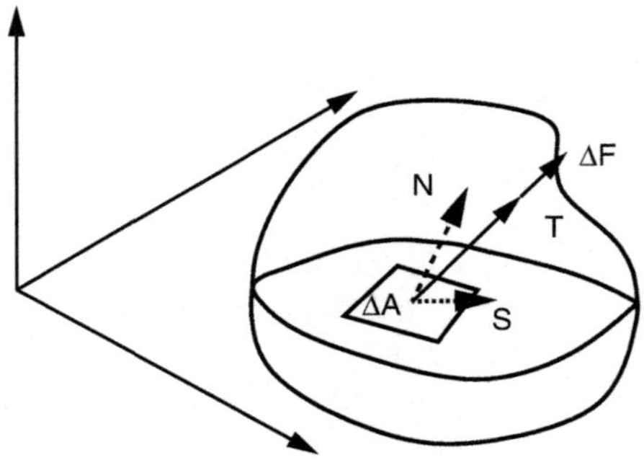

Figure 3.1 Stress vector geometry.

Coordinate surfaces, for example, in rectangular coordinates, planes *yz* with outward unit normal parallel to the x-axis, are special. Indeed, components of stress vectors, acting on coordinate planes, are stresses. The components T_x^x, T_y^x, T_z^x acting on the *yz*-plane are the stresses σ_{xx}, τ_{xy}, τ_{xz} where σ and τ mean *normal* and *shear* stress, respectively, and similarly for the other two coordinate planes. The superscript-subscript notation on the stress vector components refers to the directions of the outward normal and direction of action, respectively. In the double subscript notation, the first subscript signifies the normal direction while the second subscript signifies the direction of action. Each coordinate plane has three stress vector components, so there are nine stresses (three normal stresses, six shear stresses).

A sign convention is associated with the double subscript notation. When tension is positive, then both subscripts act in the same directional sense. Shear stress is positive according to the same convention. To make compression positive, as is often done in geomechanics, the subscripts act in opposite directions, and again, the same convention

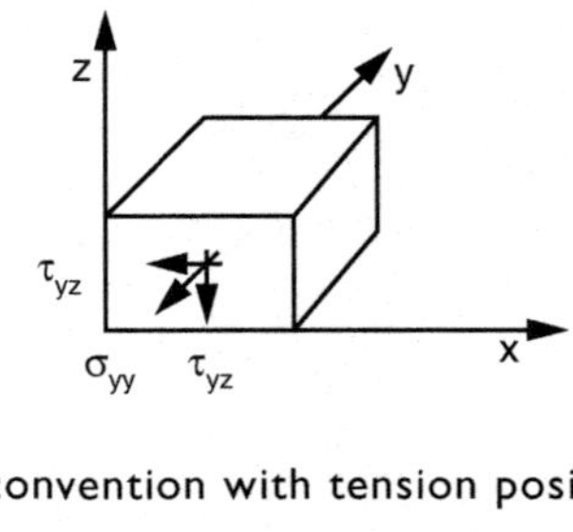

Figure 3.2 Illustration of sign convention with tension positive.

applies to shear stress. Figure 3.2 shows an example of the sign convention that makes tension positive.

A compact and quite useful notation to express the concepts of continuum mechanics is a subscript notation and summation convention that replaces coordinates x,y,z with numbers 1,2,3, so that stresses are conveniently expressed as σ_{ij} (i,j = 1,2,3). A repeated subscript implies summation and a comma denotes differentiation. For example:

$\sigma_{ii} = \sigma_{11} + \sigma_{22} + \sigma_{33} = \sigma_{xx} + \sigma_{yy} + \sigma_{zz}$ and $v_{1},_{2} = \dfrac{\partial v_1}{\partial x_2} = \dfrac{\partial v_x}{\partial y}$. Another example is:

$$\begin{aligned}\sigma &= (1/3)\sigma_{ij}\delta_{ij} \\ &= (1/3)(\sigma_{11}\delta_{11} + \sigma_{12}\delta_{21} + \sigma_{13}\delta_{13} + \sigma_{21}\delta_{21} + \sigma_{22}\delta_{22} + \sigma_{23}\delta_{23} + \sigma_{31}\delta_{31} + \sigma_{32}\delta_{32} + \sigma_{33}\delta_{33}) \\ \sigma &= (1/3)(\sigma_{11} + \sigma_{22} + \sigma_{22})\end{aligned}$$

where δ_{ij} is the Kronecker delta that has a value of 1 when $i = j$ and 0 otherwise.

Two types of external forces act on a free body: *contact* forces distributed over the surface of the body that represent the effect of the material removed on the material remaining, and *body* forces acting throughout the volume of the free body. External forces are those with origins outside the body. The force of gravity and therefore weight is an external body force. The resultant F of external forces acting on a body is:

$$F = \int_S TdS + \int_V \rho g dV = \int_S T_i dS + \int_V T_i \gamma_i dV \tag{3.2}$$

where ρ is mass density, g is the acceleration of gravity, and γ is weight density or specific weight ($\gamma = \rho g$). There are three equations in Equation (3.2).

Linear momentum P of the body is:

$$P = \int_B vdm = \int_V \rho v dV = \int_V \rho v_i dV \tag{3.3}$$

where m is mass, v is velocity, and B is the material body that occupies volume V. Mass density is continuous in the sense that $\rho = dm/dV$. The time rate of change of linear momentum is:

$$\dot{P} = \frac{D}{Dt}(P) = \int_B (\frac{D}{Dt}v)dm = \int_B adm = \int_V \rho a dV = \int_V \rho a_i dV \tag{3.4}$$

where a is acceleration and a tacit assumption is made that mass is conserved. Here, the derivative:

$$\frac{D}{Dt} = \frac{\partial}{\partial t} + (v_x \frac{\partial}{\partial x} + v_y \frac{\partial}{\partial y} v_z \frac{\partial}{\partial z}) \tag{3.5}$$

is a *material* derivative or a derivative "following the particle." The first part is a partial derivative with respect to time at a fixed point; the second part is a "convective" part attributed to particle (mass point) motion. Acceleration is the material derivative of velocity. The material derivative commutes with material integration over B but does not commute with spatial integration over V. When quantities are small, the convective part is negligible and one has:

$$\frac{D}{Dt} = \frac{\partial}{\partial t} = \frac{d}{dt} \tag{3.6}$$

where the last notation is a convenient shorthand but needs to be used with caution.

Conservation of mass simply states that the mass M of the body B is constant. Thus, $M = \int_B dm = \int_V \rho dV$ and therefore $\dot{M} = 0$ where the dot signifies material derivative with respect to time. The material derivative of the volume integral on the right of the expression for mass is not straightforward because the volume occupied by the body changes with time, so there is a part obtained at fixed volume and a part obtained from motion of the body. The latter can be visualized as sweeping mass in over a portion of the body surface and leaving mass behind elsewhere as the original volume deforms to again encapsulate the original mass of the body. Thus:

$$\frac{D}{Dt}(I) = \int_V (\frac{\partial}{\partial t}\rho)dV + \int_S \rho v_n dS = 0 \tag{3.7}$$

where I is a volume integral, $\rho v_n dS$ is the net mass swept into V per unit time, and v_n is the component of velocity normal to S, that is, $v_n = v_i n_i$. Generally, the material derivative of a volume integral of some quantity A is:

$$\frac{D}{Dt}(I) = \int_V (\frac{\partial}{\partial t}A)dV + \int_S A v_n dS = \int_V [(\frac{D}{Dt}A) + A(\frac{\partial v_x}{\partial x} + \frac{\partial v_y}{\partial y} + \frac{\partial v_z}{\partial z})]dV \tag{3.8}$$

where the last expression follows from transformation of the surface integral to a volume integral. When $A = \rho$, conservation of mass may be expressed as:

$$\frac{D}{Dt}\rho + \rho(\frac{\partial v_x}{\partial x} + \frac{\partial v_y}{\partial y} + \frac{\partial v_z}{\partial z}) = 0 \tag{3.9}$$

The integral transformation is achieved through application of the divergence theorem, a powerful tool in continuum mechanics. One form of the divergence theorem is:

$$\int_S A n_i dS = \int_V (\partial A / \partial x_i) dV \tag{3.10}$$

where A may be a scalar, a vector, or another object.

The balance of linear momentum states that the time rate of change of linear momentum is balanced by the resultant of external forces, that is, $\int_S T_i dS + \int_V \gamma_i dV = \int_V \rho a_i dV$. When applied to a "pill box" illustrated in Figure 3.3, a computation shows that $\Delta V = h\Delta S$ and as $h \rightarrow 0$, ΔV while ΔS remains finite. The momentum balance then reduces to $T_i^+ \Delta S^+ + T_i^- \Delta S^- = 0$ and therefore $T_i^+ = -T_i^-$ indicating that the action of material external to S is equal but opposite to the internal material action. This result corresponds to Newton's third law for particles (mass points) stating that for every action there is an equal but opposite reaction.

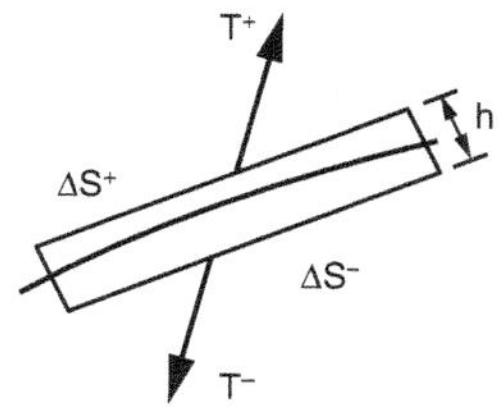

Figure 3.3 A "pillbox" and linear momentum balance.

When the balance of linear momentum is applied to the tetrahedron in Figure 3.4 where $\Delta S_1 = n_1\Delta S$, $\Delta S_2 = n_2\Delta S$, $\Delta S_3 = n_3\Delta S$, $\Delta V = (1/3)h\Delta S$ are the face areas and volume of the tetrahedron, the computation shows that in the 1-direction:

$$(-T_1^{\ 1} + \varepsilon)dS_1 + (-T_1^{\ 2} + \varepsilon)dS_2 + (-T_1^{\ 3} + \varepsilon)dS_3 + (T_1^{\ n} + \varepsilon)dS + (\gamma_1 + \varepsilon_4)(1/3)hdS$$
$$= \rho(\dot{v}_1 + \varepsilon)(1/3)hdS$$

$$(\sigma_{11} + \varepsilon)n_1 dS + (\sigma_{12} + \varepsilon)n_2 dS + (\sigma_{13} + \varepsilon)n_3 dS + T_1^{\ n} + \varepsilon)dS + (\gamma_1 + \varepsilon)(1/3)hdS$$
$$= \rho(\dot{v}_1 + \varepsilon)(1/3)hdS$$

where tractions are replaced by stress and εs are small additions because the tractions are not exactly at the origin (the point of interest). After division by *dS* and assuming a limit as *h* approaches 0, $T_1^n = \sigma_{11}n_1 + \sigma_{12}n_2 + \sigma_{13}n_3$. Similar results are obtained in the 2- and 3-directions. The final result is the famous *Cauchy stress principle*. Thus:

$$T_i = \sigma_{ij} n_j \tag{3.11}$$

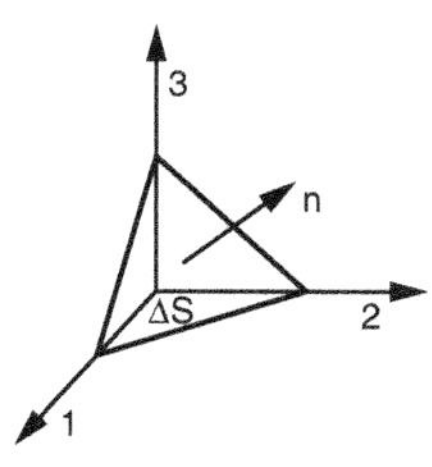

Figure 3.4 Tetrahedron with normal *n*.

When the balance of linear momentum is applied to a free body, one has:

$$\int_S T_i dS + \int_V \gamma_i dV = \frac{D}{Dt}\int_V \rho v_i dV$$

$$\int_S \sigma_{ij} n_j dS + \int_V \gamma_i dV = \int_V (\frac{D}{Dt}\rho v_i dV + \rho v_i v_{j,j})dV$$
$$= \int_V \rho a_i dV$$

$$\int_V \sigma_{ij,j} dV + \int_V \gamma_i dV = \int_V \rho a_i dV$$

in consideration of Equation (3.8), the conservation of mass (Equation 3.9), and the divergence theorem (Equation 3.10). The last equation must be true for arbitrary volumes. Hence, the equations of motions in terms of stress are:

$$\sigma_{ij,j} + \gamma_i = \rho a_i \tag{3.12}$$

The stress equations of equilibrium are obtained when acceleration is zero. Application of the balance of angular momentum leads to symmetry of stress, that is, $\sigma_{ij} = \sigma_{ji}$, so the order of subscripts is not important. For example, $\tau_{yz} = \tau_{zy}$.

Stress is a tensor quantity (of second order) that does not transform under a rotation of reference axes in the same way that a vector does (a tensor of first order). With reference to Figure 3.5, the "old" axes are denoted by *i,j* (=1,2,3); the new axes are denoted by *a,b* (=1,2,3). Direction cosines of angles between axes old and new are denoted by l_{ai}. Note that $l_{ai} = l_{ia}$, but $l_{12} \neq l_{21}$. Stresses referred to the new rotated axes are related to stresses in the old original system as:

$$\sigma_{ab} = l_{ai} l_{bj} \sigma_{ij} \tag{3.13}$$

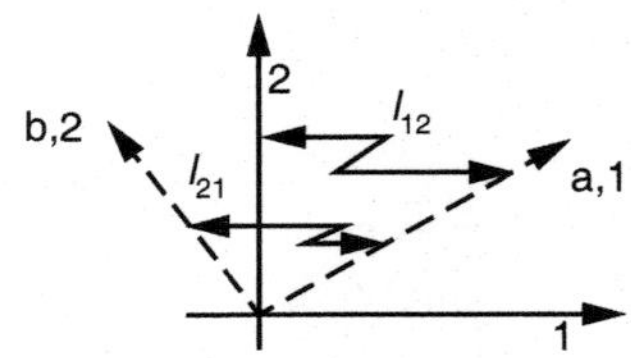

Figure 3.5 Rotation of axes.

A tabulation of direction cosines between old and new axes is given in Table 3.1.

Table 3.1 Direction cosines between old and new axes.

Old New	x	y	z
	lax	lay	laz
b	lbx	lby	lbz
c	lcx	lcy	lcz

A special rotation occurs when one of the old and new axes coincide, say the 3- or *z*-axis and the rotation is through angle θ considered positive in a counterclockwise direction. In this case, Equation (3.13) leads to the formulas in rectangular coordinates:

$$\sigma_{aa} = (\frac{\sigma_{xx} + \sigma_{yy}}{2}) + (\frac{\sigma_{xx} - \sigma_{yy}}{2})\cos(2\theta) + \tau_{xy}\sin(2\theta)$$

$$\sigma_{bb} = (\frac{\sigma_{xx} + \sigma_{yy}}{2}) - (\frac{\sigma_{xx} - \sigma_{yy}}{2})\cos(2\theta) - \tau_{xy}\sin(2\theta) \tag{3.14a}$$

$$\tau_{ab} = \tau_{ba} = -(\frac{\sigma_{xx} - \sigma_{yy}}{2})\sin(2\theta) + \tau_{xy}\cos(2\theta)$$

which is the two-dimensional part of the rotation, and use is made of double angle formulas. The remaining part is the three-dimensional part given by:

$$\begin{aligned}\sigma_{cc} &= \sigma_{zz} \\ \tau_{ac} &= \tau_{xz}\cos(\theta)+\tau_{yz}\sin(\theta) \\ \tau_{bc} &= -\tau_{xz}\sin(\theta)+\tau_{yz}\cos(\theta)\end{aligned} \tag{3.14b}$$

Shear stress rotation in this case is shown in Figure 3.6.

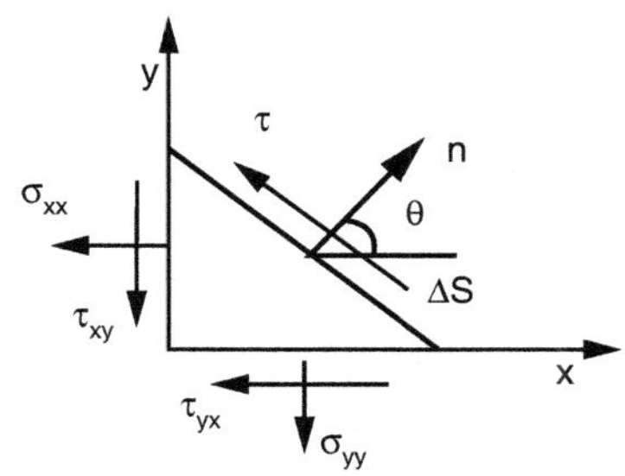

Figure 3.6 Shear stress as a function of rotation angle.

A table of direction cosines for this rotation is given in Table 3.2.

Table 3.2 Direction cosines z-axis rotation.

Old *New*	x	y	z
a	cos(θ)	sin(θ)	0
b	−sin(θ)	cos(θ)	0
c	0	0	1

When the stresses are cast into matrix form:

$$\begin{bmatrix}\sigma_{xx} & \tau_{xy} & \tau_{xz} \\ \tau_{yx} & \sigma_{yy} & \tau_{yz} \\ \tau_{zx} & \tau_{zy} & \sigma_{zz}\end{bmatrix} = \sigma(xyz) \tag{3.15}$$

and the table of direction cosines is also cast into matrix form $[R]$, transformation of stress under a rotation of axes can be expressed compactly as:

$$\sigma(abc) = [R]\sigma(xyz)[R]^t \tag{3.16}$$

where the super-t means transpose.

The 3×3 array (Equation 3.15) is a real symmetric matrix and thus amenable to diagonalization, that is, rotation to principal axes with vanishing of the off-diagonal terms (shear stress). This special rotation leads to principal values and principal directions of stress. If n_j

are the direction cosines of a unit vector that defines a principal direction, then $\sigma_{ij} n_j = \sigma\delta_{ij} n_j$. This system is homogeneous, so the determinant of the coefficients of n_j must vanish. Thus:

$$\begin{vmatrix} \sigma_{11} - \sigma & \sigma_{12} & \sigma_{13} \\ \sigma_{21} & \sigma_{22} - \sigma & \sigma_{13} \\ \sigma_{31} & \sigma_{32} & \sigma_{33} - \sigma \end{vmatrix} = 0 \tag{3.17}$$

that results in the cubic equation $\sigma^3 - \sigma^2 I_1 + \sigma I_2 - I_3 = 0$ which can be factored $(\sigma - \sigma_1)(\sigma - \sigma_2)(\sigma - \sigma_3)$, so that after expansion and comparison, one obtains:

$$\begin{aligned} I_1 &= \sigma_1 + \sigma_2 + \sigma_3 \\ I_2 &= \sigma_1\sigma_2 + \sigma_2\sigma_3 + \sigma_3\sigma_1 \\ I_3 &= \sigma_1\sigma_2\sigma_3 \end{aligned} \tag{3.18}$$

Sometimes a negative sign is used for I_2. The formulas (Equation 3.18) are the sums of the products of the roots taken one, two, and three at a time and are invariant with respect to axis orientation. The roots $\sigma_1, \sigma_2, \sigma_3$ are the desired principal values of stress. By convention, they are ordered such that $\sigma_1 \geq \sigma_2 \geq \sigma_3$. The first is the major principal stress, the next is the intermediate principal stress, and the last is the minor principal stress. The major principal stress is algebraically the largest normal stress seen at a point; the minor principal stress is the smallest normal stress at a point. Back substitution of the principal values into the original system leads to directions of the principal stresses. The principal directions are at right angles to one another. These three principal directions are normal to surfaces that are free of shear stress. When the principal stresses are not all different, then the principal directions are not unique. For example, if the principal stresses are equal, then any direction is a principal direction.

Shear stress at a point also depends on orientation. Consider the plane in Figure 3.6 where $\tau = -(\frac{\sigma_{xx} + \sigma_{yy}}{2})\cos(2\theta) + \tau_{xy}\sin(2\theta)$. A maximum occurs at the angle given by $\tan(2\theta) = -\frac{(\sigma_{xx} + \sigma_{yy})/2}{\tau_{xy}}$ with value $\tau(\max) = [(\frac{\sigma_{xx} + \sigma_{yy}}{2})^2 + (\tau_{xy})^2]^{1/2}$ which can also be expressed as $\tau\,(\max) = (1/2)(\sigma_1 - \sigma_3)$ where the principal stresses are those in the considered plane. A minimum shear stress may also be found with a value equal in magnitude but negative in sign of the maximum shear stress. The normal to the plane of maximum shear stress bisects the 90° angle between the two principal directions, while the normal to the plane of minimum shear is at 90° to the maximum shear plane normal. In a plane problem, the principal stress directions form a curvilinear orthogonal network as do the principal shear directions but at 45° to the principal stress net.

In three dimensions, the situation is somewhat similar with each pair of principal stresses associated with a maximum and minimum shear stress. The principal shear stresses in three dimensions are $\tau_1 = (1/2)|\sigma_2 - \sigma_3|$, $\tau_2 = (1/2)|\sigma_3 - \sigma_1|$, $\tau_3 = (1/2)|\sigma_1 - \sigma_2|$ where the absolute value sign implies two values, one positive and the other negative.

A useful stress concept is deviatoric stress. Deviatoric stresses are:

$s_{ij} = \sigma_{ij} - (1/3)\sigma_{kk}\delta_{ij}$. For example, $s_{11} = \sigma_{11} - (1/3)(\sigma_{11} + \sigma_{22} + \sigma_{33})$, $s_{12} = \sigma_{12}$. Deviatoric shear stresses are just the shear stresses proper. Deviatoric normal stresses are the normal stresses reduced by subtraction of the mean normal stress (sometimes referred to as the hydrostatic or spherical part of stress).

Invariants of deviatoric stress are obtained in the same way that the invariants in Equation (3.18) were obtained with the results:

$$\begin{aligned} J_1 &= s_1 + s_2 + s_3 \\ J_2 &= s_1 s_2 + s_2 s_3 + s_3 s_1 \\ J_3 &= s_1 s_2 s_3 \end{aligned} \tag{3.19}$$

where s_1, s_2, s_3 are the principal values of deviatoric stress. A negative sign is sometimes used with J_2. Also, $s_1 = \sigma_1 - (1/3)(\sigma_1 + \sigma_2 + \sigma_3 + \sigma_1 - I_1/3$ and so on.

Principal stresses can be expressed in terms of invariants. Thus (Fung, 1965; Nayak and Zienkiewicz, 1972):

$$\begin{aligned} \sigma_1 &= (I_1/3) + (2/\sqrt{3})J_2^{1/2}\cos(\alpha) \\ \sigma_2 &= (I_1/3) + (2/\sqrt{3})J_2^{1/2}\cos(\alpha + 2\pi/3), \qquad \cos(3\alpha) = \frac{3\sqrt{3}J_3}{2J_2^{3/2}} \\ \sigma_3 &= (I_1/3) + (2/\sqrt{3})J_2^{1/2}\cos(\alpha - 2\pi/3) \end{aligned} \tag{3.20}$$

3.2 Strain review

Two concepts enter into most discussions of strain: transformation of coordinates and change in length of an element of arc. Transformation of coordinates is symbolically straightforward in a three-dimensional Euclidean space where a scalar ("dot" or inner) product of two vectors is nonnegative. A transformation for a Cartesian frame x_i to some other frame of reference, say, the theta frame θ_i, is defined by the three equations:

$$\theta_i = \theta_i(x_j) \qquad (i, j = 1, 2, 3) \tag{3.21}$$

where one needs to distinguish between the function and the value of the function. The inverse of (3.21) is:

$$x_i = x_i(\theta_j) \qquad (i, j = 1, 2, 3) \tag{3.22}$$

where a one-to-one correspondence is implied. The square of an element of arc length in the Cartesian frame is:

$$(ds)^2 = dx_i dx_i = \delta_{ij} dx_i dx_j \tag{3.23}$$

From Equation (3.22), one obtains $dx_i = (\partial_i/\partial\theta_j)\, d\theta_j$. Hence:

$$(ds)^2 = \frac{\partial x_i}{\partial \theta_j}\frac{\partial x_i}{\partial \theta_k} d\theta_j d\theta_k = g_{jk}(\theta_i) dx_j dx_k \tag{3.24}$$

where g_{jk} is known as the Euclidean metric tensor.

Consider a motion that moves points in a material body initially at X_j to points currently at x_i, so $x_i = x_i(X_j)$ and $X_i = X_i\,(x_j)$ where all points are referred to the same underlying frame of reference. Elements of arc length squared in the initial and current configurations are:

$$(dS_0)^2 = G_{ij}dX_idX_j = G_{ij}\frac{\partial X_i}{\partial x_m}\frac{\partial X_j}{\partial x_n}dx_mdx_n$$
$$(dS)^2 = g_{ij}dx_idx_j = g_{ij}\frac{\partial x_i}{\partial X_m}\frac{\partial x_j}{\partial X_n}dX_mdX_n \quad (3.25)$$

A change in the arc length squared is:

$$(dS)^2 - (dS_0)^2 = (g_{ij}\frac{\partial x_i}{\partial X_m}\frac{\partial x_j}{\partial X_n} - G_{ij})dX_mdX_n = 2E_{mn}dX_mdX_n$$
$$(dS)^2 - (dS_0)^2 = (g_{ij} - G_{ij}\frac{\partial X_i}{\partial x_m}\frac{\partial X_j}{\partial x_n})dx_mdx_n = 2e_{mn}dx_mdx_n \quad (3.26)$$

The coefficients in Equation (3.26) E_{mn} and e_{mn} are the *Lagrangian* and *Eulerian* strain tensors and are defined in the undeformed (initial) and deformed (current) configurations. Both are symmetric. In a Cartesian system, one has the simplification $G_{ij} = g_{ij} = \delta_{ij}$.

Displacement u is a vector connecting a point initially at X_i now at x_i. Thus:

$$u_i = x_i - X_i \quad (3.27)$$

that can be considered as a function of either current or initial coordinates. As a function of initial coordinates or current coordinates:

$$\frac{\partial x_i}{\partial X_m} = \frac{\partial u_i}{\partial X_m} + \delta_{im}, \quad \frac{\partial X_i}{\partial x_m} = \delta_{im} - \frac{\partial u_i}{\partial x_m} \quad (3.28)$$

After substituting Equation (3.28) into expressions for strain, the result is:

$$2E_{mn} = \frac{\partial u_i}{\partial X_m}\frac{\partial u_i}{\partial X_n} + \frac{\partial u_n}{\partial X_m} + \frac{\partial u_m}{\partial X_n}$$
$$2e_{mn} = -\frac{\partial u_i}{\partial x_m}\frac{\partial u_i}{\partial x_n} + \frac{\partial u_n}{\partial x_m} + \frac{\partial u_m}{\partial x_n} \quad (3.29)$$

which express strains in terms of displacements. When displacement derivatives are small, the derivative products in Equation (3.29) are negligible and the strains become linear ("small") expressions in displacement derivatives. In this regard, one notes that:

$$\frac{\partial u_n}{\partial X_m} = \frac{\partial u_n}{\partial x_k}(\frac{\partial u_k}{\partial X_m} + \delta_{km}) \cong \frac{\partial u_n}{\partial x_m} \quad (3.30)$$

to the small strain approximation. The difference between evaluation of displacement gradients in the initial and current configurations is then negligible. Hence, in case of small strains, the strain-displacement relations are:

$$\varepsilon_{xx} = \frac{\partial u}{\partial x}, \quad \varepsilon_{yy} = \frac{\partial v}{\partial y}, \quad \varepsilon_{zz} = \frac{\partial w}{\partial z}$$
$$\gamma_{xy} = 2\varepsilon_{xy} = \frac{\partial u}{\partial y} + \frac{\partial v}{\partial x}, \quad \gamma_{yz} = 2\varepsilon_{yz} = \frac{\partial v}{\partial z} + \frac{\partial w}{\partial y}, \quad \gamma_{zx} = 2\varepsilon_{zx} = \frac{\partial w}{\partial x} + \frac{\partial u}{\partial z} \quad (3.31)$$

where γ is engineering shear strain; and u, v, and w are displacements in the x, y, and z directions, respectively. The normal strains in Equation (3.31) have the usual interpretation of change in length per unit of original length, while the engineering shear strains have the interpretation of a change in angle between two line elements that were initially orthogonal.

Particle velocity v_i is the time rate of change of position and acceleration a_i is time rate of change of velocity. Thus:

$$v_i = \frac{\partial}{\partial t} x_i(X_j, t), \quad a_i = \frac{\partial}{\partial t} v_i(X_j, t) \tag{3.32}$$

where the X_i indicate the particle of interest. When velocity is viewed as a field $v_i(x_i, t)$:

$$a_i = \frac{\partial}{\partial t} v_i(x_j, t) + v_j \frac{\partial v_i}{\partial x_j} \tag{3.33}$$

In case of displacement, one again may follow a given particle and adopt a field view. When following a particle:

$$\begin{aligned} \frac{D}{Dt}(u_i) &= \frac{D}{Dt}(x_i - X_i) \\ &= \frac{D}{Dt}(x_i) \\ &= \frac{\partial x_i(X_i, t)}{\partial t} \\ \frac{D}{Dt}(u_i) &= v_i \end{aligned} \tag{3.34}$$

where $D(X_i)/Dt = 0$ because one is following the same particle. From the field view:

$$\frac{D}{Dt}(u_i) = \frac{\partial u_i(x_j, t)}{\partial t} + v_j \frac{\partial u_i}{\partial x_j} \tag{3.35}$$

A rate of deformation that occurs in plasticity theory is defined in terms of velocities. Thus by definition:

$$2D_{ij} = v_{i,j} + v_{j,i} \tag{3.36}$$

However, consider the case of small strains and a rate computed as:

$$2\frac{D}{Dt}\varepsilon_{ij} = \frac{D}{Dt}(u_{i,j} + u_{j,i}) = v_{i,j} + v_{j,i} = 2\dot{\varepsilon}_{ij} \tag{3.37}$$

because of the small strain assumption, $D()/Dt = \partial()/\partial t$. Otherwise Equations (3.36) and (3.37) are not the same. One reason is that material differentiation and partial differentiation do not commute.

Yet another kinematic relationship is a strain increment defined in terms of displacement increments:

$$2d\varepsilon_{ij} = \frac{\partial du_i}{\partial x_j} + \frac{\partial du_j}{\partial x_i} \tag{3.38}$$

where the displacement increments are those of a particle.

3.3 Stress-strain relations

The conservation of mass, the balances of linear and angular momentum, and strain-displacement relations lead to 10 equations in one unknown mass density, three unknown displacements, six unknown stresses, and six unknown strains for a total of 16. Six more independent equations are needed. In this regard, velocities and accelerations can be computed from definitions. The six equations needed may be quite general. Thus:

$$F_i(\sigma,\varepsilon,\dot{\sigma},\dot{\varepsilon},t,T,x,y,z)=0 \qquad (i=1,2,\ldots,6) \tag{3.39}$$

where T is temperature, subscripts are dropped to focus on the concept, and powers of the variables may be present. If Equations (3.39) are independent of position, then the stress-strain relations describe a *homogeneous* material. If t does not appear explicitly in Equation (3.39), then the material response is independent of time scale change and is nonviscous. When temperature change is not important or when temperature is constant, the behavior is *isothermal*. If rates are also absent, then the law is a *total* law. However, if only rates appear in Equation (3.39), then the law is an *incremental* or *flow* law. If Equation (3.39) is invariant under rotation of axes, the material is *isotropic*, otherwise *anisotropic*.

A linearly elastic solid is characterized by generalized Hooke's law, that is:

$$\sigma_{ij}=C_{ijmn}\varepsilon_{mn}, \qquad \varepsilon_{ij}=S_{ijmn}\sigma_{mn} \tag{3.40a,b}$$

where C_{ijmn} and S_{ijmn} are elastic moduli and compliances, respectively, and are mutual inverses. Small (infinitesimal) strains are implied. In the isotropic case:

$$\begin{aligned}
\sigma_{ij} &= 2\mu\varepsilon_{ij}+\lambda\varepsilon_{kk}\delta_{ij}\\
d\sigma_{ij} &= 2\mu d\varepsilon_{ij}+\lambda d\varepsilon_{kk}\delta_{ij}\\
(d\sigma_{ij}/dt) &= 2\mu(d\varepsilon_{ij}/dt)+\lambda(d\varepsilon_{kk}/dt)\delta_{ij}
\end{aligned} \tag{3.41}$$

in total, incremental form and rate form. Again, small strains are implied. A nonlinear law would be:

$$\sigma_{ij}=2\mu\varepsilon_{ij}+\lambda\varepsilon_{kk}\delta_{ij}+2\mu\varepsilon_{ik}\varepsilon_{kj} \tag{3.42}$$

An example of a quasi-linear law is:

$$\dot{\sigma}_{ij}=H_{ijkl}(\sigma_{mn})\dot{\varepsilon}_{kl} \tag{3.43}$$

because the highest derivative is of order one. A viscous Newtonian fluid has a stress-strain rate law in the form:

$$\sigma_{ij}=-p\delta_{ij}+2\mu\dot{\varepsilon}_{ij}+\lambda\dot{\varepsilon}_{kk}\delta_{ij} \tag{3.44}$$

that is incompressible when $2\mu+3\lambda=0$. A nonviscous, incompressible fluid used in seepage analysis is Darcy's law that may be expressed as:

$$v_i=k_{ij}(p,_j+\gamma_i) \tag{3.45}$$

where γ is specific weight of the fluid flowing.

Of interest is the time rate of change of finite strain:

$$2\dot{E}_{mn} = 2\dot{\varepsilon}_{mn} + \dot{u}_{im}\dot{u}_{in} \tag{3.46}$$

that is not homogeneous in time. However, in case of small strain:

$$2\dot{\varepsilon}_{mn} = \dot{u}_{m,n} + \dot{u}_{n,m} \tag{3.47}$$

that is homogeneous in time.

3.4 Principal stress space

A function F(*x,y,z*) describes a surface in *xyz* space. Similarly, a function F(σ_{ij}) describes a hypersurface in the space of the nine stresses σ_{ij}. At any point in a material body, there are three mutually orthogonal directions associated with vanishing of shear stress. These are principal stress directions $(\vec{n}_1, \vec{n}_2, \vec{n}_3)$ that form an orthogonal right-handed triple. This triple of directions may be used as coordinate directions to plot magnitudes of the principal stresses acting parallel to the principal directions. A negative principal stress acts antiparallel to the associated direction. The physical orientation of the principal planes is found from the direction cosines of the principal directions. Although principal directions may not be unique, an orthogonal right-handed triple of directions can always be found. Figure 3.7 illustrates the construction of principal stress space (*123*) in association with physical space (*xyz*). A plot in principal stress space does *not* follow the ordering convention $\sigma_1 \geq \sigma_2 \geq \sigma_3$. For example, the principal stress parallel to $\vec{n}_2$ may be larger or smaller than σ_1 or σ_3. An ellipsoid in principal stress space has the equation:

$$\left(\frac{\sigma_1}{a}\right)^2 + \left(\frac{\sigma_2}{b}\right)^2 + \left(\frac{\sigma_3}{c}\right)^2 = 1 \tag{3.48}$$

That admits positive and negative values of stress in the 1, 2, and 3 directions.

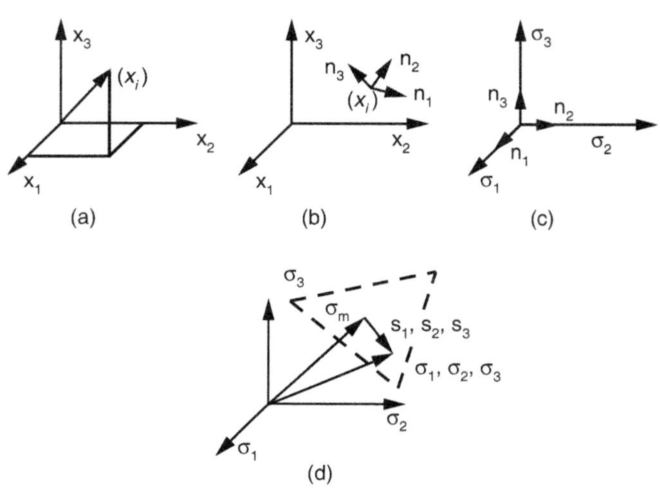

Figure 3.7 Principal directions for establishing principal stress space and locating a point. The triangle in D is in the deviatoric or pi-plane normal to the space diagonal. Addition of hydrostatic and deviatoric vectors locates a point of interest.

3.5 Yield functions, failure, and loading criteria

The initial portion of a stress-strain curve associated with a cylinder of material subjected to an axial compression or tension usually appears as a steep, approximately linear rise to a limiting point beyond which large strains become possible with further small increases in stress. This point is the point of initial yield; the stress at this point is the yield stress σ_Y as shown in Figure 3.8 and marks the limit to a purely elastic deformation. The yield stress is also the material strength. Within the strain range 1 in Figure 3.8, unloading (prior to reaching the yield point) follows the elastic line downwards towards the origin. Deformation is reversible, that is, elastic in this range.

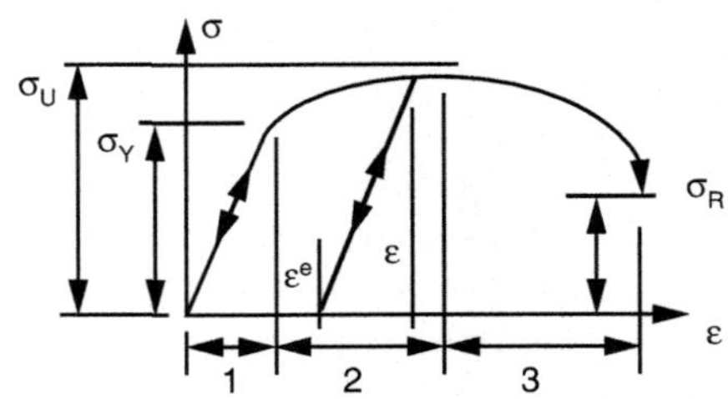

Figure 3.8 Uniaxial stress-strain plot.

If straining continues beyond the elastic limit, plastic behavior occurs. The greatly reduced slope of the stress-strain curve beyond the elastic limit indicates the potential for a large increase in strain with little increase in stress. A rising stress-strain curve beyond the elastic limit indicates *strain hardening*. Unloading from a point within the range 2 in Figure 3.8 follows a line parallel to the original elastic line. A portion ε^e of the total strain ε is elastic and therefore recoverable. The difference between the total strain and the elastic strain is a plastic strain $\varepsilon^p = \varepsilon - \varepsilon^e$. The same is true of any increment of stress and strain in this region. Thus, $d\varepsilon^p = d\varepsilon - d\varepsilon^e$.

The high point of the stress-strain curve is the ultimate strength σ_U. Reloading follows the unloading line; the response is purely elastic until the stress just prior to unloading is reached. This stress is greater than the initial yield stress in the range 2. Indeed, yield stress increases throughout the range of strain hardening. A material at yield is also at failure. Yielding and failure are synonymous with respect to stress, while yielding has a connotation of inelastic straining.

The falling portion of the stress-strain curve in region 3 is a region of *strain softening* that is terminated by rupture with rupture strength σ_R. The yield stress decreases in this region. However, strain softening is sometimes illusory and is often associated with instability as the geometry of the test specimen changes significantly. The change is usually a reduction in cross-sectional area of the test specimen that corresponds to "necking" of ductile steel tested in tension and to "hour-glassing" when testing brittle material such as rock in compression. In this regard, an ideally brittle material fails at the elastic limit where rupture strength, ultimate strength, and yield strength coincide.

If the stress-strain curve neither rises nor falls after the initial yield point is reached, then the response is *ideally* or *perfectly* plastic. Strain-hardening materials have an intrinsic stability as strength increases with strain, while perfectly plastic materials do not. When the region of yielding in a perfectly plastic material grows large enough to extend from boundary to boundary, unlimited displacement becomes possible in the form of a collapse mechanism. Here, yielding occurs at a point, while collapse relates to a global event. Of course, strain-hardening materials may also form collapse mechanisms. Strain-softening material appears intrinsically unstable.

Yield, failure

The one-dimensional concepts of yielding, elastic-plastic strain, loading, and unloading need to be generalized to three dimensions as part of an effort to establish a plasticity theory. In one dimension, strength at the elastic limit is a stress. In three dimensions, the elastic limit is a functional relationship among the stresses. Thus, $Y(\sigma_{ij}) = 0$ is an implicit statement of material strength and is a yield or failure criterion. Constants that appear in explicit expressions of strength are strength properties of the material considered. In case of uniaxial compression, $Y = \sigma - C_o$ where C_o is unconfined compressive strength. In uniaxial tension, $Y = \sigma - T_o$ where T_o is tensile strength. In geomechanics, compressive strength is generally dependent on stress, while tensile strength is often not considered to be stress dependent. If $Y(\sigma_{ij}) < 0$, then the state of stress is in the elastic domain; Y cannot be greater than 0 because stress cannot exceed strength, which is clear in case of uniaxial stress. If the term "failure" is preferred to "yielding," then a change in notation from Y as the yield function to F as the strength or failure criterion may be done; that is, the elastic limit is given by $F(\sigma_{ij}) = 0$ instead of $Y = 0$.

Yield criteria may also be expressed in terms of principal stresses σ_1,σ_2,σ_3 and direction angles, say, of the major principal stress, θ_1, θ_2, θ_3. Directions of the intermediate and minor principal stresses are known from the requirement of right-handed orthogonality of principal directions. In case of isotropic materials, yielding is independent of direction. Hence, yielding is an implicit function of principal stresses only: $Y(\sigma_{ij}) = (\sigma_1, \sigma_2, \sigma_3) = 0$. A convenient change of variables $\sigma_m = (\sigma_1 + \sigma_3) / 2$ $\tau_m = (\sigma_1 - \sigma_3) / 2$, $\sigma_1 = \sigma_m + \tau_m$, $\sigma_3 = \sigma_m - \tau_m$ allows the yield condition to be expressed as $Y(\sigma_m, \sigma_2, \tau_m) = 0$. This latter form gives rise to two distinct categories of yield criteria: (1) those that are independent of the intermediate principal stress and (2) those that are influenced by the intermediate principal stress.

When the yield condition is independent of the intermediate principal stress, an explicit form is possible. Thus, $\tau_m = f(\sigma_m)$. $\sigma_m = f^{-1}(\tau_m) = g(\tau_m)$. The simplest function form of f is a constant. Thus, $\tau_m = k$. Because the sign of the shear stress is not important for yielding, absolute values should be used. Hence, in this simple case:

$$|\tau_m| = k \tag{3.49}$$

which is the famous Tresca yield condition used mainly in metal plasticity, although occasionally in mechanics of some soft rock types. Clearly, k is the yield strength in shear.

The next increase in complexity is a linear function. Thus:

$$|\tau_m| = \sigma_m \sin(\phi) + k\cos(\phi) \tag{3.50a}$$

Alternatively:

$$|\tau| = \sigma \tan(\phi) + k \tag{3.50b}$$

where σ and τ are the normal and shear stresses acting upon a *potential* failure surface. The strength constants ϕ and k are the angle of *internal friction* and *cohesion*, respectively. The linear form (Equation 3.50b) is the famous Mohr-Coulomb yield criterion, which is especially popular in soil mechanics.

The Mohr-Coulomb criterion is shown in principal stress space in Figure 3.9b. Each of the six principal shear stresses $\pm |\sigma_1 - \sigma_2|$, $\pm |\sigma_2 - \sigma_3|$, $\pm |\sigma_3 - \sigma_1|$ forms a side of the pyramid seen in the figure. Also shown in the figure is the Tresca yield condition, which appears as a six-sided hexagon. Compression is positive in the figure.

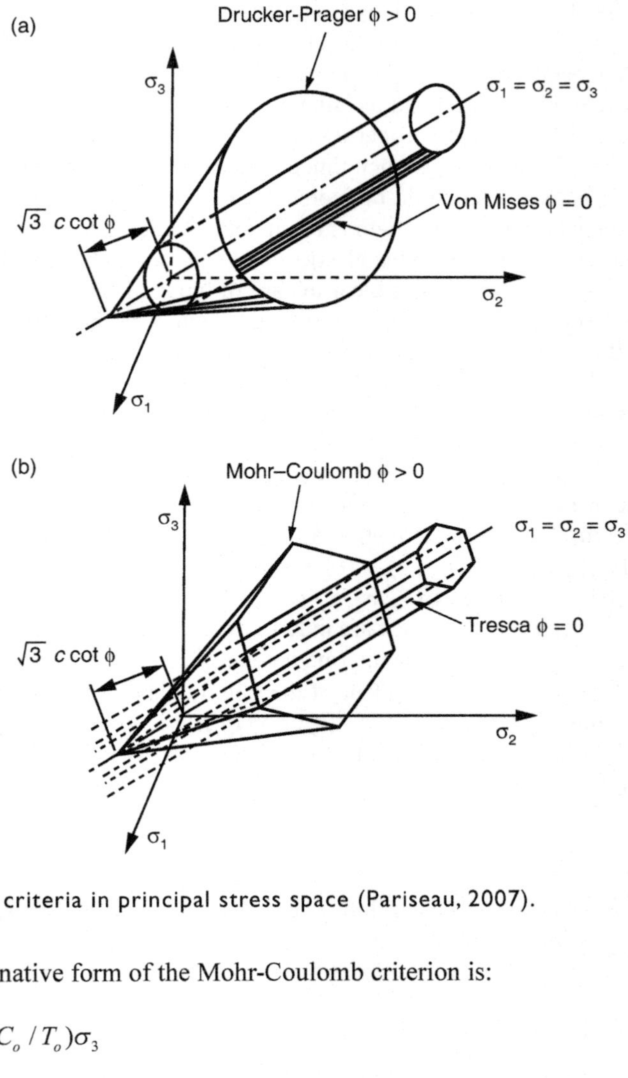

Figure 3.9 Yield criteria in principal stress space (Pariseau, 2007).

A useful alternative form of the Mohr-Coulomb criterion is:

$$\sigma_1 = C_o + (C_o / T_o)\sigma_3 \tag{3.50c}$$

where C_o and T_o are unconfined compressive and tensile strengths, respectively, $\sigma_1 = C_p$ is compressive strength under confining pressure $\sigma_1 = p$. Thus, $C_p = C_o + (C_o/T_o)p$ is also an alternative form of the Mohr-Coulomb criterion. There are only two independent strength constants. Hence:

$$\sin(\phi) = \frac{C_o - T_0}{C_o + T_0}, \quad k = \sqrt{\frac{C_o T_0}{4}}, \quad C_o = \frac{2k\cos(\phi)}{1-\sin(\phi)}, \quad T_o = \frac{2k\cos(\phi)}{1+\sin(\phi)}$$

Nonlinearity is the next step in complexity (Pariseau, 1967). Thus:

$$|\tau_m|^n = a\sigma_m + b \tag{3.51}$$

where a and b are strength constants given in terms of unconfined compressive and tensile strengths as:

$$a = \frac{|C_o/2|^n - |T_o/2|^n}{(C_o/2)+(T_o/2)}, \quad b = \frac{(T_o/2)|C_o/2|^n + (C_o/2)|T_o/2|^n}{(C_o/2)+(T_o/2)}$$

The exponent n, which defines the degree on nonlinearity, is determined from experiment, as are the unconfined compressive and tensile strengths. Experience indicates n is usually a value between 1 and 2.

Another popular nonlinear yield condition in rock mechanics is the Hoek-Brown criterion that was originally expressed in terms of the major and minor principal stresses. Thus:

$$\sigma_1 = \sigma_3 + \sqrt{a\sigma_3 + b} \tag{3.52}$$

in original form where $a = \dfrac{C_o^2 - T_o^2}{T_o}$ and $b = C_o$. After a change of variables:

$$(\tau_m + a/8)^2 = (1/8)^2[a^2 + 16(a\sigma_m + b^2)] = a'\sigma_m + b'$$

which is a parabola in a normal stress – shear stress plane. However, this parabola is not symmetric about the normal stress axis. Consequently, there is ambiguity in the use of this criterion. This ambiguity can be removed by restricting $\sigma_m \geq (-b^2/16a)$ and using an absolute value of τ_m, which makes the result symmetric with respect to the normal stress axis. In this regard, numerous yield criteria are proposed in the technical literature expressed as polynomials in terms of the major and minor principal stresses. However, only those polynomials that result in symmetry about the normal stress axis after a similar change of variables are acceptable as yield criteria (Pariseau, 2012).

Yield criteria for isotropic media that are influenced by the intermediate principal stress include the well-known Mises and Drucker-Prager criteria. These criteria are shown in Figure 3.9a and may be expressed in terms of stress invariants. The Mises criterion is simply $|J_2^{1/2}| = k$ where J_2 is the second invariant of deviatoric stress. This criterion appears as a cylinder in principal stress. The Drucker-Prager criterion has the form:

$$\left|J_2^{1/2}\right| = AI_1 + B \tag{3.53}$$

where I_1 is the first stress invariant and A and B are strength constants. These constants can be expressed in terms of unconfined compressive and tensile strengths. Thus:

$$A = \frac{1}{\sqrt{3}}\left(\frac{C_o - T_o}{C_o + T_o}\right),\quad B = \frac{2}{\sqrt{3}}\left(\frac{C_o T_o}{C_o + T_o}\right)$$

$$C_o = \frac{\sqrt{3}B}{1-\sqrt{3}A},\qquad T_o = \frac{\sqrt{3}B}{1+\sqrt{3}A}$$

Nonlinear N-type criteria that depend on all three principal stresses can be formed in much the same manner as n-type criteria were formed. Thus:

$$\left|J_2^{N/2}\right| = AI_1 + B \tag{3.54}$$

where $A = \left(\frac{1}{\sqrt{3}}\right)^N\left[\frac{C_o^N - T_o^N}{C_o + T_o}\right),\ B = \left(\frac{1}{\sqrt{3}}\right)^N\left[\frac{C_o^N T_o + C_o T_o^N}{C_o + T_o}\right]$. Again, the exponent N is determined from experiment and is usually between 1 and 2.

Another interesting failure criterion that has achieved considerable popularity in soil mechanics is the Lade criterion (Lade, 1993):

$$Y = \left(\frac{I_1'^3}{I_3'} - 27\right)\left(\frac{I_1'}{p_a}\right)^m - \eta_1 = 0 \tag{3.55}$$

where

$$I_1', (\sigma_1 + ap_a) + (\sigma_2 + ap_a) + (\sigma_3 + ap_a)$$

$$I_3', (\sigma_1 + ap_a)(\sigma_2 + ap_a)(\sigma_3 + ap_a)$$

p_a, atmospheric pressure
η, a and m are constants

The three material constants are determined by fitting Equation (3.55) to experimental test data. The geometry of this yield criterion is shown in Figure 3.10. An advantage of this criterion is in the continuously turning tangent that allows for easy differentiation. The same is true of the Mises, Drucker-Prager, and N-type yield criteria. Tresca, Mohr-Coulomb, Hoek-Brown, and n-type criteria have edges that pose a challenge to evaluating derivatives of the yield function.

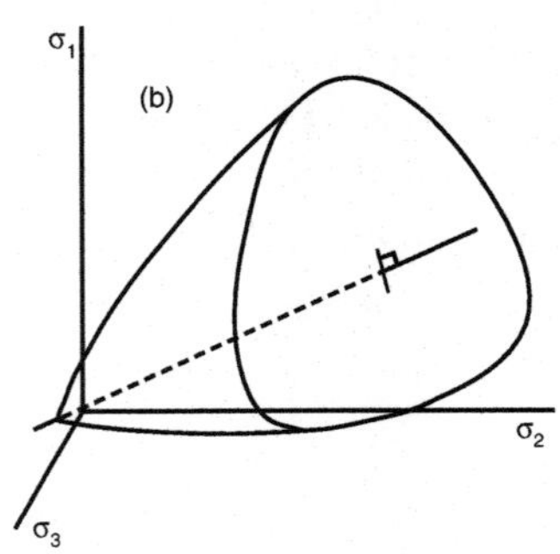

Figure 3.10 Geometry of the Lade criterion in principal stress space (Lade, 1993).

Many other yield criteria are described in the technical literature, such as "cap" models that close the compression side of a criterion such as Drucker-Prager. Cap models address a concern in plasticity theory that relates to dilatancy or relative expansion during plastic flow. The expansion may be too large in geoplasticity as a physical possibility. An alternative approach to this theoretical problem is the use of nonassociated flow rules, described later. Other yield or failure criteria are formed from extravagant fits to experimental test data, mainly triaxial test data. Some require more than a dozen fitting functions and parameters to take nonlinearities into account and raise questions of practicality in engineering application. In any case, the few well-known criteria introduced here are sufficient for an introductory account of geoplasticity.

Loading criteria

Generalization of the concept of loading to three-dimensional stress states is aided by visualizing stress in principal stress space where the yield function $Y(\sigma_{ij}) = Y(\sigma_i) = 0$ after rotation of the axes of stress to the principal axes. This function is a surface in stress space, a yield surface. A reasonable supposition is that the yield surface Y encloses the origin and is convex, in the sense that a line from the origin intersects the Y in only one point. This condition would allow for star-shaped yield surfaces. A stricter convexity requires any straight line to intersect the yield surface in two points at most. Strict convexity requires a yield surface to be piecewise continuous. Loading in stress follows a stress increment $d\sigma_{ij} = dS$ where the latter is a stress increment vector in stress space. In one dimension, if the applied stress is increased along a rising stress-strain curve, then $d\sigma/dt > 0$ and loading occurs; if decreased, then $d\sigma/dt < 0$ and unloading occurs. In three dimensions, the stress increment vector dS should point away from the elastic region when loading occurs. Unloading is indicated when dS points towards the elastic region, because during the next instant of time, the stress state will surely be in the elastic region. The gradient to the yield surface $gradY$ is a vector normal to the yield surface and allows for the determination of loading or unloading. Thus, the scalar product $(gradY) \cdot (dS) > 0$ indicates an outward pointing stress increment and therefore *loading*. *Unloading* occurs when $(gradY) \cdot (dS) < 0$. If the stress point is on the yield surface and moves to another nearby point on the yield surface, then the "loading" is considered *neutral loading*. The yield or failure criterion is thus a "loading" function as well. A yield function may depend on plastic strain, as is the case for a strain-hardening material, and on temperature and other parameters. In such cases and in general, loading occurs when $(\partial Y/\partial \sigma_{ij})d\sigma_{ij} \geq 0$ with the equality signifying neutral loading. The equality also holds in the case of ideal or perfect plasticity. As expected, unloading occurs when $(\partial Y/\partial \sigma_{ij})d\sigma_{ij} \leq 0$.

3.6 Plastic stress-strain laws

A stability postulate by Drucker (1959) provides unification of the concepts of loading, strain hardening, convexity of the yield function, and importantly, a route to plastic stress-strain laws. The postulate states that application of an increment of load results in positive work done by the increments of stress on the corresponding increments of strain during loading, and that over a cycle of loading and unloading, the incremental work is nonnegative. Thus:

$$\begin{aligned} &d\sigma_{ij}d\varepsilon_{ij} > 0 \quad (loading) \\ &d\sigma_{ij}d\varepsilon_{ij} \geq 0 \quad (cycle) \end{aligned} \tag{3.56}$$

Over a cycle of stress application and release, the elastic portion of the total strain increment is recovered. Hence:

$$d\sigma_{ij} d\varepsilon_{ij} = d\sigma_{ij}(d\varepsilon_{ij}^{e} + d\varepsilon_{ij}^{p}) = d\sigma_{ij} d\varepsilon_{ij}^{p} \geq 0 \qquad (cycle) \tag{3.57}$$

where the equality holds for an ideally plastic material. During plastic straining, the yield condition must be satisfied, that is, $Y = 0$, and for *ideally plastic* material:

$$dY = \frac{\partial Y}{\partial \sigma_{ij}} d\sigma_{ij} = 0 \qquad (ideally\ plastic) \tag{3.58}$$

where the stress increment $d\sigma_{ij}$ must be in a plane tangent to the yield surface. The increment cannot point outward because the yield surface is fixed in space, depending only on stress. An inward-pointing stress increment would result in unloading as the stress point moved off the yield surface and therefore in a purely elastic strain increment. The gradient to the yield surface is normal to the tangent plane, and so is normal to the stress increment vector as seen in Equation (3.58). The plastic strain increment vector according to Equation (3.57) is also normal to all related stress increments that lie in the tangent plane. As a consequence, the plastic strain increment and gradient to the yield surface must be parallel and therefore proportional. Thus, in case of ideally plastic materials:

$$d\varepsilon_{ij}^{p} = \lambda \frac{\partial Y}{\partial \sigma_{ij}} \tag{3.59}$$

which is the "flow rule" for ideally plastic material. Here λ is a function that expresses the proportionality between the plastic strain increment and yield surface gradient.

In case of work hardening, the argument is more subtle but leads to the same result. Work hardening implies increasing strength with plastic strain, so the yield condition is $Y(\sigma_{ij}, \varepsilon_{ij}^{p}; C_{\alpha})$ where the parameters C_{α} may also be important. Straining from one plastic state to another so the material is continuously at yield again requires:

$$dY = \frac{\partial Y}{\partial \sigma_{ij}} d\sigma_{ij} + \frac{\partial Y}{\partial \varepsilon_{ij}^{p}} d\varepsilon_{ij}^{p} + \frac{\partial Y}{\partial C_{\alpha}} dC_{\alpha} = 0 \tag{3.60}$$

Equations (3.51) and (3.53) express the Prager (1948) continuity or consistency requirement for ideally plastic and work-hardening materials.

A geometric interpretation of Equation (3.57) is that the stress and strain increment vectors in stress space form an acute angle when one imagines a strain space superposed over stress space. At a point near the current yield surface, all stress increment vectors that represent loading must lie to one side of a plane constructed normal to the strain increment vector and be outward pointing, as shown in Figure 3.11. The plane normal to the strain increment vector must therefore be parallel to a plane tangent to the yield surface. Thus, the strain increment vector is normal to the yield surface and therefore proportional to $\partial Y / \partial \sigma_{ij}$, which are direction numbers of the yield surface normal in stress space. Hence, Equation (3.59) applies.

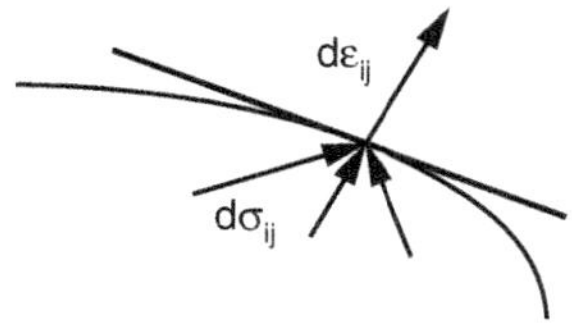

Figure 3.11 Stress and strain increment geometry.

The yield surface is on one side of the plane constructed normal to the strain increment vector at the considered point. Because this point is arbitrary and the yield surface must be piecewise continuous, the yield surface must be convex, according to the stability postulate of Drucker.

The function λ must be positive, because during loading the incremental work must be positive. However, this function does not depend on stress or strain increments, so the incremental plastic stress-strain relations (Equation 3.58) are quasi-linear. The form (Equation 3.58) indicates that Y is a potential function. Indeed, Y is a *plastic potential* that also serves as a loading function and failure or strength criterion. In this case, Equations (3.58) are *associated rules of flow*. When the parameters C_α are considered functions of plastic strain, the consistency condition (Equation 3.59) shows that λ may be expressed as:

$$\lambda = -\frac{(\partial Y / \partial \sigma_{kl}) d\sigma_{kl}}{(\partial Y / \partial \varepsilon_{ij}^{p})(\partial Y / \partial \sigma_{ij})} \tag{3.61}$$

The plastic potential and failure criterion need not be the same. In this case, the plastic potential is still Y, but the failure criterion that marks the limit to purely elastic deformation is F, and the rules of flow are *nonassociated*. After substitution into Equation (3.59):

$$d\varepsilon_{ij}^{p} = -\frac{(\partial Y / \partial \sigma_{ij})(\partial Y / \partial \sigma_{kl}) d\sigma_{kl}}{(\partial Y / \partial \varepsilon_{ij}^{p})(\partial Y / \partial \sigma_{ij})} = H_{ijkl}(\sigma_{mn,}\ \varepsilon_{mn}^{p}) d\sigma_{kl} \tag{3.62}$$

which is a quasi-linear incremental stress-strain law for work-hardening materials with coefficients that depend on stress and plastic strain, but not on increments or rates that may be obtained by dividing by dt.

Derivation of the elastic-plastic stress-strain law begins with the assumption that the total strain increment is the sum of elastic and plastic strain increments during plastic deformation. For convenience, a matrix notation is adopted such that {} denotes a 6×1 column matrix, a superscript "t" means transpose, and [ ] denotes a 6 × 6 matrix. Thus, $\{d\varepsilon\} = \{d\varepsilon^e\} + \{d\varepsilon^p\}$. Hooke's law for the elastic portion is $\{d\varepsilon^e\} = [S]\{d\sigma\}$, where $[S]$ is a matrix of elastic compliances with inverse $[C]$, a matrix of elastic moduli. Hence $\{d\varepsilon\} = [S]\{d\sigma\} + \lambda\{\partial Y/\partial\sigma)$ that can be brought into the form

$$\{\partial F / \partial \sigma\}^t [C]\{d\varepsilon\} = \{\partial F / \partial \sigma\}^t \{d\sigma\} + \lambda \{\partial F / \partial \sigma\}^t [C]\{\partial Y / \partial \sigma\} \text{ But also}$$

$$0 = \{\partial F / \partial \sigma\}^t \{d\sigma\} + \{\partial F / \partial \varepsilon^p\}\{d\varepsilon^p\} = \{\partial F / \partial \sigma\}^t \{d\sigma\} + \lambda \{\partial F / \partial \varepsilon^p\}\{\partial Y / \partial \sigma\}$$

in consideration of the consistency condition and allowing for a failure criterion F different from the plastic potential Y (yield or loading criterion). These two expressions may be solved for the scalar function λ that after back substitution results in the elastic-plastic stress-strain model:

$$\{d\sigma\} = ([C]\{d\varepsilon\} - \frac{[C]\{\partial Y / \partial\sigma\}\{\partial F / \partial\sigma\}^t [C]}{\{\partial F\}^t [C]\{\partial Y / \partial\sigma\} - \{\partial F / \partial\varepsilon^p\}^t \{\partial Y / \partial\sigma\}}\{d\varepsilon\} \tag{3.63}$$

for either ideally plastic or work-hardening plasticity. The incremental, time-independent flow rule (Equation 3.56) can be put in a more compact form. Thus:

$$\{d\sigma\} = ([C^e] - [C^p])\{d\varepsilon\} = [C^{ep}]\{d\varepsilon\} \tag{3.64}$$

where $[C^p]$ is s a correction to the matrix of elastic moduli $[C^e]$, and $[C^{ep}]$ is the sum of the two.

In case of multiple yield surfaces, Koiter (1953) suggests adding contributions from each piece of a piecewise continuous yield surface when the stress point is at an edge or corner. Thus, when there are α active yield surfaces, the plastic strain increment is given by:

$$d\varepsilon_{ij}^p = \Sigma_\alpha \lambda_\alpha \frac{\partial Y_\alpha}{\partial\sigma_{ij}} \tag{3.65}$$

There is an alternative to the Drucker stability postulate, based on an incremental loading–unloading cycle, is a postulate by Il'ushin (1961) that is based independently on a strain cycle of a concept of loading. This postulate is discussed further by Khan and Huang (1995) in a comprehensive treatment of plasticity theory.

A summary of the governing equations for a time-independent, small strain, elastic-plastic model includes:

1. Stress equations of motion (Equation 3.12)
2. Strain-displacement equations (Equation 3.37 or Equation 3.38)
3. Plastic potential and failure criterion (Y, F)
4. Hooke's law
5. Rules of flow (Equation 3.63)

In case of multiple yield functions (Equation 3.65), multiple λs may be eliminated from consideration in somewhat the same manner as in the single case.

References

Drucker, D.D. (1959) A definition of stable inelastic material. *Journal of Applied Mechanics*, 26, 101–106.

Fung, Y.C. (1965) *Foundations of Solid Mechanics*. Prentice-Hall, Englewood Cliffs, NJ.

Ilyushin, A.A. (1961) On the postulate of plasticity. *Journal of Applied Mathematics and Mechanics*, 25(3), 746–752.

Khan, A.S. & Huang, S. (1995) *Continuum Theory of Plasticity*. Wiley, New York.

Koiter, W.T. (1953) Stress-strain relations, uniqueness and variational theorems for elastic-plastic materials with a singular yield surface. *Quarterly of Applied Mathematics*, 11, 350–354.

Lade, P.V. (1993) Rock strength criteria: The theories and the evidence. In: *Comprehensive Rock Engineering*, Volume 1. Pergamon, Oxford.

Nayak, G.G. & Zienkiewicz, O.C. (1972) Convenient form of stress invariants for plasticity. *Journal of the Structural Division, ASCE*, 98(ST4), 949–954.

Pariseau, W.G. (1967) Post-yield mechanics of rock and soil. *Mineral Industries*, College of Earth & Mineral Sciences, The Pennsylvania State University, University Park, PA, 36(8).

Pariseau, W.G. (2007) *Design Analysis in Rock Mechanics*. Taylor & Francis, London.

Pariseau, W.G. (2012) A symmetry requirement for failure criteria. *International Journal of Rock Mechanics and Mining Sciences*, 52, 42–49.

Prager, W. (1948) The stress-strain laws of the mathematical theory of plasticity – A survey of recent progress. *Journal of Applied Mechanics*, 15, 226–233.

Chapter 4

Two-dimensional theory (plane strain)

Three-dimensional plasticity theory is suitable for orientation of thought, but it is too complicated for practical analyses. Some simplification must be made. A natural one is plane strain. Plane stress and axial symmetry are also possible simplifications, but a plane strain approach is more useful in geomechanics. In this approach, deformation in, say, the z-direction is nil. Plane strain plasticity theory is unusual in that a statically determinant subsystem of three equations in three stresses appears. Under stress boundary conditions, this system can be solved without reference to deformation. A solution to the stress subsystem is then a solution to a problem of "limiting equilibrium" in allusion to the fact that the equations of equilibrium and the yield condition comprise the stress subsystem. However, solutions to problems of limiting equilibrium are only part of a complete solution to a problem in plane plasticity.

Deformations are related to the stresses, so if displacement-related boundary conditions are present, then there must be some constraint on solutions to the stress subsystem problem. A complete solution thus requires a statement concerning deformation in addition to stress. Deformations may be characterized as strain increments or rates and hence by an incremental displacement field or by a velocity field. Stress and velocity fields are thus the objects of interest in two-dimensional plasticity.

Because the stress subsystem involves equilibrium and yield only, one can expect useful stress predictions from theory to the extent that the failure criterion and boundary conditions are known with acceptable accuracy. When elastic and plastic regions are present, the location of the elastic-plastic boundary becomes central to a problem. To compute the location of the elastic-plastic boundary, one must follow the internal progression of yielding from the outset. Serious difficulties are involved in the computation. In many instances, the location is a matter of conjecture. In other instances, a semi-inverse approach may be used with the assumption of the stress field and therefore the location of the elastic-plastic boundary at some moment in time. The location of the elastic-plastic boundary and the "loads" acting there constitute "internal" boundary conditions. Assumptions concerning these conditions assist in making a problem solvable. However, even when a stress field satisfies the stress subsystem of equations and boundary conditions (an *admissible stress field*), a velocity field that is compatible with the admissible stress field and satisfies velocity boundary conditions may not be found. Such a lack of uniqueness is a common feature of plane strain plasticity problems.

In plane strain theory, the stress and velocity subsystems are usually but not always hyperbolic in character and are associated with two distinct families of characteristic curves. When the yield function is smooth and the rules of flow are associated, the characteristic curves of the two subsystems coincide. In case of nonassociated flow rules, the characteristic curves of the two systems do not coincide.

4.1 Yield envelope in plane strain theory

An issue of some importance to detailed development of theory concerns the yield condition or failure criterion. Under plane strain conditions, one supposes that the z-direction stresses can be eliminated in the stress-strain equations. Certainly, this is true in the purely elastic case. The yield condition then becomes a function of three stresses, σ_{xx}, σ_{yy}, τ_{xy}, and therefore a function of the major and minor principal stress σ_1, σ_3 and orientation angle θ of σ_1 with respect to the x-axis in the plane of interest. Alternatively, one may suppose that under plane strain conditions, the intermediate principal stress is functionally dependent on the other two. Thus, in the isotropic case where independence of direction prevails:

$$Y(\sigma_i) = Y(\sigma_1, g(\sigma_1, \sigma_3), \sigma_3) = Y(\sigma_1, \sigma_3) = 0 \tag{4.1}$$

While the relationship $\sigma_2 = g(\sigma_1, \sigma_3)$ is possible in principle, in practice the relationship may be impossible to obtain explicitly. A change of variables from principal stresses to mean normal stress and maximum shear in the plane of interest is advantageous. Thus:

$$\begin{aligned} \sigma_m &= (\sigma_1 + \sigma_3)/2, \quad \tau_m = (\sigma_1 - \sigma_3) \\ \sigma_1 &= \sigma_m + \tau_m \quad\;\; \sigma_3 = \sigma_m - \tau_m \end{aligned} \tag{4.2}$$

Hence, $Y(\sigma_1, \sigma_3) = Y(\sigma_m, \tau_m) = 0$ at yield. Explicitly, one has:

$$\tau_m = f(\sigma_m), \quad \sigma_m = g(\tau_m) \tag{4.3a,b}$$

where g is the inverse of f. Now a Mohr's circle has the representation:

$$(\sigma - \sigma_m)^2 + \tau^2 = \tau_m^2 \tag{4.4}$$

in a normal stress shear stress plane as shown in Figure 4.1. Also shown in Figure 4.1 are segments of two expressions of yield, Equation (4.3a) and an *envelope* of Mohr circles representing stress states at yield or failure.

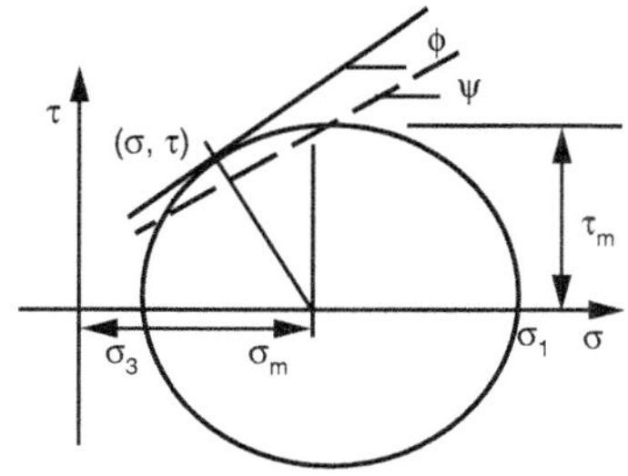

Figure 4.1 Mohr's circle and envelope.

The stresses (σ, τ) are normal and shear stresses acting on potential failure surface inclined with *normal* at an angle θ to the major principal stress σ_1 as shown in Figure 4.2. This angle is related to the yield condition and failure envelope. Inspection of the geometry of Figure 4.1

shows that $\theta = \pi/4 + \phi/2$ where ϕ is the inclination of the failure envelope to σ_1. The failure surface is inclined $\pi/4 - \phi/2$ to the direction of σ_1.

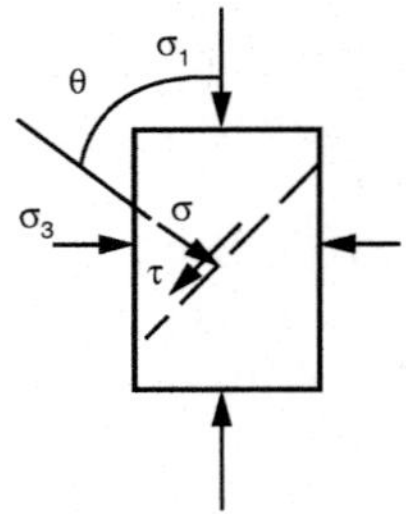

Figure 4.2 Failure surface in relation to principal stresses.

Equations (4.3) and (4.4) depend on the parameter σ_m that locates the center of a circle. Differentiation of Equation (4.4) with respect to σ_m and then eliminating it from these two equations leads to the yield or failure envelope. After substitution of Equation (4.3a) into Equation (4.4), one has:

$$(\sigma - \sigma_m)^2 + \tau^2 = f^2(\sigma_m)$$

After differentiation:

$$-(\sigma - \sigma_m) + 0 = ff'$$

where f' is differentiation with respect to σ_m. In principle, this last result allows for solution of σ_m in terms of σ, so $\sigma_m = h(\sigma)$. Thus, $f(\sigma) = f(h(\sigma))$, and the equation of the envelope is:

$$\begin{aligned} \tau^2 &= f^2[1-(f')^2] \\ \tau &= f(\sigma) \end{aligned} \tag{4.5a,b}$$

where now $f = f(\sigma)$. Equation (4.5) is the failure envelope and is an alternative expression of the yield condition (Equation 4.3) when an envelope exists.

Inspection of Equation (4.5) shows that the derivative f' is restricted for the failure envelope to exist, that is, $|f'| \le 1$. If ψ is the inclination of the plot of (Equation 4.3a) to the normal stress axis, then $-1 \le \tan(\psi) \le 1$. There is a relationship between the slope of the failure envelope and the failure criterion (Equation 4.3). Differentiation of Equation (4.4) shows that:

$$\begin{aligned} \frac{d\tau}{d\sigma} &= -\frac{\sigma - \sigma_m}{\tau} = \tan(\phi) \\ \frac{d\tau_m}{d\sigma_m} &= -\frac{\sigma - \sigma_m}{\tau} = \tan(\psi) \end{aligned} \tag{4.6}$$

But also from rotation of the reference axis, $\tau = \tau_m \cos(\theta)$, and since $\theta = \pi/4 + \phi/2$, one has from Equation (4.6) the general result:

$$\sin(\phi) = \tan(\psi) \tag{4.7}$$

This result restricts existence of the failure envelope according to values of $-\pi/4 \le \psi \le \pi/4$ obtained before. As a consequence, the failure criterion (Equation 4.3a) is more general than the envelope form of the failure criterion (Equation 4.5b), which does not exist for $|\psi| > \pi/4$.

Figure 4.1 has a mirror image in the lower half plane for negative shear stress. The reason is that the sign of the shear stress is a matter of convention and has no importance to failure. An implication is that not any function is suitable for describing yield and failure. Only functions that are symmetric with respect to the normal stress axis in Figure 4.1 are physically acceptable. Many functions $f(\sigma_1, \sigma_3)$ have been proposed based on good fits to laboratory test data that do not meet this simple symmetry requirement when transformed to the usual normal stress – shear stress plane (Pariseau, 2012). The yield condition or failure criterion, either terminology is applicable, should be stated using the absolute value of shear stress to guarantee symmetry. Thus:

$$\begin{aligned} |\tau_m| &= f(\sigma_m), \quad \sigma_m = g(\pm\tau_m) \\ |\tau| &= f(\sigma), \quad \sigma = g(\pm\tau) \end{aligned} \tag{4.8}$$

where the envelope form is also restricted to functions symmetric with respect to the normal stress axis.

An example illustrates the relationship between the functions $|\tau_m| = f(\sigma_m)$ and $|\tau| = f(\sigma)$ in the quadratic case when $\tau_m^2 = A\sigma_m + B$ with the unconfined compressive strengths $C_o = 18$, $T_o = 2$ in arbitrary units. The strength constants in the yield criterion are $A = (C_o - T_o)/2$ and $B = C_oT_o/4$. Slope of the yield criterion is $d\tau_m/d\sigma_m = A/2\tau_m = \tan(\psi)$. Hence, an envelope exists according to Equation (4.5) only if $\tau_m \geq 4$ or $\tau_m \geq 5/8$. The envelope is expressed by the formula $\tau^2 = A\sigma + (A^2/4 + B)$ and has the slope $d\tau/d\sigma = A/2\tau$. Figure 4.3 illustrates the criterion in the quadratic case with associated envelope. Unconfined compressive strength is associated with the envelope, but tensile strength is not as seen in the figure. The dotted Mohr half-circle is at the lower limit of the mean normal stress Sm (and maximum shear stress Tm) for the envelope existence. Stress states with smaller mean normal stress satisfy the failure criterion but are not encompassed by the envelope, thus illustrating the possibility of stress states satisfying the yield criterion $\tau_m^2 = A\sigma_m + B$ but are not indicated by an envelope criterion $\tau^2 = A\sigma + (A^2/4 + B)$. An example of such is uniaxial tensile strength as seen in the figure. Interestingly, in the linear case when $n = 1$ and $|\tau_m| = A\sigma_m + B$ with $A = (C_o - T_o)/(C_o + T_0)$ and $B = C_oT_o/(C_o + T_o)$, an envelope exists for all A such that $\tau = \sigma A/\sqrt{1-A^2} + (B/\sqrt{1-A^2})$ or in the more familiar form $\tau = \sigma\tan(\phi) + k$, where ϕ and k are the angle of internal friction and cohesion, respectively. Note: $A = \sin(\phi)$ in the linear case.

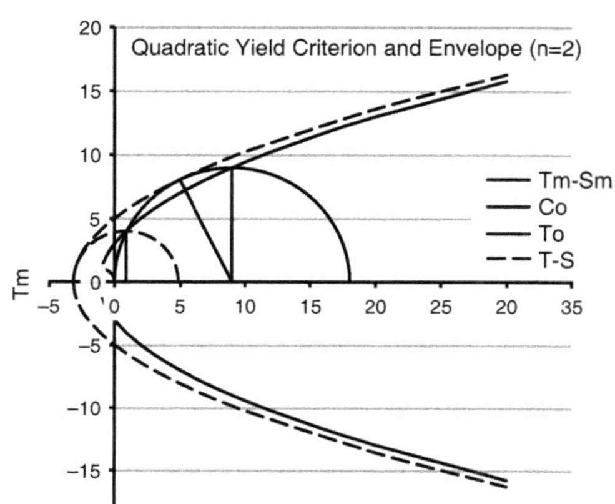

Figure 4.3 Illustration of yield criterion and associated envelope in the quadratic case. Yield criterion is a solid curve; the envelope is a dotted curve. Unconfined compressive and tensile strength Mohr half-circles are shown. Also shown is a Mohr half-circle at the lower limit to Sm for which the envelope exists (dotted).

4.2 The stress subsystem of equations

In plane strain, the stress subsystem of equations is composed of two equations of motion or equilibrium, as the case may be, and a yield condition. Thus:

$$\begin{aligned} &\sigma_{xx,x} + \tau_{xy,y} = \gamma_x - \rho\dot{v}_x \\ &\tau_{yx,x} + \sigma_{yy,y} = \gamma_y - \rho\dot{v}_y \\ &Y(\sigma_{xx}, \sigma_{yy}, \tau_{xy}) = 0 \end{aligned} \tag{4.9}$$

where positive velocities and components of specific weight act in the positive coordinate directions and *compression* is positive.

The Cartesian stresses may be expressed in terms of the principal stresses through a rotation of reference axis. Thus, with respect to the principal axes that are rotated through a counterclockwise angle θ from the x axes:

$$\begin{aligned} &\sigma_{xx} = \sigma_m + \tau_m \cos(2\theta) \\ &\sigma_{yy} = \sigma_m - \tau_m \cos(2\theta) \\ &\tau_m = \tau_m \sin(2\theta) \end{aligned} \tag{4.10a}$$

The rotation (Equation 4.10) is actually from the 1–3 axes through a negative, clockwise angle to the x axes. In consideration of the failure criterion (Equation 4.3b), one has:

$$\begin{aligned} &\sigma_{xx} = g(\tau_m) + \tau_m \cos(2\theta) \\ &\sigma_{yy} = g(\tau_m) - \tau_m \cos(2\theta) \\ &\tau_m = \tau_m \sin(2\theta) \end{aligned} \tag{4.10b}$$

Thus, use of the yield condition has reduced the number of unknowns to two τ_m, θ. The yield condition is satisfied; the question of equilibrium (motion) remains.

Differentiation of Equation (4.10) with $g' = dg/d\tau_m = d\sigma_m/d\tau_m$ results in:

$$\begin{aligned} &\sigma_{xx,x} = (g' + \cos(2\theta))\tau_{m,x} - (2\tau_m \sin(2\theta))\theta,_x \\ &\sigma_{yy,y} = (g' - \cos(2\theta))\tau_{m,y} + (2\tau_m \sin(2\theta))\theta,_y \\ &\tau_{xy,x} = \sin(2\theta)\tau_{m,x} + 2\tau_m \cos(2\theta)\theta,_x \\ &\tau_{yx,y} = \sin(2\theta)\tau_{m,y} + 2\tau_m \cos(2\theta)\theta,_y \end{aligned} \tag{4.11}$$

The equations of motion are now:

$$\begin{aligned} &(g' + \cos(2\theta))\tau_{m,x} + \sin(2\theta)\tau_{m,y} - (2\tau_m \sin(2\theta))\theta,_x + (2\tau_m \cos(2\theta))\theta,_y \\ &= \rho\dot{v}_x - \gamma_x \\ &\sin(2\theta)\tau_{m,x} + (g' - \cos(2\theta))\tau_{m,y} + (2\tau_m \cos(2\theta))\theta,_x + (2\tau_m \sin(2\theta))\theta,_y \\ &= \rho\dot{v}_y - \gamma_y \end{aligned} \tag{4.12}$$

The variables τ_m, θ are functions of x and y and have differentials:

$$\begin{aligned} &\tau_{m,x} dx + \tau_{m,y} dy = d\tau_m \\ &\theta,_x dx + \theta,_y dy = d\theta \end{aligned} \tag{4.13}$$

Equations (4.12) and (4.13) can be combined into a single system that has the form:

$$\begin{aligned}
&A_1\tau_{m,x} + B_1\tau_{m,y} + C_1\theta,_x + D_1\theta,_y = E_1 \\
&A_2\tau_{m,x} + B_2\tau_{m,y} + C_2\theta,_x + D_2\theta,_y = E_2 \\
&\tau_{m,x}dx + \tau_{m,y}dy + 0 + 0 = d\tau_m \\
&0 + 0 + \theta,_x dx + \theta,_y dy = d\theta
\end{aligned} \tag{4.14}$$

This system of four equations in the four unknown derivatives can be solved provided the determinant of the coefficients does not vanish. The solution in determinant form is:

$$\tau_{m,x} = \frac{|N_1|}{\Delta}, \quad \tau_{m,y} = \frac{|N_2|}{\Delta}, \quad \theta,_x = \frac{|N_3|}{\Delta}, \quad \theta,_y = \frac{|N_4|}{\Delta}, \tag{4.15}$$

The determinants are:

$$\Delta = \begin{vmatrix} A_1 & B_1 & C_1 & D_1 \\ A_2 & B_2 & C_2 & D_2 \\ dx & dy & 0 & 0 \\ 0 & 0 & dx & dy \end{vmatrix}, \quad |N_1| = \begin{vmatrix} E_1 & B_1 & C_1 & D_1 \\ E_2 & B_2 & C_2 & D_2 \\ d\tau_m & dy & 0 & 0 \\ .d\theta & 0 & dx & dy \end{vmatrix}, etc \tag{4.16}$$

If circumstances are such that Δ vanishes, then the determinants in the numerators of Equation (4.15) must also vanish for a solution to be possible. Setting $\Delta = 0$ and expanding leads to a quadratic in *dy*/*dx*. Thus:

$$(dy/dx)^2\,[AC] + (dy/dx)(-[BC]-[AD]) + [BD] = \rightarrow (dy/dx)^2 \tag{4.17}$$

The bracket notation [XY] means $(X_1Y_2\text{-}X_2Y_1)$, and:

$$\begin{aligned}
&[AC] = 2\tau_m(g'\cos(2\theta)+1) \quad &&[BC] = 2\tau_m g'\sin(2\theta) \\
&[AD] = 2\tau_m g'\sin(2\theta) &&[BC] = 2\tau_m(1-g'\cos(2\theta))
\end{aligned} \tag{4.18}$$

Solution of Equation (4.17) results in the directions that lead to $\Delta = 0$. Thus:

$$dy/dx = \frac{\sin(2\theta) \pm \sqrt{1-\tan^2(\psi)}}{\cos(2\theta)+\tan(\psi)} \tag{4.19}$$

where tan(ψ) is the slope of the failure criterion (Equation 4.3a). Equations (4.19) are differential equations of two families of curves associated with vanishing of the determinant of the coefficients in Equation (4.15). These curves are *characteristic curves* of the stress subsystem of equations. Inspection of Equation (4.19) shows three possibilities:

$$\begin{aligned}
&(1)\ |\tan(\psi)| > 1 \\
&(2)\ |\tan(\psi)| = 1 \\
&(3)\ |\tan(\psi)| < 1
\end{aligned} \tag{4.20}$$

In the first case, the characteristics are imaginary; there are no real characteristics and the stress subsystem is *elliptic*. In the second case, there is one family of real characteristic curves and the system is *parabolic*. In the third case, there are two real families of characteristic curves and the system is *hyperbolic*. The third case is of most interest and implies the existence of a failure envelope. In this case, recall, sin $(\phi) = \tan(\psi)$ that after substitution in Equation (4.19) gives:

$$dy\,/\,dx = \frac{\sin(2\theta) \pm \cos(\phi)}{\cos(2\theta) + \sin(\phi)} \tag{4.21}$$

The form of Equation (4.21) is more informative after the substitution of $(\phi) = (\pi/2) - 2\mu$ in Equation (4.21). The result after some use of trigonometric identities is:

$$dy\,/\,dx = \tan(\theta \pm \mu) \quad : \begin{cases} C_1 \\ C_2 \end{cases} \tag{4.22}$$

According to Equation (4.22), the characteristic curves form angles of $\pm\mu$ with the direction of σ_1 which is inclined to the x-axis at an angle θ. When a + sign is associated with the first or C_1 family of characteristic curves and a − sign with the second or C_2 family, then the geometry of characteristic curves has the form shown in Figure 4.4.

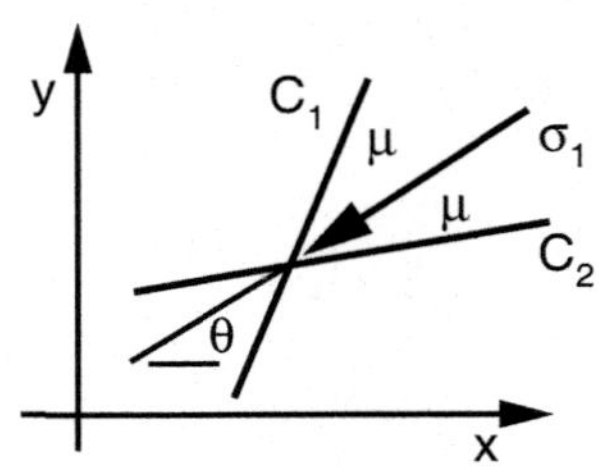

Figure 4.4 Characteristic curves in relation to σ_1.

The requirement that the determinants in the numerators of Equation (4.15) vanish when the determinant of the coefficients vanishes leads to:

$$d\tau_m\{[BC]-[BD](dy\,/\,dx)\} - d\theta[CD] + [CE]dy - [DE]dx = 0 \tag{4.23}$$

where dx and dy are increments along the characteristic curves (Equation 4.22), and (dx/dy) is the reciprocal of Equation (4.22). The new terms in square brackets are:

$$\begin{aligned} [CD] &= -4(\tau_m)^2 \\ [CE] &= 2\tau_m(E_2\sin(2\theta) + E_1\sin(2\theta)) \\ [DE] &= 2\tau_m(E_2\cos(2\theta) - E_1\sin(2\theta)) \end{aligned} \tag{4.24}$$

Reduction of Equation (4.23) leads to differential relationship along the characteristic curves:

$$\pm\cot(\phi)\,d\tau_m + 2\tau_m\,d\theta + \{E_1\sin(\theta\mp\mu) - E_2\cos(\theta\mp\mu)ds = 0 \quad (4.25)$$

where ds is an element of arc length along a characteristic curve with:

$$dx = \cos(\theta\mp\mu)ds, \quad dy = \sin\pm(\theta\mp\mu)ds \quad (4.26)$$

If the variable σ_m were used instead of τ_m, the result would be:

$$\pm\cos(\phi)\,d\sigma_m + 2f(\sigma_m)\,d\theta + \{E_1\sin(\theta\mp\mu) - E_2\cos(\theta\mp\mu)\}ds = 0 \quad (4.27a)$$

where $d\tau_m/d\,\sigma_m = \tan(\psi) = \sin(\phi)$, $\tau_m = f(\sigma_m)$ as before.

In the event $\phi = 0$, then with neglect of body forces, weight, and inertia, Equation (4.25) shows that $d\tau_m = 0$. Hence, τ_m = constant = $2k$, which is the well-known Tresca yield condition. Equation (4.27a) shows that:

$$d\sigma_m \pm 2kd\theta = 0 \quad (4.27b)$$

along the characteristic curves. Hence:

$$\sigma_m \pm 2k\theta = \sigma_m^o \pm 2k\theta_o \quad (4.27c)$$

where the "o" signifies a starting value obtained from boundary conditions.

Again, with $\phi = 0$, the characteristic curves have slopes given by $dy/dx = \tan(\theta \pm \pi/4)$ according to Equation (4.22), since $\mu = \pi/4$ in this case, and coincide with directions of maximum and minimum shear stress, which are orthogonal. These relations appear in metal plasticity and are a special case of the more general development.

The stress subsystem of equations in characteristic form is given by Equations (4.22) and (4.25) or Equation (4.23):

$$dy/dx = \tan(\theta\pm\mu) \quad : \begin{cases} C_1 \\ C_2 \end{cases} \text{and}$$

$$\pm\cos(\phi)\,d\tau_m + 2\tau_m\,d\theta + \{E_1\sin(\theta\pm\mu) - E_2\cos(\theta\pm\mu)\}ds = 0 \quad \begin{cases} C_1 \\ C_2 \end{cases}$$

where $\mu = \pi/4 - \phi/2$ and a failure envelope is assumed to exist. The characteristics form an acute angle of 2μ between C_2 and C_1 curves according to Equation (4.22). In case $\phi = 0$, the characteristics are orthogonal. Equations (4.22) and (4.23) are invariant with respect to rotation of the reference axis. Moreover, expansions of the remaining determinants in the numerators of Equation (4.15) lead to the same results in Equation (4.25). The transformation of the original system of the equations of motion and yield to this system is advantageous because of the appearance of total differentials (derivatives) instead of partial derivatives. The two systems are equivalent; a solution of one satisfies the other.

Nothing in the preceding development restricted the yield condition $\tau_m = f(\sigma_m)$ to be linear, such as the well-known Mohr-Coulomb condition $\tau_m = \sigma_m \tan(\psi) + k = \sigma_m \sin(\phi) + c\cos(\phi)$, where c and ϕ are cohesion and angle of internal friction, respectively. In case of an n-type yield criterion $\tau_m^n = a\sigma_m + b$, $d\sigma_m / d\tau_m = n\tau_m^{n-1} / a = 1/\sin(\phi)$, and the angle of internal friction is no longer constant but rather stress dependent. The characteristic subsystem of stress (Equations 4.22 and 4.25) has the same form, but ϕ and μ are now stress dependent.

Geometry of the characteristic curves is of critical importance. If one knows this geometry, integrals of Equation (4.22), then one knows $\theta(x,y)$, which may be used in Equation (4.25) to find τ_m. The yield condition in the form $g(\tau_m)$ leads to σ_m. The combination of σ_m, τ_m, and θ allows for computation of the Cartesian stresses, σ_{xx}, σ_{yy}, τ_{xy}, through the equations of transformation (rotation of reference axes). The normal stress $\sigma_{zz} = \sigma_2$ is then computed through the stress-strain relations, in principle, and thus the stress system is solved in the plastic region of a plane plastic strain problem.

4.3 Regions of constant state, radial shear, and radial stress

Three very useful stress fields arise in connection with rather simple geometric configurations of the two families of characteristic curves (Equation 4.22). These are regions where (1) both families of characteristic curves are straight lines, (2) one family is composed of straight lines (rays) and the other exponential spirals, and (3) both families are exponential spirals. These special cases apply to isotropic, weightless material that has a constant ϕ and therefore a linear failure envelope. Thus, $d\tau_m/d\sigma_m = \tan(\psi) = \sin(\phi)$ and:

$$\tau_m = a\sigma_m + b, \qquad a = \sin(\phi), \qquad b = c\cos(\phi) \tag{4.28}$$

which has the form of famous Mohr-Coulomb (MC) failure criterion with envelope:

$$|\tau| = \sigma\tan(\phi) + c \tag{4.29}$$

where c is the *cohesion* of the material and ϕ is the *angle of internal friction*. The absolute value sign is used to include the envelope in the lower half of the σ,τ plane as well as the upper half. The stress subsystem of equations is now:

$$\begin{aligned} dy &= dx\tan(\theta \pm \mu) \\ d\tau_m &\pm 2\tau_m \tan(\phi)d\theta = 0 \end{aligned} \tag{4.30a,b}$$

With $S = \sigma_m + (b/a)$ so $\tau_m = S\sin(\phi)$, Equation (4.30b) may also be expressed as:

$$dS \pm 2S\tan(\phi)d\theta = 0 \tag{4.31}$$

Region of constant state

If a characteristic curve is straight, then the stresses along that curve are constant as can be seen in Equation (4.30). A straight characteristic curve requires a constant slope and therefore a constant θ (μ is a constant because ϕ is constant). Consequently, $d\theta = 0$, and according to Equation (4.30b), τ_m = constant that, in consideration of the yield condition, implies σ_m is a constant. The converse is also true. If the stresses are constant along a straight line, then $d\theta = 0$, according to Equation (4.30b), and according to Equation (4.30a), the line must be a straight characteristic or embedded in a region of straight characteristics.

There are three possibilities of one family of characteristics being straight. These are shown in Figure 4.5. The first possibility shows parallel straight lines; the second shows rays emanating from a point. The third possibility shows straight lines tangent to a curve that is an envelope of straight characteristics.

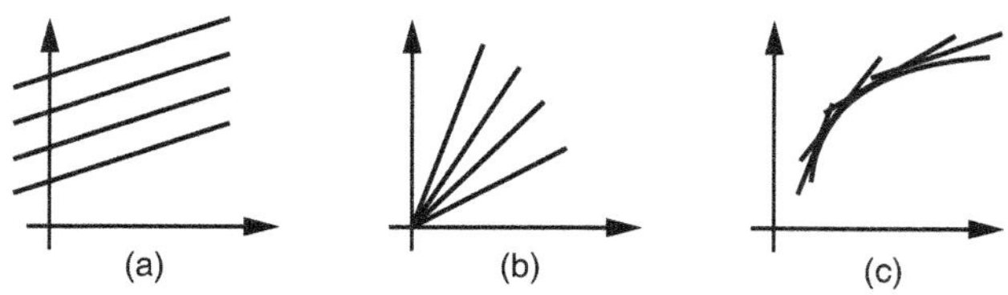

Figure 4.5 One family of straight characteristics.

A region where the stresses are constant is a region of constant state. Both families of characteristic curves are straight in a region of constant state, and again, the converse is true. If both families of characteristics are straight in a region, then the stresses are constant in the region. The analysis result is a theorem that holds under the assumption of an isotropic, weightless material with strength described by a Mohr-Coulomb criterion. Isotropy, weightlessness, and the Mohr-Coulomb criterion are reasonable assumptions for relatively strong materials at yield because of the high stresses and implied loads necessary for failure. Stresses caused by weight would be negligible. However, neglect of weight would not be reasonable for weak materials under low stress, such as a sand bank.

Region of radial shear

In polar coordinates $x = r\cos(\omega)$, $y = r\sin(\omega)$, and the equations of characteristics (Equation 4.30a) have the form:

$$dr / r = d\omega \cot(\theta - \omega \pm \mu) = d\omega \cot(\delta \pm \mu) \tag{4.32}$$

where $\delta = \theta - \omega$ is the angle from a ray to the direction of major principal stress as shown in Figure 4.6.

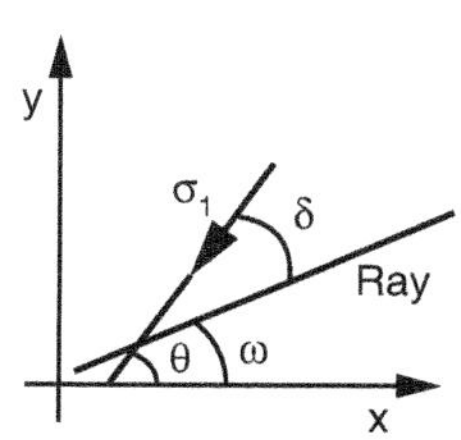

Figure 4.6 Ray and principal stress directions in polar coordinates.

If one of the families of characteristics is straight and passes through the origin, then $\omega = \theta \pm \mu$. Suppose $\omega = \theta + \mu$, then the family of first characteristics are rays and $dr/r = d\omega \cot(0)$, which implies ω = constant that, in turn, implies θ = constant *along a given characteristic.* This characteristic is therefore straight, as assumed, and the stresses are constant along the given characteristic. However, as ω is varied, so is the considered ray and so are the stresses, even though they are constant along any particular ray. The pole or origin is a stress singularity.

The second family of characteristics is given by $dr/r = d\omega \cot(-2\mu)$. Integration gives $\ln(r/r_0) = -(\omega - \omega_0)$, where $\omega - \omega_0$ defines the ray $r = r_0$. Thus, the second family of characteristics is composed of exponential spirals:

$$(r / r_0) = \exp[-(\omega - \omega_0)\tan(\phi)] \tag{4.33}$$

As ω increases counterclockwise in the positive sense, the radial distance r to a point on a given spiral decreases as shown in Figure 4.7.

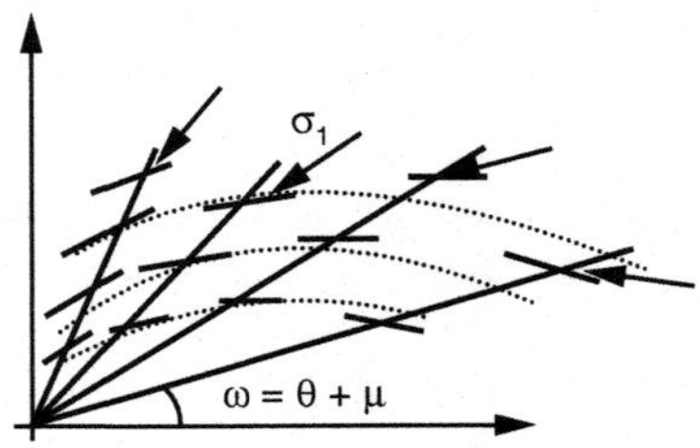

Figure 4.7 Rays and exponential spirals as characteristics.

The stress field associated with the geometry of the characteristics is obtained by integrating Equation (4.30b). In this case, $d\theta = d\omega$. Hence, along the second family of characteristics $d\tau_m - 2\tau_m \tan(\phi)d\omega = 0$. But $dr/r = -d\omega \tan(\phi)$, so $d\tau_{m/}\tau_m + 2dr/{=}0$. Thus:

$$(\tau_m / \tau_m^o) = (r_o / r)^2 = \exp[2(\omega - \omega_o)\tan(\phi)] = \exp[2(\theta - \theta_o)\tan(\phi)] \tag{4.34}$$

along a characteristic curve of the second family where "starting" values (boundary conditions) are denoted by "o." The stresses are constant along the first family of characteristics, which are straight lines ($\omega = \omega_o$). A similar analysis applies when $\omega = \theta - \mu$ with suitable changes.

In the case when $\phi = 0$, the first family of characteristics are still rays ($\omega = \omega_o$), but the second family is given by integration of $dr = 0$, which follows from $dr/r = d\omega \cot(-2\mu)$ with $2\mu = \pi / 2$. The second family of characteristics are the circles $r = r_o$. Equations (4.27b) apply along the characteristics. Thus, $\sigma_m = \sigma_m^o \pm 2k(\omega_o - \omega)$ with $\omega = \theta + \mu = \theta + \pi / 4$.

Region of radial stress

Rays are principal stress directions in a region of radial stress. When the rays are directions of the major principal stress σ_1, then $\theta = \omega$ and the differential equations of characteristics are $dr/r = d\omega \cot(\pm\mu)$ from (4.32). After integration:

$$(r / r_o) = \exp[\pm(\omega - \omega_o)\cot(\mu) \quad : \begin{cases} C_1 \\ C_2 \end{cases} \tag{4.35}$$

These curves are exponential spirals that form acute angles with the rays. The C_1 curves spiral outward with increasing ω, while the C_2 characteristics spiral outward with decreasing ω (clockwise is negative), as shown in Figure 4.8a.

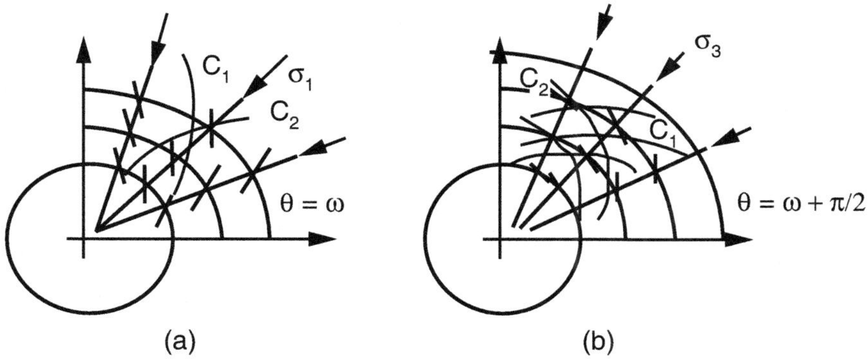

Figure 4.8 Regions of radial stress.

Use of $\theta = \omega$ in Equation (4.30b) leads to $d\tau_m \pm 2\tau_m \tan(\phi)\, d\omega = d\tau_m + 2\tau_m \tan(\phi)\, (dr/r) \tan(\mu) = 0$ along a characteristic curve in either family. Hence:

$$(\tau_m / \tau_m^o) = (r / r)^{-2\tan(\phi)\tan(\mu)} = \exp[\pm 2(\omega - \omega_o)\tan(\phi)] \tag{4.36}$$

The result (Equation 4.36) shows that maximum shear stress diminishes with increasing r as the characteristic curves spiral outward. This result is in keeping with the label of "radial stress."

When the rays are directions of minor principal stress, $\theta = \omega + \pi/2$, a similar analysis leads to characteristics described by $dr / r = \pm d\omega \tan(\mu)$, so:

$$(r / r_o) = \exp[\pm(\omega - \omega_o)\tan(\mu) \; : \begin{cases} C_1 \\ C_2 \end{cases} \tag{4.37}$$

and:

$$(\tau_m / \tau_m^o) = (r / r)^{+2\tan(\phi)\cot(\mu)} = \exp[\pm 2(\omega - \omega_o)\tan(\phi)] \tag{4.38}$$

In this case, the maximum shear stress increases with increasing r. The exponential spirals in this case are shown in Figure 4.7b.

In the case when $\phi = 0$, the characteristics are still given by Equation (4.35) when rays $\omega = \omega_o$ coincide with the direction of σ_1 and by Equation (4.37) when the rays coincide with the direction of σ_3. Equations (4.27b) apply along the characteristics. Thus, $\sigma_m - \sigma_m^o = \pm 2k(\omega - \omega_o)$. Thus, when the rays coincide with the direction of $\sigma_{1,}$ $\sigma_m - \sigma_m^o = 2k \ln[(r_o - r)]$, and when the rays coincide with the direction of σ_3, $\sigma_m - \sigma_m^o = 2k \ln[(r - r_o)]$. In both instances, $\tau_m = k$ throughout the plastic region. The angle θ is known by choice, so the principal stresses and Cartesian stresses may be computed, as desired.

4.4 The velocity subsystem of equations

Although the plasticity here is time-independent, the use of rates is convenient. Small strains are assumed, so the strain rate is given in terms of velocities by:

$$2\dot{\varepsilon}_{ij} = v_{i,j} + v_{j,i} \tag{4.39}$$

Further, the strain rate is composed of elastic and plastic parts, so:

$$\dot{\varepsilon}_{ij} = \dot{\varepsilon}^e_{ij} + \dot{\varepsilon}^p_{ij} \tag{4.40}$$

The elastic part is related to stress rates by Hooke's law. Thus:

$$\dot{\varepsilon}^e_{ij} = S_{ijmn}\dot{\sigma}_{mn} \tag{4.41}$$

The plastic part is obtained through normality. Hence:

$$\dot{\varepsilon}^p_{ij} = \lambda \frac{\partial Y}{\partial \sigma_{ij}} \tag{4.42}$$

In plane strain theory, the velocity subsystem may be obtained by elimination of the function λ from Equation (4.42). Thus, in plane strain:

$$\frac{\dot{\varepsilon}_{xx} - \dot{\varepsilon}^e_{xx}}{\partial Y / \partial \sigma_{xx}} = \frac{\dot{\varepsilon}_{yy} - \dot{\varepsilon}^e_{yy}}{\partial Y / \partial \sigma_{yy}} = \frac{\dot{\varepsilon}_{xy} - \dot{\varepsilon}^e_{xy}}{\partial Y / \partial \sigma_{xy}} = \lambda \tag{4.43}$$

Equations (4.43) lead to:

$$\begin{aligned} &\dot{\varepsilon}_{xx}(\partial Y / \partial \sigma_{xy}) - \dot{\varepsilon}_{xy}(\partial Y / \partial \sigma_{xx}) = \dot{\varepsilon}^e_{xx}(\partial Y / \partial \sigma_{xy}) - \dot{\varepsilon}^e_{xy}(\partial Y / \partial \sigma_{xx}) \\ &\dot{\varepsilon}_{xy}(\partial Y / \partial \sigma_{yy}) - \dot{\varepsilon}_{yy}(\partial Y / \partial \sigma_{xy}) = \dot{\varepsilon}^e_{xy}(\partial Y / \partial \sigma_{yy}) - \dot{\varepsilon}^e_{yy}(\partial Y / \partial \sigma_{xy}) \end{aligned} \tag{4.44}$$

If u and v are the x and y components of velocity, then:

$$\begin{aligned} &u,_x (\partial Y / \partial \sigma_{xy}) - (1/2)(u,_y + v,_x)(\partial Y / \partial \sigma_{xx}) = E_1 \\ &(1/2)(u,_y + v,_x)(\partial Y / \partial \sigma_{yy}) - v,_y (\partial Y / \partial \sigma_{xy}) = E_2 \end{aligned} \tag{4.45}$$

where the right-hand-side terms are just those in Equation (4.44). This system has the form:

$$\begin{aligned} &A_1 u,_x + B_1 u,_y + C_1 v,_x + D_1 v,_y = E_1 \\ &A_2 u,_x + B_2 u,_y + C_2 v,_x + D_2 v,_y = E_2 \\ &u,_x\, dx + u,_y\, dy + 0 + 0 = du \\ &0 + 0 + v,_x\, dx + v,_y\, dy = dv \end{aligned} \tag{4.46}$$

where differentials of u and v have been added to Equation (4.45). The system (Equation 4.46) can be solved by the method of determinants. Thus:

$$u_{,x} = \frac{|N_1|}{\Delta}, \quad u_{,y} = \frac{|N_2|}{\Delta}, \quad v_{,x} = \frac{|N_3|}{\Delta}, \quad v_{,y} = \frac{|N_4|}{\Delta}, \tag{4.47}$$

where:

$$\Delta = \begin{vmatrix} A_1 & B_1 & C_1 & D_1 \\ A_2 & B_2 & C_2 & D_2 \\ dx & dy & 0 & 0 \\ 0 & 0 & dx & dy \end{vmatrix}, \quad |N_1| = \begin{vmatrix} E_1 & B_1 & C_1 & D_1 \\ E_2 & B_2 & C_2 & D_2 \\ du & dy & 0 & 0 \\ dv & 0 & dx & dy \end{vmatrix}, etc \tag{4.48}$$

Expanding Δ and setting the results to zero leads to the expression:

$$(dy/dx)^2 [AC] + (dy/dx)(-[BC]-[AD]) + [BD] = 0 \tag{4.49}$$

Expanding the second of Equation (4.48) results in:

$$du\{[BC]-[BD](dx/dy)\} - dv[CD] + [CE]dy - [DE]dy = 0 \tag{4.50}$$

The coefficients in Equations (4.49) and (4.50) are:

$$\begin{aligned}
[AC] &= (1/2)(\partial Y/\partial\sigma_{xy})(\partial Y/\partial\sigma_{yy}) \\
[BC] &= 0 \\
[AD] &= (\partial Y/\partial\sigma_{xx})(\partial Y/\partial\sigma_{xy}) \\
[BD] &= (1/2)(\partial Y/\partial\sigma_{xx})(\partial Y/\partial\sigma_{xy}) \\
[CD] &= (1/2)(\partial Y/\partial\sigma_{xx})(\partial Y/\partial\sigma_{xy}) \\
[CE] &= -(1/2)(\partial Y/\partial\sigma_{xx})E_2 - (\partial Y/\partial\sigma_{yy})E_1 \\
[DE] &= \partial Y/\partial\sigma_{xy})E_1
\end{aligned} \tag{4.51}$$

The condition that $\Delta = 0$ is therefore:

$$dy/dx = -\frac{\partial Y/\partial\sigma_{xy}}{\partial Y/\partial\sigma_{yy}} \pm \left[\left(\frac{\partial Y/\partial\sigma_{xy}}{\partial Y/\partial\sigma_{yy}}\right)^2 - \left(\frac{\partial Y/\partial\sigma_{xy}}{\partial Y/\partial\sigma_{yy}}\right) \right]^{1/2} \tag{4.52}$$

Alternatively:

$$dy/dx = -\frac{\dot{e}_{xy}}{\dot{e}_{yy}} \pm \left[\left(\frac{\dot{e}_{xy}}{\dot{e}_{yy}}\right)2 - \left(\frac{\dot{e}_{xx}}{\dot{e}_{yy}}\right)^2 \right]^{1/2} \tag{4.53}$$

where e is used for the plastic part of strain.

The plastic strain rates are obtained from the rules of flow (Equation 4.42). Recall the yield condition (strictly, the plastic potential) may be expressed in the forms $Y(\sigma_m,\tau_m) = 0$, $\tau_m - f(\sigma_m)$, $\sigma_m - g(\tau_m)$. The major principal plastic strain rate is then:

$$\dot{e}_1 = \lambda(\partial Y / \partial \sigma_1) = \lambda[(\partial Y / \partial \sigma_m)(\partial \sigma_m / \partial \sigma_1) + (\partial Y / \partial \tau_m)(\partial \tau_m / \partial \sigma_1)$$

by the chain rule for differentiation. Hence, the principal plastic strain rates are:

$$\begin{aligned} \dot{e}_1 &= -\lambda(1-g')/2 \\ \dot{e}_1 &= -\lambda(1+g')/2 \end{aligned} \tag{4.54}$$

where $g' = 1/\tan(\psi)$ as before. The Cartesian plastic strain rates are then:

$$\begin{aligned} \dot{e}_1 &= -\lambda(1-g'\cos(2\theta))/2 \\ \dot{e}_1 &= -\lambda(1+g'\cos(2\theta))/2 \\ \dot{e}_{xy} &= -\lambda g'\sin(2\theta))/2 \end{aligned} \tag{4.55}$$

Using Equation (4.55) in Equation (4.53) leads to the equations of velocity characteristics. Thus:

$$dy / dx = \frac{\sin(2\theta) \pm \sqrt{1-\tan^2(\psi)}}{\cos(2\theta) + \tan(\psi)} \tag{4.56}$$

which is the same as for the characteristics of stress, provided the plastic potential and yield (failure) criterion are the same. Again, use of trigonometric identities leads to:

$$dy / dx = \tan(\theta \pm \mu) \quad : \begin{cases} C_1 \\ C_2 \end{cases} \tag{4.57}$$

If the plastic potential and failure criterion are not the same, so the rules of flow are not associated, then a distinction between ψ's in the stress and velocity subsystems is needed. Corresponding distinctions between μ's and ϕ's are also required.

The velocity relations along the velocity characteristics (Equation 4.57) are obtained from Equations (4.50) and (4.51). Thus, after some algebra:

$$\cos(\theta \pm \mu)\, du + \sin(\theta \pm \mu)\, dv = [\, \dot{\varepsilon}_1^e \;(1 + \sin(\phi) + \dot{\varepsilon}_3^e \;(1 - \sin(\phi)] \tag{4.58}$$

If the elastic strain rates are negligible or if the strain rates are entirely plastic (ideally plastic material, non-hardening), of if they are constant in relation to the plastic strain rates, then the velocity subsystem is:

$$\begin{aligned} dy / dx &= \tan(\theta \pm \mu) \quad : \begin{cases} C_1 \\ C_2 \end{cases} \\ du + dv \tan(\theta \pm \mu) & \quad : \begin{cases} C_1 \\ C_2 \end{cases} \end{aligned} \tag{4.59a,b}$$

Suppose the x-axis is tangential to a first characteristic curve ($\theta + \mu = 0$). In this case $du = 0$, so no change in velocity occurs along the characteristic. Hence, the characteristic is inextensible. A similar argument holds when the x-axis is tangential to a second characteristic curve ($\theta - \mu = 0$). Thus, the characteristic curves form a network of inextensible curves.

Again, consider the case when the x-axis is tangent to a C_1, so the companion C_2 curve is inclined at -2μ to the x-axis as shown in Figure 4.9. According to Equation (4.59b), $du + dv \tan(-2\mu) = du + dv\cot(\phi)$. Hence, $du/dv = \tan(\phi)$. A similar analysis shows that $dv/du = -\tan(\phi)$ when the x-axis is coincident with a C_2 curve. Thus:

$$dv / du = \pm \tan(\phi) \quad : \begin{cases} C_1 \\ C_2 \end{cases} \tag{4.60}$$

where v and u have the interpretation of normal and tangential velocity components on the characteristic curves.

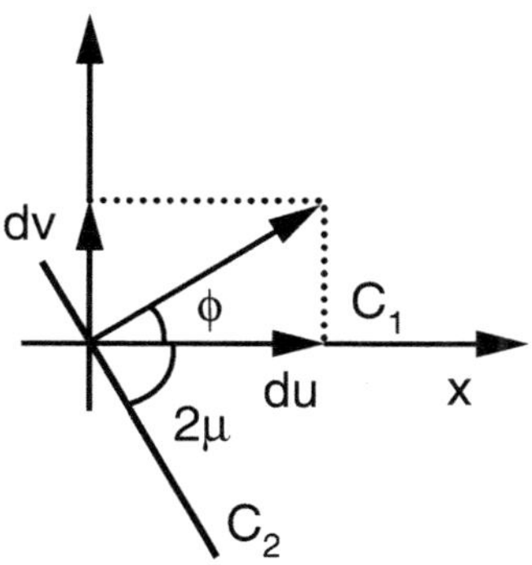

Figure 4.9 Velocity along a characteristic curve.

Another interesting form of the velocity equations is obtained by rotation of the reference axes to a system (s, n) where s is parallel to a characteristic and n is normal. The rotation expresses velocities components v_s and v_n in terms of the Cartesian components u and v with inverse relations:

$$\begin{aligned} u &= v_s \cos(\theta \pm \mu) - v_n \sin(\theta \pm \mu) \\ v &= v_s \sin(\theta \pm \mu) - v_n \cos(\theta \pm \mu) \end{aligned} \tag{4.61}$$

After differentiation and substitution into Equation (4.60), one obtains:

$$dv_s - v_n\, d(\theta \pm \mu) = 0 : \begin{cases} C_1 \\ C_2 \end{cases} \tag{4.62}$$

where a variable μ is allowed.

A more general rotation of axes is:

$$\begin{aligned} u &= v_s \cos(\alpha) - v_n \sin(\alpha) \\ v &= v_s \sin(\alpha) - v_n \cos(\alpha) \end{aligned} \tag{4.63}$$

where α is an arbitrary angle of axes rotation, positive counterclockwise, from (x, y) to (s, n). Under the rotation (Equation 4.63), the velocity relations (Equation 4.60) assume the form:

$$(dv_s - v_n\, d\alpha)\cos(\theta - \alpha \pm \mu) + (dv_n - v_s\, d\alpha)\sin(\theta - \alpha \pm \mu) = 0 \tag{4.64}$$

4.5 A special plane strain velocity field (streaming flow)

A velocity field that is compatible with the three special stress fields is one of "streaming flow" where the particle paths coincide with streamlines. If α is interpreted as the angle from the x-axis to a streamline in Equation (4.63), then the normal component of velocity v_n vanishes by definition of a streamline, while the tangential velocity v_s is the velocity V. From Equation (4.63), $(dv_s)\cos(\theta - \alpha \pm \mu) + v_s \sin(\theta - \alpha \pm \mu)\, d\alpha = 0$. Hence:

$$dV / V + \tan(\theta - \alpha \pm \mu)\, d\alpha = 0 \tag{4.65}$$

If θ-α is a constant, then Equation (4.65) leads to:

$$V / V_o = \exp[-(\alpha - \alpha_o)\tan(\theta_o - \alpha_o \pm \mu) \ : \begin{cases} C_1 \\ C_2 \end{cases} \tag{4.66}$$

where the subscript "o" is a boundary or "starting" value.

In a region of constant state, θ is a constant and the characteristics are straight lines. The assumption that θ-α is a constant then implies α is a constant. Thus, the left side of Equation (4.66) is also a constant. The motion is then a rigid body motion that clearly satisfies the original velocity relations (Equation 4.61).

In a region of radial shear, one family of characteristics is straight; θ and α are constant along a straight characteristic. According to Equation (4.66), the velocity is also constant along a straight characteristic (rays in polar coordinates) but changes from one straight characteristic to another.

In a region of radial stress, the rays are principal directions, and either $\theta = \omega$ or $\theta = \omega + \pi/2$. If the flow is radial, then $\omega = \alpha$ also. From Equation (4.65), when $\theta = \omega$ and the rays are directions of σ_1, then:

$$V / V_o = \exp[\pm(\omega - \omega_o)\tan(\mu) \qquad : \theta = \omega \tag{4.67a}$$

When the rays are directions of σ_3, then:

$$V / V_o = \exp[\pm(\omega - \omega_o)\cot(\mu) \qquad : \theta = \omega + \pi / 2 \tag{4.67b}$$

In consideration of the characteristic curve geometry (Equation 4.35), one has from Equation (4.67a):

$$V / V_o = (r_o / r)^{\tan^2 \mu} \tag{4.68a}$$

and from Equations (4.37) and (4.67b):

$$V / V_o = (r_o / r)^{\cot^2 \mu} \tag{4.68b}$$

which indicate velocity decreases with r in both cases, although at different rates, as one might expect.

4.6 Example problems

An important problem that is often encountered in engineering practice is a hollow cylinder under internal and external pressure. This problem makes use of the third special stress and velocity fields. A number of interesting features of elastic-plastic problems in general are revealed during the problem-solving process. An appreciation of the cascading complexities of elastic-plastic analysis of Coulomb material is also developed. Central to the solution of this problem are conditions regarding the boundary between an entirely elastic region and the elastic-plastic zone. The shape of the boundary is known to be circular because of symmetry, but the position varies with the applied pressures and must be computed as part of the solution. During the process of locating the elastic-plastic boundary, the meaning of "velocity" in time-independent plasticity is clarified.

Another problem of considerable engineering importance concerns wedge indentation. In this problem, one assumes location of the elastic-plastic boundary and proceeds to find stress and velocity fields that satisfy boundary conditions. The concept of a rigid-plastic response is also introduced, and with this concept, discontinuities in stress and velocity as well. Foundation-bearing capacity is a problem of some importance and is considered as a special case of the wedge problem, a problem with many variants and much associated literature.

A third example problem involves retaining walls where weight of material may not be negligible as in the preceding examples. The difficulty that arises when weight is important is a fundamental one and leads to consideration of numerical methods for integrating the governing differential equations from characteristic form. Integration gives rise to several types of boundary value problems with a structure clearly revealed in the characteristic form of equations describing plane strain plasticity.

Hollow cylinder under external pressure

The basic assumption in the problem of a hollow cylinder under internal pressure p_a and external pressure p_b is one of radial displacement u. The plane strain condition requires the axial displacement and strain to be zero. In the purely elastic case, the solution is $u = Cr + D/r$ where r is a radial coordinate, and C and D are constants to be determined from boundary conditions (e.g., Love, 1944). In this example, $p_a = 0$ at $r = a$, while p_b is specified at $r = b$; compression is positive as shown in Figure 4.10.

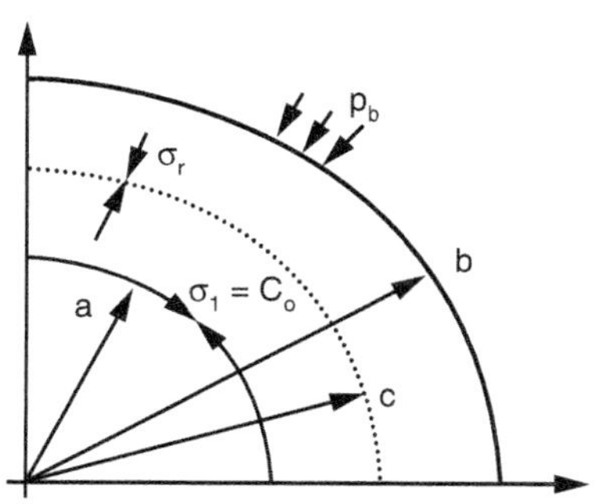

Figure 4.10 Elastic-plastic hollow cylinder under external pressure.

The strains in cylindrical coordinates are where the negative sign is introduced into the strain-displacement equations, because compression is positive and, in consideration of Hooke's law:

$$\begin{aligned}
&\varepsilon_r = -\frac{\partial u}{\partial r} = -(C - \frac{D}{r^2}), \quad \varepsilon_\theta = -\frac{u}{r} = -(C - \frac{D}{r^2}) \\
&\varepsilon_r = -(C - \frac{D}{r^2}) = (\frac{1}{E})(\sigma_r - v\sigma_\theta - v\sigma_z) \\
&\varepsilon_\theta = -(C + \frac{D}{r^2}) = (\frac{1}{E})(\sigma_\theta - v\sigma_r - v\sigma_z) \\
&\varepsilon_z = 0 = (\frac{1}{E})(\sigma_z - v\sigma_r - v\sigma_\theta) \\
&\sigma_z = v(\sigma_r - \sigma_\theta)
\end{aligned} \tag{4.69}$$

where E and v are Young's modulus and Poisson's ratio, respectively. Application of the boundary condition on $r = b$ leads to $CE = (1 - 2v)DE/r^2 - p_b(1 + v)(1-2v)$ and:

$$\begin{aligned}
&\sigma_r = p_b - (DE/(1+v))(\frac{1}{b^2} - \frac{1}{r^2}) \\
&\sigma_\theta = p_b - (DE/(1+v))(\frac{1}{b^2} + \frac{1}{r^2})
\end{aligned} \tag{4.70}$$

that apply in the elastic region defined loosely as $r \le b$.

Application of the second boundary condition on $r = a$, $p_a = 0$, leads to:

$$\begin{aligned}
&\sigma_r = p_b - p_b(\frac{1}{b^2} - \frac{1}{r^2}) / (\frac{1}{b^2} - \frac{1}{a^2}) \\
&\sigma_\theta = p_b - p_b(\frac{1}{b^2} + \frac{1}{r^2}) / (\frac{1}{b^2} + \frac{1}{a^2})
\end{aligned} \tag{4.71}$$

that apply to the entire cylinder in the purely elastic case.

In case $p_a \neq 0$, then:

$$\begin{aligned}
&\sigma_r = p_b - (p_b - p_a)(\frac{1}{b^2} - \frac{1}{r^2}) / (\frac{1}{b^2} - \frac{1}{a^2}) \\
&\sigma_\theta = p_b - (p_b - p_a)(\frac{1}{b^2} + \frac{1}{r^2}) / (\frac{1}{b^2} + \frac{1}{a^2})
\end{aligned} \tag{4.71a}$$

As the applied pressure $p_b(t)$ is increased in time, at some point the cylinder will reach the elastic limit. Yielding will occur under the pressure p_b^* when the yield condition is first satisfied. In this example, a Mohr-Coulomb failure criterion is assumed in the form $\tau_m = A\sigma_m + B$.

By symmetry of the problem and Equation (4.71), $\sigma_1 = \sigma_\theta$, $\sigma_3 = \sigma_r$ that in the yield condition results in:

$$-p_b(1-r^2) = A[-p_b(1/a^2)] + B(1/b^2 - 1/a^2) \tag{4.72}$$

This result indicates that yielding first occurs where r is least, that is, at $r = a$, under the pressure:

$$-p_b^* = (\frac{B}{1-A})(1-\frac{a^2}{b^2}) = [\frac{c\cos(\phi)}{1-\sin(\phi)}](1-\frac{a^2}{b^2}) = (\frac{C_o}{2})(1-\frac{a^2}{b^2}) \tag{4.73}$$

Inspection of Equation (4.73) shows that for a given cylinder geometry, first yield depends only on the unconfined compressive strength of the material. The radial stress at $r = a$ is zero, while the circumferential stress is a compression such that:

$$\sigma_\theta = \sigma_1 = C_o = (\frac{2B}{1-A}) = \frac{2c\cos(\phi)}{1-\sin(\phi)} \tag{4.74}$$

Indeed, and as a check:

$$C_o = (\frac{2p_b^*}{1-(a/b)^2}) \tag{4.75}$$

As b becomes indefinitely large, one sees that the stress concentration at the hole wall is just 2, which is the known stress concentration about a circular hole excavated in a hydrostatic stress field.

Similar results occur when $p_a \neq 0$, that is, first yielding occurs at $r = a$. A pressure difference at first yield analogous to Equation (4.73) is:

$$-p_b^* = p_a^*(\frac{B}{1-A})(1-\frac{a^2}{b^2}) = [\frac{c\cos(\phi)}{1-\sin(\phi)}](1-\frac{a^2}{b^2}) = (\frac{C_o}{2})(1-\frac{a^2}{b^2}) \tag{4.73a}$$

provided $p_b^* > p_a^*$, so certainly $\sigma_\theta = \sigma_1 > \sigma_r = \sigma_3$. A result analogous to Equation (4.74) is then:

$$\sigma_\theta(a) = \sigma_1 = (\frac{2B}{1-A}) = \frac{1+A}{1-A}\sigma_r(a) \tag{4.74a}$$

$$= \frac{2c\cos(\phi)}{1-\sin(\phi)} + (\frac{1+A}{1-A})\sigma_r(a)$$

$$\sigma_\theta(a) = \sigma_1 = C_o + (\frac{C_o}{T_o})\sigma_r(a)$$

where relationships among the Mohr-Coulomb strength properties are used. If $p_b^* < p_a^*$, so the internal pressure exceeds the external pressure, then there is a possibility of failure in tension at the inner wall of the cylinder.

Once yielding is initiated, the elastic-plastic boundary moves into the cylinder with further increase in the external pressure p_b. Within the plastic zone, the maximum shear stress is given by $\tau_m = \tau_m^o (r/r)^{-2\tan(\phi)\cot(\mu)}$; the mean normal stress is $\sigma_m = (\tau_m / A) - (B/A)$, and $\tau_m^o = C_o / 2 = B/(1-A)$. Hence:

$$\begin{aligned} \sigma_r &= (B/A)[(r/a)^{-2\tan(\phi)\cot(\mu)} - 1] \\ \sigma_r &= (B/A)[(\frac{1+A}{1-A})(r/a)^{-2\tan(\phi)\cot(\mu)} - 1] \quad (a \le r \le c) \end{aligned} \tag{4.75a}$$

where c is the radius to the elastic-plastic boundary.

The pressure required to extend the plastic zone to the outer wall of the cylinder and therefore to cause collapse is given by the first of Equation (4.75) after setting $r = c = b$. Thus, at collapse:

$$-p_b^{**} = (B/A)[(b/a)^{-2\tan(\phi)\cot(\mu)} - 1] \tag{4.76}$$

Stresses in the elastic region need to be recomputed because of yielding. At the elastic-plastic boundary, material is just at yield. The stresses given by Equation (4.70) satisfy the yield condition, so:

$$-(\frac{DE}{1+v})(\frac{1}{c^2}) = A[p_b - (\frac{DE}{1+v})(\frac{1}{b^2})] + B$$

Hence now:

$$-(\frac{DE}{1+v}) = (Ap_b + B)(\frac{A}{b^2} - \frac{1}{c^2}) \tag{4.77}$$

After back substitution into Equation (4.70), one obtains:

$$\begin{aligned} \sigma_r &= p_b (Ap_b + B)(\frac{1}{b^2} - \frac{1}{r^2}) / (\frac{A}{b^2} - \frac{1}{r^2}) \\ \sigma_\theta &= p_b - (Ap_b + B)(\frac{1}{b^2} + \frac{1}{r^2}) / (\frac{A}{b^2} - \frac{1}{r^2}) \qquad (c \le r \le b) \end{aligned} \tag{4.78}$$

that clearly show the dependence of stress in the elastic region on the location of the elastic-plastic boundary.

The location of the elastic-plastic boundary remains to be determined from boundary conditions. While one may reasonably suppose material at the elastic-plastic boundary is just at yield, equality of the circumferential stresses cannot be guaranteed. However, equilibrium requires equality of the radial stresses at the elastic-plastic boundary. Thus:

$$(\frac{B}{A})[(\frac{c}{a})^{-2\tan(\phi)\cot(\mu)} - 1] = p_b - (Ap_b + B)(\frac{1}{b^2} + \frac{1}{c^2}) / (\frac{A}{b^2} - \frac{1}{c^2}) \tag{4.79}$$

that is an implicit function for c in terms of p_b. Solving Equation (4.79) results in:

$$(\frac{P_b}{C_o}) = (1/2)[\frac{c}{a}]^{-2\tan(\phi)\cot(\mu)}](\frac{1}{A} - \frac{c^2}{b^2}) - (1/2)(\frac{1-A}{A}) \quad (4.80)$$

that is normalized relative to unconfined compressive strength C_o. Although explicit solution for c is not possible, Equation (4.80) allows for easy graphical solution.

Use of Equation (4.80) in Equation (4.78) gives normalized expressions in the elastic region, and normalization of Equation (4.75) in the plastic region leads to:

$$\left.\begin{matrix}(\sigma_r / C_o)\\(\sigma_\theta / C_o)\end{matrix}\right\} = (1/2)(c/a)^{-2\tan(\phi)\cot(\mu)}][(1/A) \pm (c^2/r^2)] - (1-A)/(2A) \quad (c \le r \le b) \quad (4.81)$$

$$\left.\begin{matrix}(\sigma_r / C_o)\\(\sigma_\theta / C_o)\end{matrix}\right\} = (1/2)(r/a)^{-2\tan(\phi)\cot(\mu)}][(1/A) \pm 1)] - (1-A)/(2A) \quad (a \le r \le c) \quad (4.82)$$

Substitution of $r = c$ in the first of Equation (4.81) and comparison with the first of Equation (4.82) shows that equilibrium is indeed insured by equality of radial stress at the elastic-plastic boundary. Of interest is the equality $\tan(\phi)\cot(\mu) = A/(1 - A)$ where $A = \sin(\phi)$. The case of $\phi = 0$ must be solved as a separate problem.

Calculations of radial (SR) and tangential (ST) stress in the case of a hollow cylinder with inner radius $a = 1$ and outer radius $b = 3$ are shown in Figure 4.11. An angle of internal friction (phi) of 35° is assumed. Stresses in Figure 4.11 are normalized by division by the unconfined compressive strength of the cylinder. The evolution of the stress field is in response to an increasing external pressure. Purely elastic stress ends with first yield at $r = a$ under a normalized external pressure $(p_b^* / C_o) = 0.444$. The stresses at this point are indicated with open symbols. The circumferential stress at first yield at $r = a$ is just the unconfined compressive strength of the cylinder material. With increasing pressure, the plastic zone grows; position of the plastic zone $r = c$ as function of the external pressure is shown in Figure 4.12. The approach of the plot in Figure 4.12 towards the vertical with increased external pressure is a precursor to collapse. The plastic zone reaches the outer radius when $(p_b^* / C_o) = 6.769$. Thus, the collapse pressure is about 6.8 times unconfined compressive strength. The circumferential stress at the outer wall $r = b$ at collapse is almost 26 times unconfined compressive strength.

A mistake that is sometimes made is to locate the elastic-plastic boundary by equating the circumferential stresses given by the purely elastic formulas (Equation 4.71) and the plastic formulas (Equation 4.75). The procedure is illustrated graphically in Figure 4.13, where the open symbols relate to Equation (4.71) and the solid symbols relate to Equation (4.75). The intersection points of the purely elastic and plastic curves are well away from the peak stresses that occur at the actual elastic-plastic boundary. Serious underestimation of the peak circumferential stress is associated with this deceptive and erroneous solution procedure. The proper procedure is to determine the location of the elastic-plastic boundary as a function of the external boundary conditions. Any attempt to equate radial stresses to one another shows the fallacy quite clearly as well. Equality occurs only at the cylinder walls regardless of where the actual elastic-plastic boundary is located.

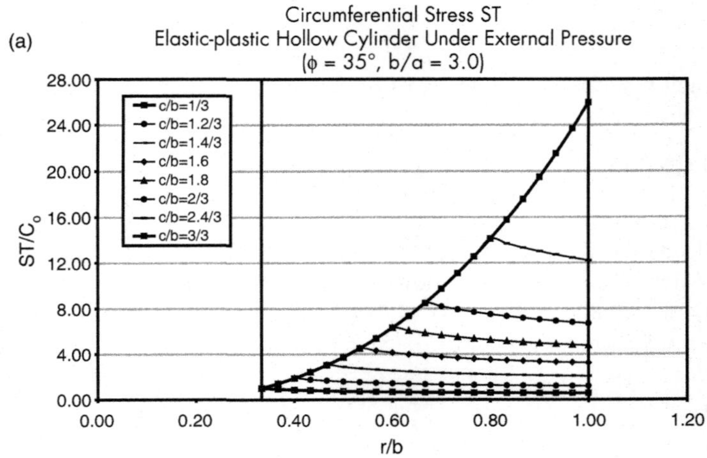

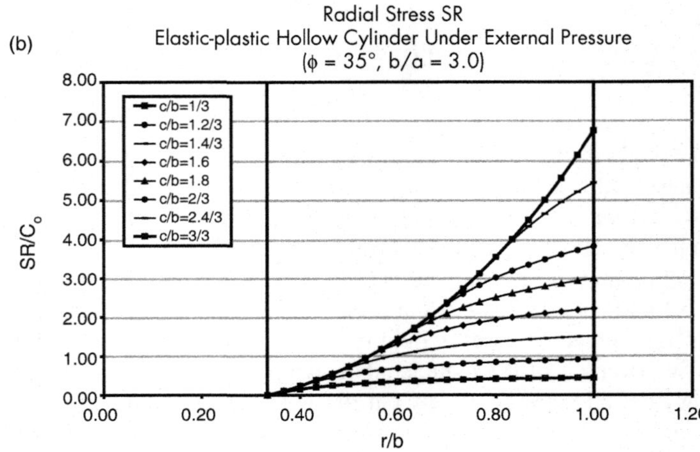

Figure 4.11 Stress distributions in a hollow cylinder under external pressure. (a) Circumferential stress. (b) Radial stress.

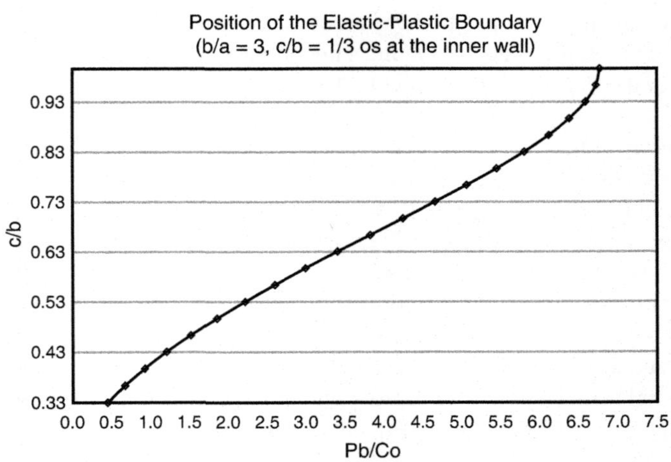

Figure 4.12 Position of the elastic-plastic boundary as a function of applied pressure.

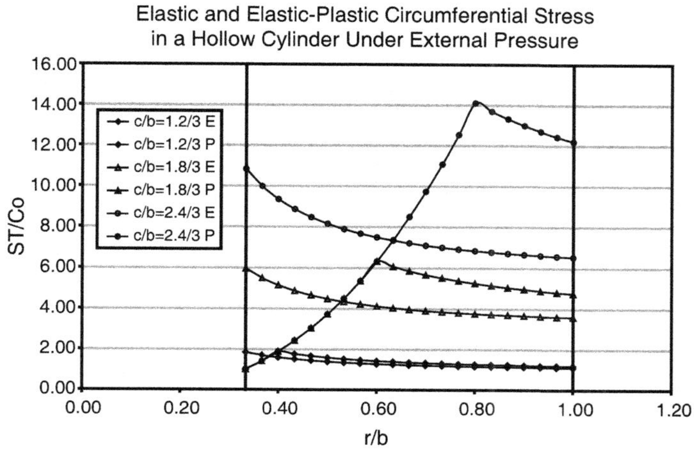

Figure 4.13 Elastic and elastic-plastic circumferential stress at various external pressures.

All shear stresses are nil, but the axial stress σ_z remains to be determined. In principle, the process of eliminating σ_z from the stress-strain relations at the outset would lead to a formula that would allow computation of σ_z as it does in the purely elastic case. However, no analogous algebraic formula exists for the elastic-plastic region. Moreover, just as there is a possibility of a discontinuity in circumferential stress at the elastic-plastic boundary, there is also a possible discontinuity in the axial stress. If the axial load were known, then the average axial stress in the plastic region could be computed from knowledge of the axial stress in the elastic region. Recourse to numerical analysis becomes necessary to determine σ_z.

A velocity field for the hollow cylinder is radial is given by Equation (4.68b)

$$(V / V_o + = (r_o / r)^{\cot^2(\mu)}$$

In this example, $r_o = a$, $V_o = V(a)$. Velocity usually refers to a time rate of change, but in time-independent plasticity, any time-like, monotonically increasing variable will do. The distance c to the elastic-plastic boundary is a convenient time variable in this example. Thus, $V = dr/dc$, $V(a) = da/dc$, and:

$$dr / dc, = (da / dc)(a / r)^{\cot^2(\mu)} \tag{4.83}$$

Integration of Equation (4.83) gives:

$$r^{1+\cot^2(\mu)} - r_o^{1+\cot^2(\mu)} = a^{1+\cot^2(\mu)} - a_o^{1+\cot^2(\mu)} \tag{4.84}$$

where a particle originally at r_o is now at r. During this motion, the inner cylinder wall originally at a_o is now at a. Radial displacement $u = r\text{-}r_o$ and $u(a) = a\text{-}a_o$. After substitution into Equation (4.84) and expansion in series form, one has with neglect of second order and higher terms:

$$u = u_a(a / r)^{\cot^2(\mu)} \tag{4.85}$$

that gives the displacement of a particle currently at r in terms of the displacement and radius on the inner surface of the cylinder. Equation (4.85) applies only in the plastic zone and was obtained with neglect of the elastic strains implying a rigid-plastic model. Of course, the elastic strains in the plastic region are only considered negligible after achieving a solution for stress. Consistency would seem to require neglect of the elastic strains altogether. However, if the strains in the elastic region are not neglected, then continuity of displacement at the elastic-plastic boundary requires $u = u_a (a/c)^{\cot^2(\mu)} = Cc + D/c$ and therefore:

$$u = (Cc + D/c)(c/r)^{\cot^2(\mu)} \tag{4.86}$$

The constants C and D are known from solution for stress in the elastic region. The parameter c is known from the analysis of stress as well, so the objective of determining displacements in terms of boundary conditions is achieved, but with some inconsistency.

An alternative is to continue with the rigid-plastic assumption. In this case, a velocity discontinuity must occur at the elastic-plastic, that is, rigid-plastic boundary. Material just on the rigid side of the boundary must undergo a sudden jump in velocity across the rigid-plastic boundary. However, velocity discontinuities can occur only across velocity characteristics. But the rigid-plastic boundary $r = c$ is not a velocity characteristic or an envelope of velocity characteristics. Seemingly, an alternative solution does not exist. Resolution of the situation was first given by Pariseau (1972) where additional variations on the theme of the hollow cylinder, including anisotropy, are also presented.

Wedge indentation

The main objective in solving this problem is to find the relationship between force and penetration depth as described by a force-displacement curve. Extent of damage and chip velocity are also of interest. Wedge geometry, notation, and characteristics are shown in Figure 4.14. The wedge is considered smooth and perfectly lubricated, with no shear stress acting on the wedge face that has slope length a. The wedge acts in a machined notch. Wedge angle is 2β, wedge force is F, penetration depth is h, and surface damage extends from O to D in Figure 4.14. Plane strain conditions are assumed in consideration of the small penetration depth h relative to wedge length (into the page) of one.

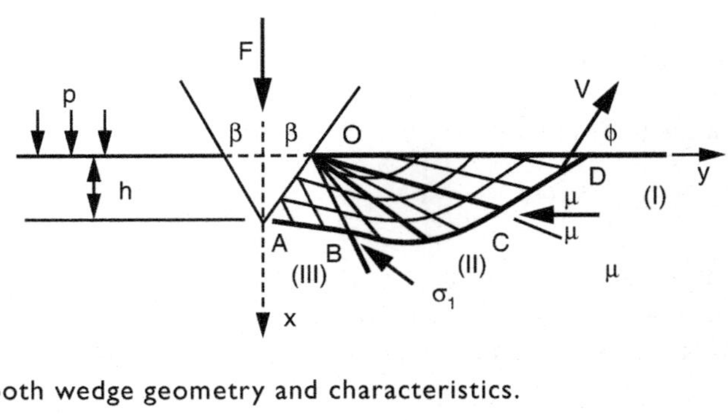

Figure 4.14 Smooth wedge geometry and characteristics.

Location of the elastic-plastic boundary is assumed to be ABCD. With neglect of elastic strains, ABCD is a rigid-plastic boundary. Regions of constant state (I, III) that are separated by a region of radial shear (II) are shown in Figure 4.14. Weight is considered negligible in this example, so a tacit assumption is made that the material is relatively strong and obeys a Mohr-Coulomb yield condition $\tau_m = A\sigma_m + B$. The wedge itself is considered rigid and so strong that breakage is not a factor.

Boundary conditions for the problem require stress normal to the wedge face to be a principal stress because no shear stress acts on the wedge face. The plane $x = 0$ is under a uniform pressure p. In consideration of the characteristics in (I) and the direction of σ_1, the pressure $p = \sigma_3$ and $\theta = \pi / 2$ in (I). From the yield condition, $\sigma_1 = C_o + (C_o/T_o)p$ along OD. Hence, $\sigma_m = p/(1 - A) + B/(1 - A)$ and $\tau_m = pA/(1 - A) + B/(1 - A)$ along OD, in (I), and in particular, along OC at the beginning of the region of radial shear (II).

In (II), the rays are C_2 lines, while the C_1 curves are exponential spirals. From Equation (4.32) after taking the origin at O and setting $\theta - \mu = \omega$, one obtains the equation for the spiral characteristics:

$$(r / r_o) = \exp[(\omega - \omega_o)\tan(\phi)] \tag{4.87}$$

From Equation (4.30), the variation of the maximum shear stress along Equation (4.87) is:

$$(\tau_m / \tau_m^o) = (r_o / r) = \exp[-2(\omega - \omega_o)\tan(\phi)] = \exp[2(\theta - \theta_o)\tan(\phi)] \tag{4.88}$$

where "o" signifies a starting or boundary value. The change in principal stress direction between regions (I) and (III) is $\theta - \theta_o = (\pi / 2 - \beta) - (\pi / 2) = -\beta$ which indicates a decrease in angle θ. The angle β is an important parameter in this problem and is known as the *fan angle*. The maximum shear stress in (III) is then given by:

$$(\tau_m / \tau_m^o) = \exp[2\beta \tan(\phi)] \tag{4.89}$$

where the starting stress τ_m^o is in (I) and the angle β is the change in θ from (I) to (III). Thus:

$$\tau_m = [pA / (1 - A) + B / (1 - A)]\exp[2\beta \tan(\phi) \text{ in (III)} \tag{4.90}$$

The mean normal stress in (I) is obtained from the yield condition. Thus, $\sigma_m = (\tau_m - B)/A$, while the major principal stress acting normal to the wedge face is simply $\sigma_1 = \sigma_m + \tau_m$. Hence, along the wedge face:

$$\sigma_1 = [-B / A) + [1 + A)[pA / (1 - A)] + B / \exp[2\beta \tan(\phi)$$

that may be brought into a normalized form:

$$(\frac{2\sigma_1}{C_o}) = -(\frac{1 - A}{A}) + [(\frac{1}{A}) + (\frac{2p / C_o}{1 - A})](1 + A)\exp[2\beta \tan(\phi)] \tag{4.91}$$

where a 2 is introduced to take into account both wedge faces.

The normal stress acting on the wedge faces is a constant, so integration is simple. Slope distance of a wedge face $a = h/\cos(\beta)$ The vertical force from Equation (4.91) in normalized form is therefore:

$$(\frac{F}{C_o h\tan(\beta)}) = -(\frac{1-A}{A}) + [(\frac{1}{A}) + (\frac{2p/C_o}{1-A})](1+A)\exp[2\beta\tan(\phi)] \tag{4.92}$$

Equation (4.92) describes a linear relationship between force and penetration depth as a smooth-faced wedge penetrates a material following a Mohr-Coulomb yield criterion. The wedge length is one measured normal to the plane of analysis (normal to the page) in Equation (4.92).

In the case where the wedge angle is $\pi/2$ and the indenter is flat, then the force is:

$$(\frac{F}{aC_o}) = -(\frac{1-A}{A}) + [(\frac{1}{A}) + (\frac{2p/C_o}{1-A})](1+A)\exp[\pi\tan(\phi)] \tag{4.93}$$

that has an interpretation of bearing capacity of strip foundation $2a$ units wide (per unit) of length normal to the plane of analysis founded on a Mohr-Coulomb material.

When the material lacks cohesion, so the angle of internal friction is zero, and the yield condition is, $\tau_m = k$, then the problem requires use of Equation (4.27b). Combining Equation (4.27b) with the yield condition and wedge angle gives:

$$\sigma_1 = p + 2k + 2k\beta \tag{4.94}$$

When no pressure acts adjacent to wedge and the wedge is flat $\beta = \pi/2$, the unit-bearing capacity is simply $\sigma_1 = k(2+\pi) = 5.14k$, a value often cited in soil mechanics texts.

The streaming flow velocity field is useful for the wedge problem. The main requirement is equality of velocity components normal to the wedge face. With reference to Figure 4.15, the downward motion of the wedge is V with component normal to the wedge face $V\sin(\beta)$. Velocity in (III) is constant and assumed to act at right angle to the C_2 lines in (III) as shown at B. The normal component of V_{II} is $V_{II}\sin(\mu)$. Thus, at B, $V_{II} = V\sin(\beta)/\sin(\mu)$. Velocity in (II) is constant along a ray, but changes from ray to ray along the C_1 exponential spirals according to Equation (4.65), where α is the angle from the x-axis to the velocity direction. For reasons given later, this angle is $\omega+\pi/2$, where $\omega = \theta - \mu$ in consideration of the fact that the C_2 lines are rays in the region of radial shear. The difference θ-α is constant in streaming flow. Along a C_1 curve then in passage from B to C, $(V_c/V_{II}) = \exp[\beta\tan(\phi)]$. Region (I) is a region of constant state, so the velocity in region (I) is also constant. In particular, velocity at D is just:

$$V_D = [V\sin(\beta)/\sin(\mu)]\exp[\beta\tan(\phi)] \tag{4.95}$$

that is sometimes referred to as the "chip" velocity.

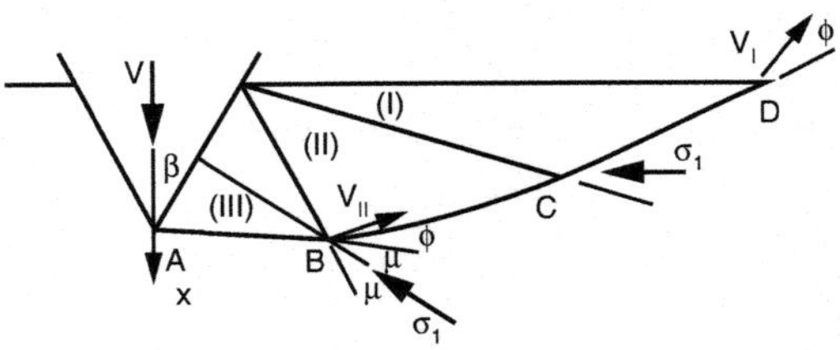

Figure 4.15 Velocity geometry for a smooth wedge.

The velocity "jumps" across the rigid-plastic boundary. In the rigid region where the strains would be purely elastic but are neglected, velocity is zero everywhere. In the plastic region, the velocity is finite and includes a singularity at the pole of the radial shear region. The velocity jump across the rigid-plastic boundary is a "strong" discontinuity that is discussed later. At this juncture, the stress field satisfies equilibrium, yield, and stress boundary conditions, while the velocity field is continuous in the plastic region and satisfies velocity boundary conditions at the wedge face and the rigid-plastic boundary.

Extent of damage is the distance (OD) in Figure 4.18, which can be computed from the geometry of the characteristic regions. Length (OD) = 2(OC)cos(μ) and (OC) is given by $(OC) = (OB)\exp[\beta\tan(\phi)]$. From wedge geometry, $(OB) = [h/\cos(\beta)][1/2\sin(\mu)]$. Hence:

$$(OD) = [h/\cos(\beta)][1/\tan(\mu)]\beta\tan(\phi)]. \tag{4.96}$$

In case of zero angle of internal friction, (OD) $h/\cos(\beta)$.

A modification of the wedge problem is to suppose "lips" form a raised region along OD as shown in Figure 4.16. If the material were incompressible, lip volume would equal the volume displaced by the wedge. Lip formation is at the heart of a more realistic wedge penetration problem and is critical to the solution of a wedge problem.

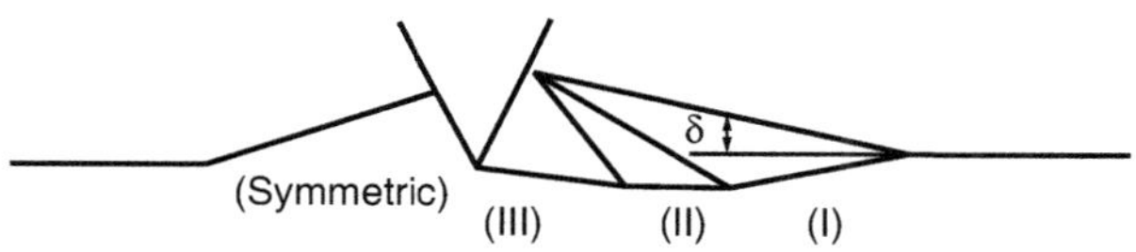

Figure 4.16 Wedge indentation with lip formation angle δ.

There are many other variations on the wedge indentation problem that include the case where friction acts at the wedge-rock interface, penetration of anisotropic material and oblique penetration (Pariseau and Fairhurst, 1967; Pariseau, 1970).

Retaining walls

An example of a problem where weight is important is the case of a smooth retaining wall shown in Figure 4.17. An objective of retaining wall analysis is estimation of the horizontal force and moment acting on the wall at the point of imminent collapse. Two collapse mechanisms are shown in Figure 4.17, the first involves downward motion of material behind the wall that is pushed outward. The second involves inward motion of the wall and upward motion of material behind the wall.

The characteristics of stress in Figure 4.17 are straight lines, because the angle θ is considered constant, and according to an assumption of a Mohr-Coulomb material, μ is also a constant. In Figure 4.17a, the major principal stress is vertical (compression is positive), while in Figure 4.17b, the major principal stress is horizontal. No shear stress is present along the wall; a surcharge p acts at the surface along the y-axis.

In either case, an element of arc length along the characteristic curves is:

$$\begin{aligned} dx &= \cos(\theta \pm \mu)ds \\ dy &= \sin(\theta \pm \mu)ds \end{aligned} \tag{4.97}$$

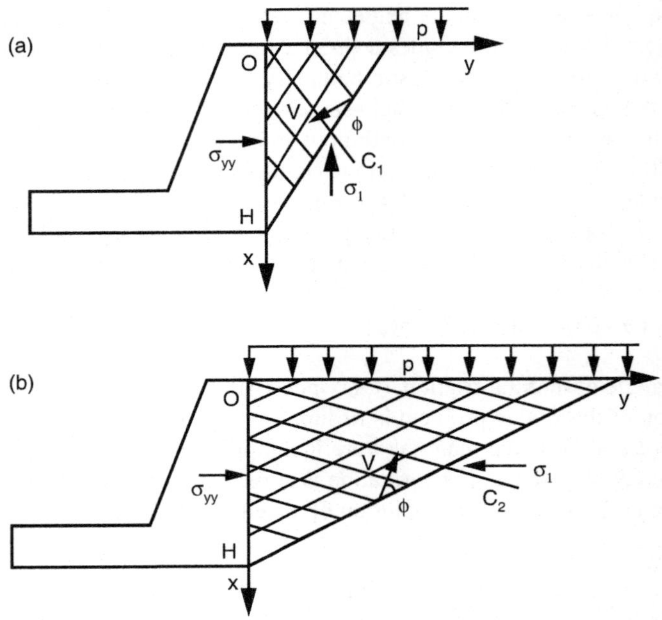

Figure 4.17 Vertical retaining walls: (a) smooth, (b) smooth wall with surcharge.

where the (+) refers to C_1 curves and the (−) refers to C_2 curves. With neglect of accelerations in Equation (4.25) and substitution from Equation (4.97), one obtains:

$$d\tau_m \pm \gamma \tan(\phi)\frac{\sin(\theta \mp \mu)}{\cos(\theta \pm \mu)}dx = 0 \qquad \begin{cases} C_1 \\ C_2 \end{cases} \tag{4.98}$$

In the first case (a) in Figure 4.16, $\theta = 0$; in the second case (b), $\theta = \pi / 2$. After substitution in Equation (4.98) and integration, one obtains:

$$\tau_o = \tau_m^o + \gamma \tan(\phi) = \begin{Bmatrix} \tan(\mu) \\ \cot(\mu) \end{Bmatrix} x, \qquad \begin{cases} (a) \\ (b) \end{cases} \tag{4.99}$$

where τ_m^o is a boundary value on $x = 0$.

Determination of τ_m^o follows from boundary and yield conditions. In case (A):

$$\sigma_3 = p(\frac{1-A}{1+A}) - \frac{2B}{1+A}, \quad \sigma_1 = p, \quad :(a) \tag{4.100}$$

where yield is expressed as $\tau_m = A\,\sigma_m + B$, and in case (B):

$$\sigma_1 = p(\frac{1-A}{1+A}) - \frac{2B}{1+A}, \quad \sigma_3 = p, \quad :(b) \tag{4.101}$$

Thus:

$$\tau_m^o = (\sigma_1 - \sigma_3)\ /2 = \begin{cases} (Ap+B)/(1+A), & :(a) \\ (Ap+B)/(1-A), & :(b) \end{cases} \tag{4.102}$$

and:

$$\begin{aligned} \tau_m &= (Ap + B) / (1{+}A) + \gamma x \tan(\phi)\tan(\mu), \quad &:(a) \\ \tau_m &= (Ap + B) / (1-A) + \gamma x \tan(\phi)\cot(\mu), \quad &:(b) \end{aligned} \tag{4.103}$$

The principal stresses may be obtained from Equation (4.103) and the yield condition, so:

$$\begin{aligned} \sigma_1 &= \sigma_m + \tau_m = (\tau_m - B)/A + \tau_m \\ \sigma_3 &= \sigma_m - \tau_m = (\tau_m - B)/A - \tau_m \end{aligned} \tag{4.104}$$

Use of Equation (4.104) in Equation (4.103) and a variety of trigonometric identities leads to:

$$\begin{aligned} \sigma_1 = \sigma_v = p + \gamma x, \qquad & \sigma_3 = \sigma_h = \left(\frac{1-\sin(\phi)}{1+\sin(\phi)}\right)(p - C_o + \gamma x), \quad &:(a) \\ \sigma_3 = \sigma_v = p + \gamma x, \qquad & \sigma_1 = \sigma_h = \left(\frac{1-\sin(\phi)}{1+\sin(\phi)}\right)(p + T_o + \gamma x), \quad &:(b) \end{aligned} \tag{4.105}$$

where C_o and T_o are unconfined compressive and tensile strength, respectively, and "v" and "h" refer to vertical and horizontal. The distribution of vertical stress is clearly an equilibrium distribution caused by gravity loading augmented by the surface load. The distribution of horizontal stress is illustrated in Figure 4.18.

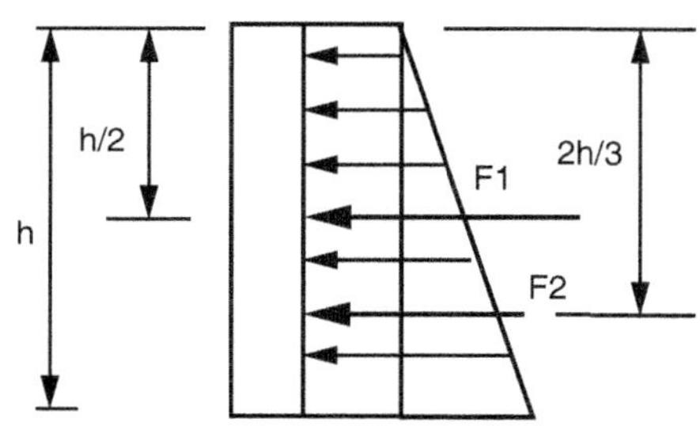

Figure 4.18 Resultant forces from rectangular and triangular distributions of stress.

In traditional soil mechanics treatments of sand, which lacks cohesion, and where no surcharge or surface load is present, Equation (4.105) reduces to:

$$\begin{aligned} \sigma_v = \gamma x, \qquad & \sigma_h = \left(\frac{1-\sin(\phi)}{1+\sin(\phi)}\right)(\gamma x), \quad &:(a) \\ \sigma_v = \gamma x, \qquad & \sigma_h = \left(\frac{1-\sin(\phi)}{1+\sin(\phi)}\right)(\gamma x), \quad &:(b) \end{aligned} \tag{4.105a}$$

Coefficients of γx (vertical stress) in expressions for the horizontal stress in Equation (4.105a) are known as coefficients of *active* and *passive* earth pressure at rest, K_a and K_b, respectively.

Integration of horizontal stress over height h of the retaining wall gives the resultant force. Thus:

$$F=\left(\frac{1-\sin(\phi)}{1+\sin(\phi)}\right)[(p-C_o+\gamma h/2)h, \quad :(a)$$
$$F=\left(\frac{1+\sin(\phi)}{1-\sin(\phi)}\right)[(p+T_o\gamma h/2)h, \quad :(b) \qquad (4.106)$$

The moment about the base of the retaining wall associated with F and the location of action of F follows from:

$$M=\int_0^h \sigma_h(x)x\,dx = Fx_o \qquad (4.107)$$

Thus:

$$M=\left(\frac{1-\sin(\phi)}{1+\sin(\phi)}\right)[(p-C_o)\left(\frac{h^2}{2}\right)+\gamma\left(\frac{h^3}{3}\right)], \quad :(a)$$
$$M=\left(\frac{1+\sin(\phi)}{1-\sin(\phi)}\right)[(p+T_o)\left(\frac{h^2}{2}\right)+\gamma\left(\frac{h^3}{3}\right)], \quad :(b) \qquad (4.108)$$

Use of Equations (4.108) and (4.106) in Equation (4.107) gives:

$$x_o=[(p-C_o)\left(\frac{h^2}{2}\right)+\gamma\left(\frac{h^3}{3}\right)]/[(p-C_o)h+\gamma\left(\frac{h^2}{2}\right)], \quad :(a)$$
$$x_o=[(p-T_o)\left(\frac{h^2}{2}\right)+\gamma\left(\frac{h^3}{3}\right)]/[(p-T_o)h+\gamma\left(\frac{h^2}{2}\right)], \quad :(b) \qquad (4.109)$$

The resultant force F may be considered as a sum of a uniform (rectangular) distribution of stress and a uniformly increasing distribution (triangular) of gravity-induced stress that act through the centers of their respective distributions as shown in Figure 4.17.

Velocity fields associated with the collapse mechanisms are shown in Figure 4.16 that depicts essentially rigid body translations with a velocity discontinuity present at the rigid-plastic boundary. Magnitude of the velocity V would be determined by a boundary condition.

4.7 Discontinuities in velocity and stress

Discontinuities in velocity and stress are allowed by physical requirements but restrictions apply. The general subject is replete with possibilities. However, only strong discontinuities are considered here. A strong discontinuity is a sudden change in a variable, a "jump." If F^- is the value of F below a discontinuity surface and F^+ is the value above, then the jump in F is simply the difference $F^+ - F^-$. Square brackets denote a jump, so $[F] = F^+ - F^-$. Weak discontinuities are sudden changes in derivatives. A sudden change in strain, a displacement derivative, is an example of a weak discontinuity. Features of discontinuities that are of particular interest include magnitude, location, and propagation.

Velocity discontinuities

Conservation of mass across a discontinuity surface only requires the product of density and velocity normal to the surface to be continuous. Thus:

$$[\rho V_n] = 0 \tag{4.110}$$

where ρ is mass density and V_n is the velocity component normal to the discontinuity surface relative to the velocity of the surface in the normal direction. This result may be derived from the conservation of mass requirement when applied to a "pill box" shown in Figure 4.19. By writing the sums of the mass entering and leaving the pill box per unit of time and then shrinking the thickness of the box to zero while keeping the surfaces finite, one arrives at the jump condition (Equation 4.110). If the material were incompressible, then the mass density is constant and velocity normal to the surface is continuous. However, any velocity component tangential to the surface may jump. In this case, the velocity discontinuity is a *sliding* discontinuity. If mass density is not constant, then normal and tangential components may be discontinuous as the mass density changes across a discontinuity surface.

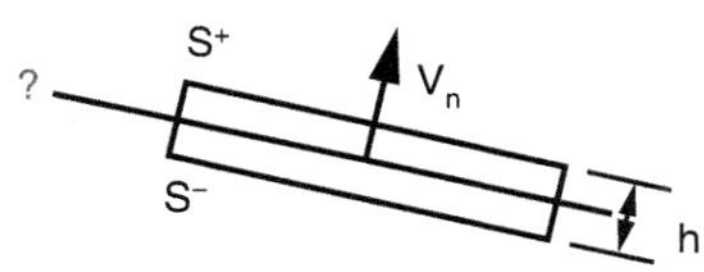

Figure 4.19 A "pill box" for mass conservation.

In plane plasticity, a surface of velocity discontinuity is seen as a curve in the considered plane. Reasoning from the strain rate velocity equations shows that a velocity discontinuity can only occur along a velocity characteristic. With neglect of the elastic contribution to the strain rates, one has:

$$\begin{aligned}
-\frac{\partial u}{\partial x} &= -(\frac{1}{2})\lambda[1-g'\cos(2\theta)] \\
-(\frac{1}{2})(\frac{\partial x}{\partial y}+\frac{\partial v}{\partial x}) &= -(\frac{1}{2})\lambda g'\sin(2\theta) \\
-\frac{\partial x}{\partial y} &= -(\frac{1}{2})\lambda[1+g'\cos(2\theta)]
\end{aligned} \tag{4.111}$$

where the plastic potential is assumed in the form of Equation (4.3b) and $g' = dg / d\tau_m$. The negative signs on the left indicate compression is positive. Choose the x-axis tangential to the discontinuity curve, so the y-axis is normal to the discontinuity as shown in Figure 4.20. Derivatives along the discontinuity are continuous. However, the velocity v normal to the discontinuity may jump. A jump in v implies the derivative in the normal direction $(\partial v/\partial y)$ becomes indefinitely large. Consequently, λ must become large and so must the derivative $(\partial u/\partial x)$. But this derivative is continuous, hence $[1 - g'\cos(2\theta)]$ must approach zero and thus. $(1 / g') = \cos(2\theta) = \sin(\phi) = \cos(2\mu)$ Consequently, $2\theta = \pm 2\mu$, and therefore $\theta \pm \mu = 0$, which implies the x-axis coincides with a velocity characteristic. Thus, a velocity discontinuity can occur only on a velocity characteristic and has the interpretation of a zone of intense shearing.

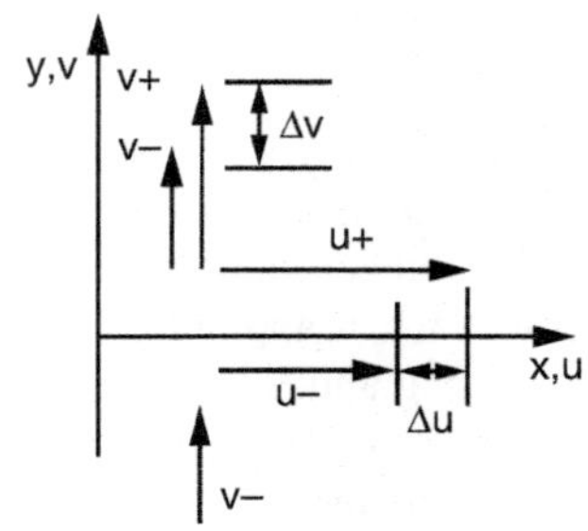

Figure 4.20 Jumps in velocity components.

When the rules of flow are associated, stress and velocity characteristics coincide, and velocity discontinuities may appear across stress characteristics. In fact, in the wedge and retaining wall problems, the boundaries between the plastic and rigid regions were characteristic curves of stress across which velocity discontinuities were used to generate velocity fields that met velocity boundary conditions in the form of jumps that are necessarily present at a rigid-plastic boundary. In the case of the hollow cylinder problem, the rigid-plastic boundary is not coincident with a characteristic curve, although a discontinuity in velocity must occur there in apparent contradiction to the result just obtained. The resolution of this puzzle is shown later.

According to Equation (4.60), when the x-axis coincides with a characteristic curve, $dv = \pm du\tan(\phi)$. For example, if velocity is discontinuous across a C_1 curve, then on a C_2 curve, $dv = du\tan(\phi)$. At a discontinuity in velocity, the differentials of velocity may be interpreted as increments:

$$\Delta v_n = \pm \Delta v_s \tan(\phi) \quad : \begin{cases} C_1 \\ C_2 \end{cases} \tag{4.113}$$

where v and u are interpreted as normal and tangential velocity components along a characteristic curve.

Transformation of the velocity equations (Equation 4.59) to an (s,n) system where s is along a characteristic and n is normal to a characteristic leads to Equation (4.61), and:

$$dv_s - v_n d\theta = 0 \tag{4.114}$$

where μ is considered a constant and θ is continuous at a velocity discontinuity. Equation (4.112) holds for both families of velocity characteristics and on either side of a velocity discontinuity. If (+) represents the upper side of a discontinuity and (−) the lower side, then across a discontinuity:

$$d(v_s^+ - v_s^-) - (v_n^+ - v_n^-)d\theta = 0 \tag{4.115}$$

that is obtained by subtraction of Equation (4.112) that holds on either side of a velocity discontinuity. Denoting a jump by Δ, one obtains:

$$d(\Delta v_n) - (\Delta v_n)d\theta = 0 \tag{4.116}$$

that states the jump in a derivative is equal to the derivative of the jump. In consideration of Equation (4.113), $d(\Delta v_s) \pm (\Delta v_s)\tan(\phi)$ After integration, one obtains the propagation equations of velocity discontinuities. Thus:

$$\begin{aligned} \Delta v_s / \Delta v_s^0 &= \exp[\pm(\theta - \theta_o)\tan(\phi) \\ \Delta v_n / \Delta v_n^0 &= \exp[\pm(\theta - \theta_o)\tan(\phi), \\ \Delta V_n / \Delta V_o &= \exp[\pm(\theta - \theta_o)\tan(\phi) \end{aligned} \quad : \begin{cases} C_1 \\ C_2 \end{cases} \tag{4.117}$$

where ΔV is the velocity jump vector magnitude with tangential and normal components indicated by "s" and "n," respectively. A velocity jump grows with increasing θ along a first characteristic, but diminishes along a second characteristic.

When jumps of equal magnitude occur across intersecting characteristics along a line of symmetry as shown in Figure 4.21, the net effect is to produce a jump along the line of symmetry. This phenomenon often appears during plastic flow of granular materials in bins and hoppers. In the hollow cylinder problem, each ray is a line of symmetry, with the result that a velocity discontinuity appears in the radial direction across the rigid-plastic boundary, which is not a characteristic. However, the double discontinuity meets the requirement that velocity discontinuities occur only across velocity characteristics.

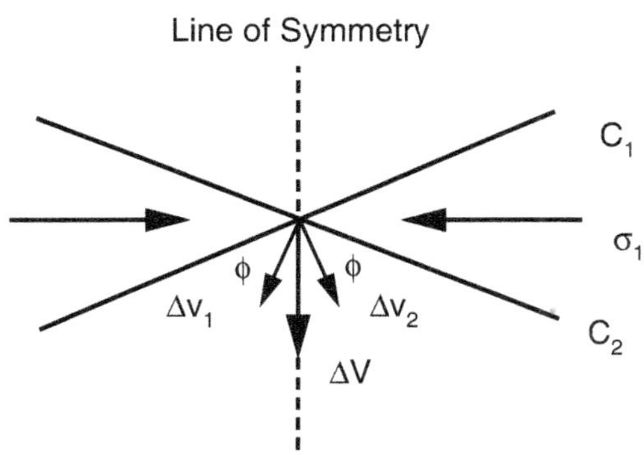

Figure 4.21 Double velocity discontinuity at a line of symmetry.

Stress discontinuities

Equilibrium requires tractions to be continuous across a surface of interest. Thus, the jump restrictions are:

$$[\sigma_{nn}] = 0, \quad [\tau_{nt}] = 0 \tag{4.118}$$

In the plane problem, illustrated in Figure 4.22, the normal stress σ_{ss} acting tangential to the trace of a discontinuity surface may be discontinuous. Stresses σ_{zz}, τ_{zs} may also be discontinuous. Equilibrium and yielding requirements restrict the location and magnitude of any stress jump. With the assumption of a Mohr-Coulomb failure criterion, yield condition, $\tau_m = A\sigma_m + B$, and with reference to Figure 4.21, relative to the principal axes:

$$\begin{aligned} \sigma'_{nn} &= (-B/A) + (\tau'_m/A)\{1-\sin(\phi)\cos[2(\theta'-\alpha)]\} \\ \sigma'_{nn} &= (-B/A) + (\tau''_{nn}/A)\{1-\sin(\phi)\cos[2(\theta''-\alpha)]\} \\ \sigma'_{ns} &= (\tau'_m)\sin[2(\theta'-\alpha)] \\ \sigma''_{ns} &= (\tau''_m)\sin[2(\theta''-\alpha)] \end{aligned} \tag{4.119}$$

By the jump conditions (Equation 4.118):

$$\begin{aligned} (\tau'_m/A)\{1-\sin(\phi)\cos[2(\theta'-\alpha)]\} &= (\tau''_m/A)\{1-\sin(\phi)\cos[2(\theta''-\alpha)]\} \\ (\tau'_m)\sin[2(\theta'-\alpha)] &= (\tau''_m)\sin[2(\theta''-\alpha)] \end{aligned} \tag{4.120}$$

After eliminating the stresses from Equation (4.120) and using some trigonometric identities, one obtains:

$$\sin(\theta''-\theta')\cos(\theta''+\theta'-2\alpha) - \sin(\phi)\sin(\theta''-\theta'')\cos(\theta''-\theta') = 0$$

If $\sin(\theta''-\theta')$ is not zero, then:

$$\cos(\theta''+\theta'-2\alpha) - \sin(\phi)\cos(\theta''-\theta') = 0 \tag{4.121}$$

otherwise θ is continuous and so are the stresses.

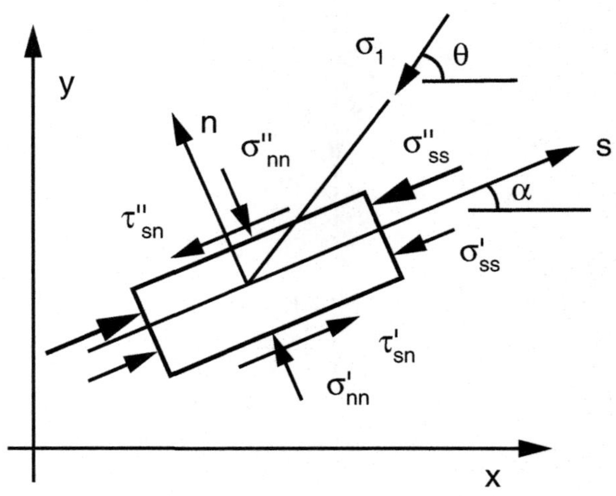

Figure 4.22 A stress discontinuity

A discontinuity in stress *cannot* occur across a stress characteristic. Suppose this assertion were not true, then $\alpha = \pm\mu$. According to Equation (4.121), after some reduction:

$$-\cos(\theta'' - \theta')\sin(\phi) \pm \sin(\theta'' - \theta')\cos(\phi) + \sin(\phi)\cos(\theta'' - \theta') = 0 \qquad (4.122)$$

Hence, $\sin(\theta'' - \theta') = 0$. But this result contradicts the original assumption. Thus, a stress discontinuity cannot occur across a stress characteristic. The situation is easy to see in graphical form as illustrated in Figure 4.23. When associated flow rules are in force, stress and velocity discontinuities cannot occur across the same surface because of the coincidence of stress and velocity characteristics.

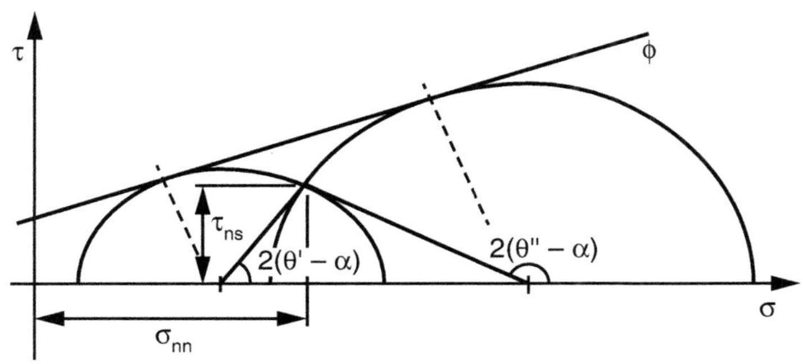

Figure 4.23 Mohr's circles for a stress discontinuity

Problems that use stress discontinuities are often regions of constant state that are joined at an angle. An example is shown in Figure 4.24, where a uniform pressure is applied to one side of a wedge. The wedge tip fails when a plastic region extends across the tip. One solution for stress of this limiting equilibrium problem is given in the figure where p is the maximum pressure that can be applied. In this example, the boundary conditions dictate the stress states, so the discontinuity angle α is not needed but can still be found and should lie within the wedge angle.

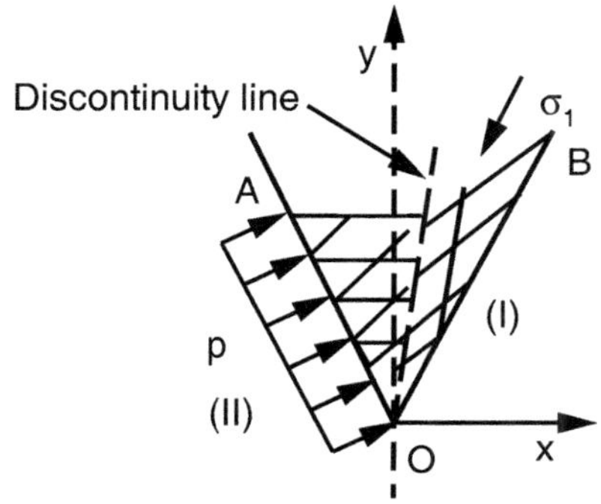

Figure 4.24 Stress discontinuity in a wedge under side pressure.

Frictional boundary conditions may be associated with stress discontinuities. Figure 4.25 illustrates a combination of wall friction, a stress discontinuity, and regions of straight characteristics where weight may be important. Wall friction constrains the direction of principal stress at a wall, but not the magnitude. A graphical representation of stress is shown in Figure 4.26. Two possibilities of principal stress orientation at the wall are present in Figure 4.26 that correspond to points B and B′. Point A corresponds to the continuity requirement for normal and shear stress across the discontinuity where the small circle represents stress in the region (I) and the large circle represents stress in region (II). Point C corresponds to the surcharge p shown in Figure 4.25. In (I), the quantities, σ_m, τ_m, θ, are known. The yield condition, the jump conditions, and a wall friction condition provide four equations for solution of the same three quantities and the unknown angle α that determines the orientation of the stress discontinuity as well.

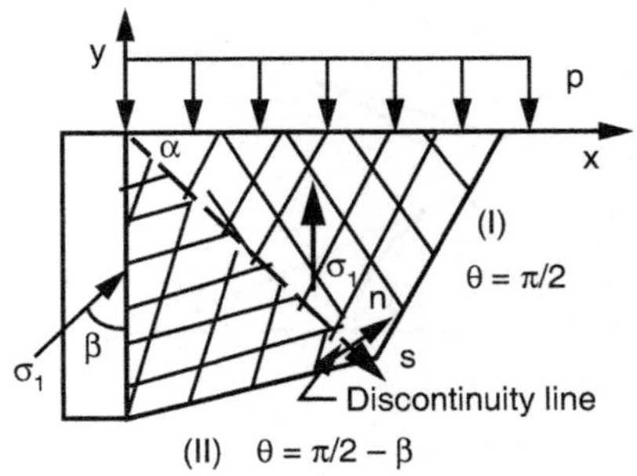

Figure 4.25 Wall friction and stress discontinuity.

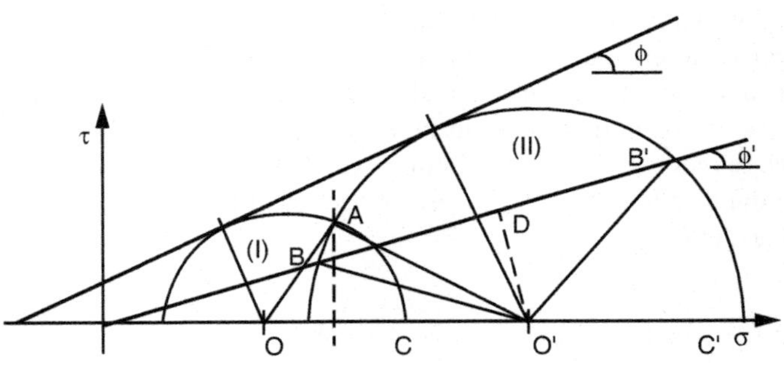

Figure 4.26 Wall friction and discontinuous stress.

4.8 Envelope solutions

In the derivation of the characteristic form of the stress subsystem of equations, the condition was used that required the determinant Δ of the coefficients of the partial derivatives of stress in the original system to vanish. Whenever $\Delta = 0$, a companion requirement was the determinant $|N| = 0$ where $|N|$ was the determinant in the numerator of the solution formulas that had the form $|N|/\Delta = 0$. For example, one had as a formula $\partial\theta/\partial x = \theta, x = |N_3|/\Delta = 0$. If $\Delta = 0$, then $|N_3| = 0$ also. The result of these two requirements was the characteristic form of the stress subsystem of equations.

Suppose, however, that $\Delta = 0$, but $|N| \neq 0$. This situation would imply that some derivative of a stress variable (τ_m or θ) became very large and that in the limit, a discontinuity of a stress would occur. Now the condition that $\Delta = 0$ was $dy/dx = \tan(\theta \pm \mu)$. Thus, a discontinuity would appear along a stress characteristic. But stress discontinuities cannot occur on a stress characteristic as shown previously, that is, a stress discontinuity cannot occur on a stress characteristic *within* a plastic domain. One concludes that a discontinuity across a curve defined by $dy/dx = \tan(\theta \pm \mu)$ must be at the *boundary* of a plastic domain (i.e., an elastic-plastic boundary), physically and mathematically.

Such boundaries are *natural* boundaries because of the mathematical and physical implications. Mathematically one has a condition that precludes extension of the plastic domain into the adjacent rigid-plastic or elastic-plastic domain. Physically, the plastic domain appears self-limiting in extent. Beyond the natural boundary, the material remains rigid or elastic. The implication is that an *envelope* of characteristic curves exists under these conditions.

An envelope of characteristics is a singular solution to the differential equations of the stress subsystem, that is, of $dy/dx = \tan(\theta \pm \mu)$. Recall that a differential equation $dy/dx = f(x,y)$ defines a one-parameter family of integral curves $F(x,y{:}C) = 0$ where the arbitrary constant of integration C is the parameter. A singular solution such as an envelope, if one exists, cannot be obtained by specializing the parameter C. Instead, to find the equation of the envelope, one must eliminate C from the system:

$$F(x, y : C) = 0 \text{ and } \frac{\partial \mathrm{F(x,y;C)}}{\partial \mathrm{C}} = 0$$

One could retain the system in parametric form to avoid eliminating C, of course. Once a singular solution is obtained, it must be tested to see that it does actually satisfy the original differential equation. The reason is that the procedure may introduce extraneous results.

As an example, consider the family of straight lines at a distance d from the origin. The normal form of one of these lines is:

$$x\sin(C) + y\cos(C) = d \tag{4.122}$$

where the parameter C is the line slope and d is the distance from the origin to the considered line. With d fixed, Equation (122) is a one-parameter family of lines at unit distance from the origin as illustrated in Figure 4.27 (otherwise one has a two-parameter family of lines).

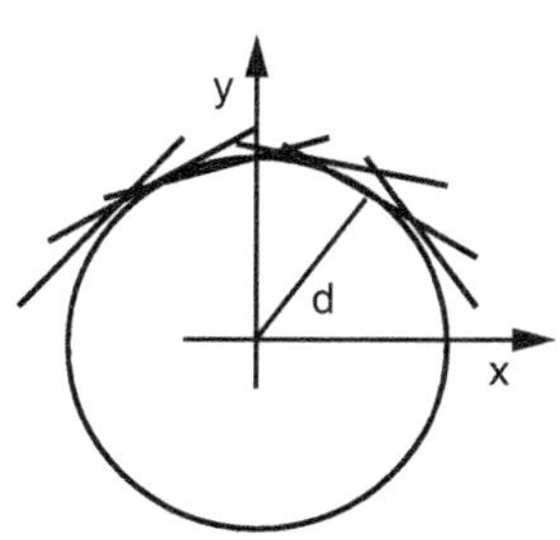

Figure 4.27 A family of straight lines at distance *d* from the origin. The circle is an envelope of the tangent lines with radius *d*.

Differentiating Equation (122) with respect to C results in:

$$x\cos(C) - y\sin(C) = 0 \quad (4.123)$$

Solving Equations (122) and (123) gives $x = d\sin(C)$ and $y = d\cos(C)$ that after substituting back into Equation (122) results in:

$$x^2 + y^2 = d^2 \quad (4.124)$$

which is the equation of a circle with radius d centered at the origin as shown in Figure 4.27.

Another example of an envelope is the singular point associated with the stress field of radial shear, where one family of characteristics consists of straight lines passing through the origin of characteristics. (The other family of characteristics are exponential spirals.) The singular point is actually a degenerate envelope of the family of straight characteristics. In the more general case, the enveloping curve would have finite length. Both cases are illustrated in Figure 4.28. Envelopes of one family of characteristics are also loci of cusps of the other family of characteristics as seen in Figure 4.28b.

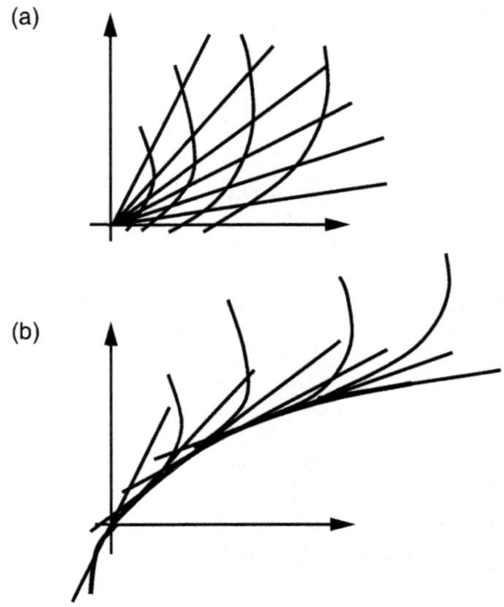

Figure 4.28 Envelopes of straight characteristics. (a) Degenerate case, envelope has zero length, (b) finite length case.

The simplest case arises when the stress variable θ is constant along any given curve of one family of characteristics, although it may be different along any particular characteristic. In this case, the characteristics are given by:

$$y = x\tan(\theta \pm \mu) + \text{constant} \quad (4.125)$$

where $\theta = \theta_o$ along a given characteristic and either the + sign or the − sign applies. A constant θ implies a straight characteristic. Thus, one obtains a one-parameter family of characteristics, which may have an envelope that can be determined by the procedure outlined previously.

The next step in complexity is to assume that θ is a function of one of the space variables, x or y. After integration by parts of the characteristic equations, one obtains:

$$y\cot(\theta \pm \mu) - x + \int \frac{y\theta' dy}{\sin^2(\theta \pm \mu)} = 0 \tag{4.126}$$

The integral in Equation (126) is a function of θ in consideration of $y = f^{-1}(\theta)$ and $\theta' dy = d\theta$. Thus:

$$y - x\tan(\theta \pm \mu) + g(\theta) = 0 \tag{4.127}$$

After differentiation of Equation (127) with respect to θ, one obtains:

$$x - g'\cos^2(\theta \pm \mu) = 0 \tag{4.128}$$

Following substitution of Equation (128) into Equation (127), the result is:

$$y - g'\cos(\theta \pm \mu)\sin(\theta \pm \mu) + g(\theta) = 0 \tag{4.129}$$

Equations (128) and (129) are the parametric expressions for the envelope of characteristics, if one exists. If $g(\theta)$ were an arbitrary constant, then one would be considering an envelope of straight characteristics as before. However, the assumption is $\theta = f(y)$.

The simplest function $g(\theta)$ compatible with $g(\theta) = 2m\theta$, is a linear function where m is a true constant and the 2 is for convenience. Thus, the envelope equations are now:

$$x - 2m\cos^2(\theta \pm \mu) = 0, \; y - 2m\cos(\theta \pm \mu)\sin(\theta \pm \mu) + 2m\theta = 0 \tag{4.130a,b}$$

provided an envelope exits.

To ensure that extraneous loci are not present, the expressions (Equation 130) must satisfy the characteristic equations. Thus:

$$\begin{aligned} dy/d\theta &= -2m + 2m\cos[2(\theta \pm \mu)] \\ dx/d\theta &= -2m\sin[2(\theta \pm \mu)] \\ dy/dx &= \frac{1 - \cos[2(\theta \pm \mu)]}{\sin[2(\theta \pm \mu)]} = \tan(\theta \pm \mu) \end{aligned} \tag{4.131a,b,c}$$

Thus, the singular solutions (Equation 130) do in fact satisfy the characteristic equations.

Explicit forms of the characteristic equations may be obtained from Equation (131) with the aid of trigonometric identities. Three example cases are:

$$\begin{aligned} &\text{Case 1.} \quad dy/d\theta = 2m[\sin(2\theta) \pm \sin(2\mu)] \\ &\qquad \text{and } dx/rd\theta = 2m[\cos(2\theta) + \cos(2\mu)] \\ &\text{Case 2.} \quad dy/d\theta = 2m[-\cos(2\theta) + \cos(2\mu)] \\ &\qquad \text{and } dx/rd\theta = 2m[\sin(2\theta) \pm \sin(2\mu)] \\ &\text{Case 3.} \quad dy/d\theta = 2m(1 - \cos[2(\theta \pm \mu)]) \\ &\qquad \text{and } dx/rd\theta = 2m\sin[2(\theta \pm \mu)] \end{aligned} \tag{4.132}$$

Other possibilities may occur, although they may also lead to the same final results.
Similarly in polar coordinates (r, ω) with $\psi = \theta - \omega$ where we recall that:

$\theta =$ counter – clockwise angle from the x – axis to the direction of the major principal stress σ_1
$\omega =$ counter – clockwise angle from the x – axis to a ray r
$\psi = \theta - \omega$ counter – clockwise angle from a ray r to the direction of the major principal stress σ_1

as illustrated in Figure 4.29:

Case 1. $d\omega / d\psi = 2m[\sin(2\mu) \pm \sin(2\mu)]$ and $dr / rd\psi = 2m[\cos(2\psi) + \cos(2\mu)]$
Case 2. $d\omega / d\psi = 2m[-\cos(2\mu) + \cos(2\mu)]$ and $dr / rd\psi = 2m[\sin(2\psi) \pm \sin(2\mu)]$
Case 3. $d\omega / d\psi = 2m(1 - \cos[2(\psi \pm \mu)])$ and $dr / rd\psi = 2m\sin[2(\psi \pm \mu)]$

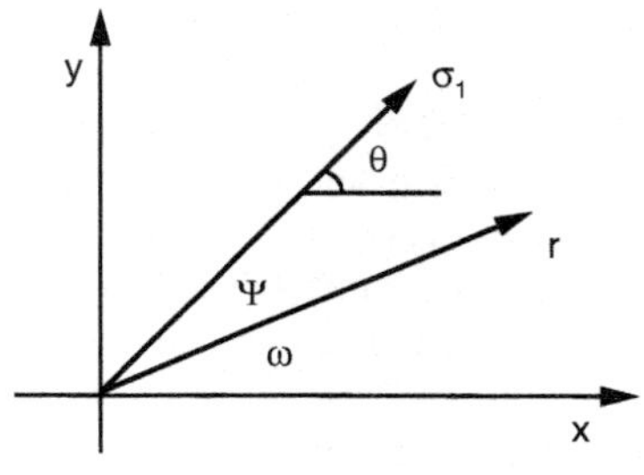

Figure 4.29 Angles in polar coordinates.

Again, these are just three possibilities. Use of other trigonometric identities may lead to other solutions, as they may simply be different starting expressions. Integration of any of these three cases introduces arbitrary constants or arbitrary functions depending upon the initial assumption of whether θ depends on x (or y) or both. Additionally, other coordinate systems may be used to seek other solutions. Indeed, a systematic survey could prove useful.

An indication of the presence of an envelope occurs when a derivative of x or y with respect to θ tends to zero. In polar form, the derivative of ω or r with respect to the parameter Ψ tends to zero. Depending upon the case used, either one or two envelopes of characteristics may be present or possible. This feature serves as a guide in deciding which to choose initially.

As an example of the use of these equations, consider Case 1 in rectangular coordinates. After integration:

$$\begin{aligned} x &= m\sin(2\theta) + 2m\theta\cos(2\mu) + \text{constant} \\ y &= m\cos(2\theta) \pm 2m\theta\sin(2\mu) + \text{constant} \end{aligned} \tag{4.133}$$

As boundary conditions, one may suppose $x = +h$ is an envelope of C_2 characteristic curves and that $x = -h$ is an envelope of C_1 characteristic curves. The y-axis ($x = 0$) is a plane of symmetry as shown in Figure 4.30a. Figure 4.30b is a more accurate plot. Specifically, the boundary conditions are:

On $x = -h$, $\theta = +(\pi/2 - \mu)$ and therefore

$$x = -h = m\sin(\pi - \mu) + 2m(\pi - \mu) + \text{constant}$$

On $x = +h$, $\theta = -(\pi/2 - \mu)$ and therefore

$$x = +h = -m\sin(\pi - \mu) + 2m(\pi - \mu)\cos(2\mu) + \text{constant}$$

On $x = 0$, $\theta = 0$ and therefore $x = 0 = m\sin(0) + 2m\,0\cos(2\mu) + \text{constant}$

One has from these boundary conditions the result that the constant is zero and that:

$$m = (-h)/[\sin(\pi - 2\mu) + 2(\pi - 2\mu)\cos(2\mu) \text{ and}$$

$$x = \left(\frac{-h}{[\sin(\pi - 2\mu) + 2(\pi - 2\mu)\cos(2\mu)]}\right)[\sin(2\theta) + 2\cos(2\mu)] \tag{4.134}$$

$$y = \left(\frac{-h}{[\sin(\pi - 2\mu) + 2(\pi - 2\mu)\cos(2\mu)}\right)[\cos(2\theta) \mp \cos 2\theta\sin(2\mu) - 1] + y_o$$

where the constant $= y_o$ and serves to identify any particular characteristic of interest and simply shifts the characteristic curves along the y-axis. Note that the first y with the minus sign corresponds to C1 curves, while the plus sign corresponds to C2 curves.

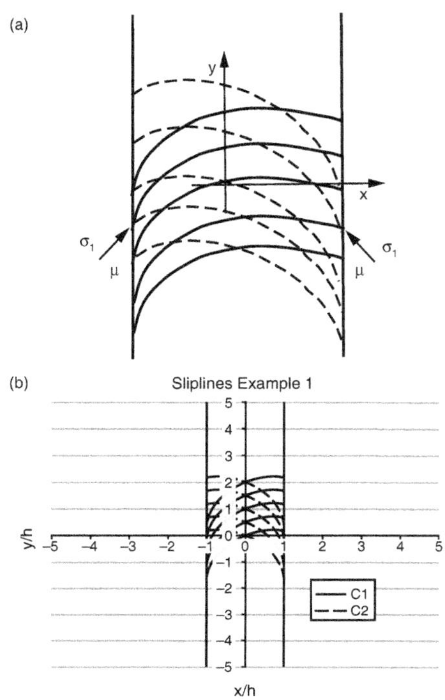

Figure 4.30 (a) Characteristic envelopes and loci of cusps between parallel plates (Case 1), (b) characteristic envelopes and loci of cusps between parallel plates (Case 1).

As a second example of the use of these equations, consider Case 2 in rectangular coordinates. After integration:

$$\begin{aligned} y &= 2m\theta\cos(2\mu) - m\sin(2\theta) + \text{constant} \\ x &= m\cos(2\theta) \pm 2m\sin(2\mu) + \text{constant} \end{aligned} \tag{4.135}$$

As boundary conditions, one may suppose $y = +h$ is an envelope of C_1 characteristic curves and that $y = -h$ is an envelope of C_2 characteristic curves. The x-axis ($y = 0$) is a plane of symmetry as shown in Figure 4.31a. Figure 4.31b is a more accurate plot. Specifically, the boundary conditions are:

$$\begin{aligned} &\text{On } y = +h,\ \theta = -\mu \text{ and therefore } y = +h = -2m\mu\cos(2\mu) - m\sin(-2\mu) + \text{constant} \\ &\text{On } y = -h,\ \theta = +\mu \text{ and therefore } y = -h = +2m\mu\cos(2\mu) - m\sin(+2\mu) + \text{constant} \\ &\text{On } y = 0\ \theta = 0 \text{ and therefore } y = 0 = -2m0\cos(0) - m\sin(-0) + \text{constant} \end{aligned}$$

One has from these boundary conditions the result that the constant is zero and that:

$$\begin{aligned} m &= (-h)/[2\mu\cos(2\mu) - \sin(2\mu)] \\ y &= \left(\frac{-h}{2\mu\cos(2\mu) - \sin(2\mu)}\right)[2\theta\cos(2\mu) - \sin(2\theta)] \\ x &= \left(\frac{-h}{2\mu\cos(2\mu) - \sin(2\mu)}\right)[\mp 2\theta\sin(2\theta) - \cos(2\mu)] + \text{constant}(x_o) \end{aligned} \tag{4.136}$$

where the constant $= x_0$ and serves to identify any particular characteristic of interest and simply shifts the characteristic curves along the x-axis. Note that the first x with the minus sign corresponds to C1 curves, while the plus sign corresponds to C2 curves.

Along the upper and lower boundaries ($y = \pm h$), the condition $\theta = \mp\mu$ holds. Using the Mohr-Coulomb yield condition $|\tau_m| = A\sigma_m + B$ and the relationship $\mu = \pi/4 - \phi/2$, one obtains the Cartesian stresses:

$$\begin{aligned} \sigma_{xx} &= (-B/A) + \tau_m[(1/A) + \sin(\phi)] \\ \sigma_{yy} &= (-B/A) + \tau_m[(1/A) + \sin(\phi)] \\ \tau_{xy} &= \tau_m\cos(\phi) \end{aligned} \tag{4.137}$$

One may also use the yield condition in the form $|\tau| = \sigma\tan(\phi) + k$, where ϕ and k are the angle of internal friction and cohesion, respectively. Thus:

$$\tau_m\cos(\phi) = \sigma_{yy}\tan(\phi) + k \text{ or } \ddot{A}_f = \sigma_n\tan(\phi) + k \tag{4.138}$$

where τ_f and σ_n on a stress characteristic (recall $A = \sin(\phi)$ and $B = k\cos(\phi)$).

In another example in polar coordinates, shown in Figure 4.32, one has after integration:

$$\begin{aligned} \omega &= m\sin(2\psi) + 2m\psi\cos(2\mu) + \text{constant} \\ \ln(r) &= -m\cos(2\psi) \pm 2m\psi\sin(2\mu) + \text{constant} \end{aligned} \tag{4.139}$$

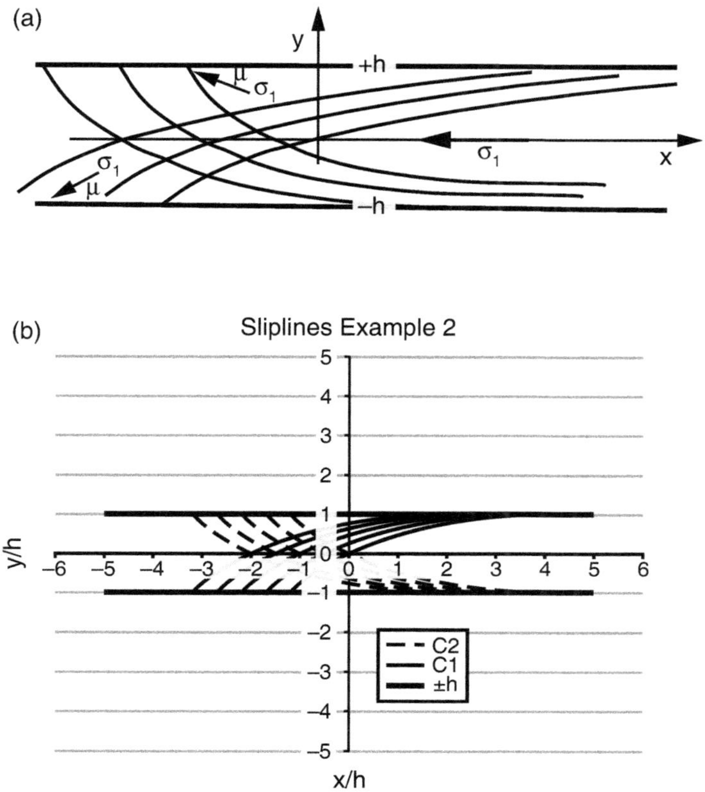

Figure 4.31 (a) Characteristic envelopes and loci of cusps between parallel plates (Case 2), (b) Graphical plot of sliplines for Example 2.

where the inclined lines are envelopes. For boundary conditions:

$$\begin{aligned}
&\omega = -\omega_o,\ \theta = -\omega_o + \mu,\ \theta - \omega = \mu = \psi, \\
&\omega = 0,\ \theta = \pi/2,\ \theta - \omega = \pi/2 = \psi, \\
&\omega = \omega_o, \theta = \omega_o + \pi - \mu, \theta - \omega = \pi - \mu = \psi
\end{aligned} \tag{4.140}$$

Using the boundary conditions (Equation 140) in the integrals (Equation 139), one obtains:

$$\begin{aligned}
&\omega = \left(\frac{-\omega_o}{\sin(2\mu) + (\pi - 2\mu)\cos(2\mu)}\right)[\sin(2\psi) + (\pi - 2\psi)\cos(2\mu)] \\
&\ln(r) = \left(\frac{-\omega_o}{\sin(2\mu) + (\pi - 2\mu)\cos(2\mu)}\right)[\cos(2\psi) \pm 2\psi\sin(2\mu)] + \text{constant}
\end{aligned} \tag{4.141}$$

where the constant in the expression for $\ln(r)$ is an identifier for a particular characteristic curve.

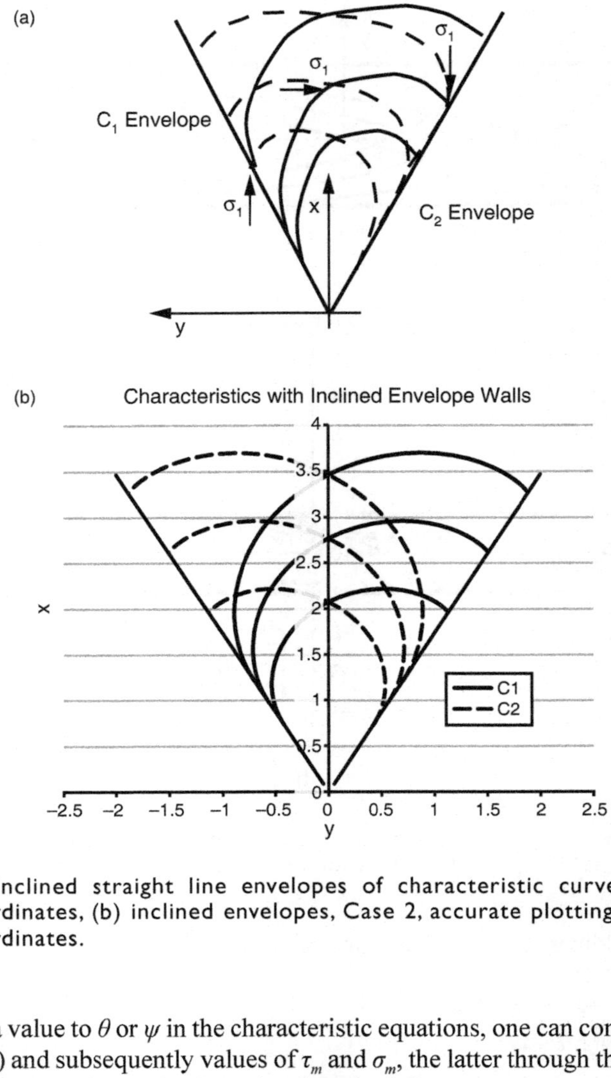

Figure 4.32 (a) Inclined straight line envelopes of characteristic curves, Case 2, polar coordinates, (b) inclined envelopes, Case 2, accurate plotting, analysis in polar coordinates.

By assigning a value to θ or ψ in the characteristic equations, one can compute the location of that value (x,y) and subsequently values of τ_m and σ_m, the latter through the yield condition. The former by the stress integrals along the characteristics:

$$\tau_m = \tau_m^o \exp[\mp 2(\theta - \theta_o)\tan(\phi)] \tag{4.142}$$

provided weight is negligible. When weight is important, numerical methods become necessary not only in case of envelope solutions but in the general case. The Cartesian stresses are with the aid of a Mohr-Coulomb yield condition:

$$\begin{aligned}
\sigma_{xx} &= \sigma_m + \tau_m \cos(2\theta) = \tau_m^o\{\exp[\mp 2(\theta - \theta_o)\tan(\phi)]\} \\
&\quad [1/\sin(\phi) + \cos(2\theta)] - k\cot(\phi) \\
\sigma_{yy} &= \sigma_m - \tau_m \cos(2\theta) = \tau_m^o\{\exp[\mp 2(\theta - \theta_o)\tan(\phi)]\} \\
&\quad [1/\sin(\phi) - \cos(2\theta)] - k\cot(\phi) \\
\tau_{xy} &= \tau_m \sin(2\theta) = \tau_m^o\{\exp[\mp 2(\theta - \theta_o)\tan(\phi)]\}\sin(2\theta)
\end{aligned} \tag{4.143}$$

where τ_m^o and θ_o are determined from boundary conditions at walls where the normal and shear stresses are perpendicular and parallel to the envelope characteristics (σ,τ).

Additional relationships of some use in case of Mohr-Coulomb yield are:

$$\begin{aligned}\sigma_1 &= \sigma_m + \tau_m \\ \sigma_3 &= \sigma_m - \tau_m, \\ \sigma_1 &= C_o + (C_o / T_o)\sigma_3\end{aligned} \tag{4.144}$$

where C_0 and T_0 are unconfined compressive and tensile strengths. Wall friction cannot be a factor when a wall is an envelope. The reason is in the conflicting requirements $\tau_o = \tau_m^o \cos(\phi') = \sigma_o \tan(\phi) + k$ and $\tau_o = \sigma_o \tan(\phi') + k'$ where ϕ' and k' are wall friction angle and adhesion respectively.

Cases of one or two circular envelopes of characteristics could be developed by using boundary conditions on r instead of ω. Other interesting possibilities are mentioned by Nadai (1963). Also, functions could replace constants of integration which would lead to more complex envelope solutions.

4.9 Numerical solution of boundary value problems

Simple finite difference procedures for three types of boundary value problems are developed. An iterative process is necessary because of the nonlinearity of the characteristic system. As a reminder, the stress subsystem of equations in characteristic form is

$$dy / dx = \tan(\theta \pm \mu) \quad : \begin{cases} C_1 \\ C_2 \end{cases}$$

$$d\tau_m \pm 2\tau_m \tan(\phi) d\theta = 0: \begin{cases} C_1 \\ C_2 \end{cases}$$

where $\mu = \pi / 4 - \phi / 2$ and a failure envelope is assumed to exist. The characteristics form an acute angle of 2μ between C_2 and C_1 curves. If weight and inertia are important, then the second pair of equations becomes:

$$d\tau_m \pm 2\tau_m \tan(\phi) d\theta \pm \{E_1 \sin(\theta \mp \mu) - E_2 \cos(\theta \mp \mu)\} \tan ds = 0 \quad : \begin{cases} C_1 \\ C_2 \end{cases}$$

where ds is an element of arc length along a characteristic curve. There are four unknowns in the characteristic system: x, y, θ, and τ_m. As a reminder, $E_1 = \rho \dot{v}_x - \gamma_x$ and $E_2 = \rho \dot{v}_y - \gamma_y$.

With reference to Figure 4.33, data along a given curve can be used to estimate the unknowns at the point (x_p, y_p) by following a C1 characteristic from point 1 and a C2 characteristic from point 2 to the unknown point P(x,y).

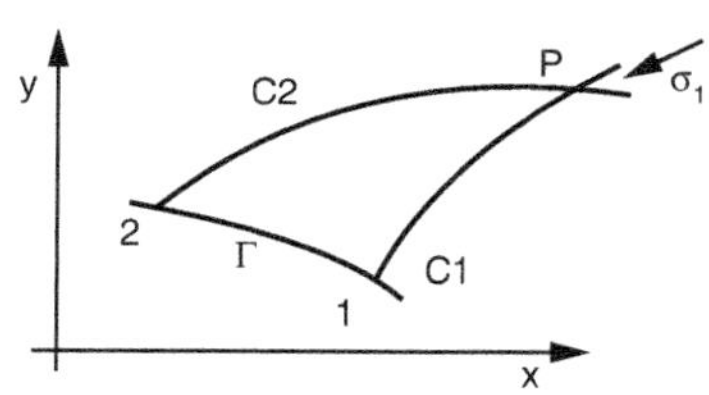

Figure 4.33 Beginning of a numerical solution to the characteristic system of equations.

An algebraic system of equations is first formed by replacing differentials with increments. Thus:

$$dx \doteq \Delta x = x_p - x_1,\ dy \doteq \Delta y = y_p - y_1,\ d\tau_m \doteq \tau^p_m - \tau^1_m,\ d\theta \doteq \theta_p - \theta_1,\ :C_1$$

$$dx \doteq \Delta x = x_p - x_2,\ dy \doteq \Delta y = y_p - y_2,\ d\tau_m \doteq \tau^p_m - \tau^2_m,\ d\theta \doteq \theta_p - \theta_1,\ :C_2$$

where the superscripts on τ_m are *not exponents* but rather indicate a value on a characteristic curve. After substitution into the characteristic system and solving for unknowns, one obtains:

$$x_p = \frac{(y_2 - y_1) - [x^2 \tan(\theta_2 - \mu)]}{[\tan(\theta_1 + \mu] - \tan(\theta_2 - \mu)]} \qquad (4.145a,b)$$

$$y_p = y_1 + (x_p - x_1)\tan(\theta_1 + \mu) \text{ or } y_p = y_2 + [x_p - x_2)\tan[(\theta_2 - \mu]$$

$$\tau^p_m = \frac{2\tau^1_m\tau^2_m[1 + (\theta_1 - \theta_2)\tan(\phi)]}{(\tau^1_m + \tau^2_m)} \qquad (4.145c,d)$$

$$\theta_p = \theta_2 + \frac{(\tau^p_m - \tau^2_m)}{2\tau^2_m \tan(\phi)} \text{ or } \theta_p = \theta_1 + \frac{(\tau^p_m - \tau^2_m)}{2\tau^2_m \tan(\phi)}$$

where the subscripts indicate starting points 1 and 2. The superscripts on τ_m are also indicative of the associated characteristic curve C_1 or C_2; they are not exponents. Successive substitution of Equation (143a) into Equation (143b) and Equation (143c) into Equation (143d) is used. Equations (143) are used for a first estimate of the unknowns. Second and successive estimates should use averages. Thus:

$$\theta^{n+1}_p = \frac{\theta_1 + \theta^n_p}{2}, \theta^{n+1}_2 = \frac{\theta_2 + \theta^n_p}{2} \qquad (4.146)$$

$$\tau^{n+1}_p = \frac{\tau_1 + \tau^n_p}{2}, \tau^{n+1}_2 = \frac{\tau_1 + \tau^n_p}{2}$$

which are averages of the most recent estimate at the unknown point and the given values, and n is an iteration number. The process continues until $\theta^{n+1}_p - \theta^n_p \le$ tolerance, for $n = 1,2,...N$ where N is a maximum number of iterations allowed and "tolerance" is some small number much less than one.

BVP I All data given along a non-characteristic curve

A systematic labeling of characteristic curves illustrated in Figure 4.34 shows preliminary work that is helpful in programming digital solutions. A curvilinear triangle ABC above the curve with all given data is amenable to solution. The dotted triangle ACD below the curve with the given data is also amenable to solution. In this regard, the characteristic curves can be transformed into curvilinear, oblique coordinates, but there is no particular advantage in doing so. The network of characteristic curves is an isogonal network, meaning the angle between characteristics is constant ($2\mu = \pi/2\text{-}\phi$). In the event that $\phi = 0$, the network is orthogonal.

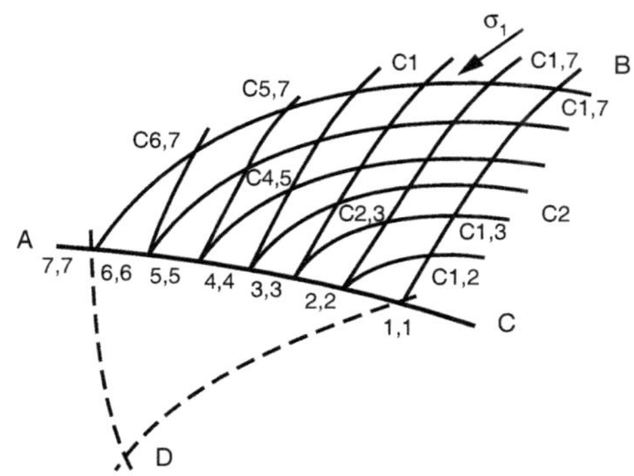

Figure 4.34 A region amenable to solution from all data given along a non-characteristic curve.

When weight or inertia is important, the numerical form (Equations 4.145c,d) has additional terms:

$$\bar{K}_1 = \{E_1 \sin(\theta - \mu) - E_2 \cos(\theta - \mu)\} \tan(\phi) dx / \cos(\theta + \mu)$$
$$\bar{K}_2 = \{E_1 \sin(\theta + \mu) - E_2 \cos(\theta + \mu)\} \tan(\phi) dx / \cos(\theta + \mu) \quad (4.147)$$
and $ds = dx / \cos(\theta \pm \mu) = dy / \sin(\theta \pm)$ on C_1 and C_2, respectively.

After replacing differentials with increments:

$$(\tau_m^p - \tau_m^1) + 2\tau_m^1 \tan(\phi)(\theta_p - \theta_1) + \bar{K}_1(x_p - x_1) = 0$$
$$(\tau_m^p - \tau_m^2) - 2\tau_m^2 \tan(\phi)(\theta_p - \theta_2) + \bar{K}_2(x_p - x_2) = 0 \quad (4.148a,b)$$

and solving for a first estimate:

$$\tau_m^p = \frac{2\tau_m^1\tau_m^2[1 + (\theta_1 - \theta_2) + \tan(\phi)] + \tau_m^2\bar{K}_1(x_p - x_1) - \tau_m^1\bar{K}_2(x_p - x_2)}{(\tau_m^1 + \tau_m^2)}$$
$$\theta_p = \theta_2 + \frac{(\tau_m^p - \tau_m^2) + \bar{K}_2(x_p - x_2)}{2\tau_m^2 \tan(\phi)} = -\theta_1 + \frac{(\tau_m^p - \tau_m^1) + \bar{K}_1(x_p - x_1)}{2\tau_m^1 \tan(\phi)} \quad (4.149a,b)$$

Again, averages (Equation 4.144) should be used after the first estimate, and again iteration continues until $\theta_p^{n+1} - \theta_p^n \leq$ tolerance, for $n = 1,2,...$N where N is a maximum number of iterations allowed and "tolerance" is some small number much less than one.

Alternatively, one can solve for θ first and then τ_m. Thus:

$$\theta_p = \frac{\tau_m^1 - \tau_m^2 + 2\tau_m^1\theta_1 \tan(\phi) + 2\tau_m^2\theta_2 \tan(\phi) - \bar{K}_1(x_p - x_1) + \bar{K}_2(x_p - x_2)}{2(\tau_m^2 + \tau_m^1)\tan(\phi)}$$
$$\tau_m^p = \tau_m^1 + 2\tau_m^1(\theta_p - \theta_2)\tan(\phi) - \bar{K}_1(x_p - x_1) \text{ or} \quad (4.150c,d)$$
$$\tau_m^p = \tau_m^2 + 2\tau 3(\theta_p - \theta_2)\tan(\phi) - \bar{K}_2(x_p - x_2)$$

and again averages should be used after the first estimate or iteration.

BVP II All data given along two characteristic curves

When all four unknowns are given along to adjoining characteristics as illustrated in Figure 4.35, a solution over the curvilinear quadrilateral shown in the figure is possible.

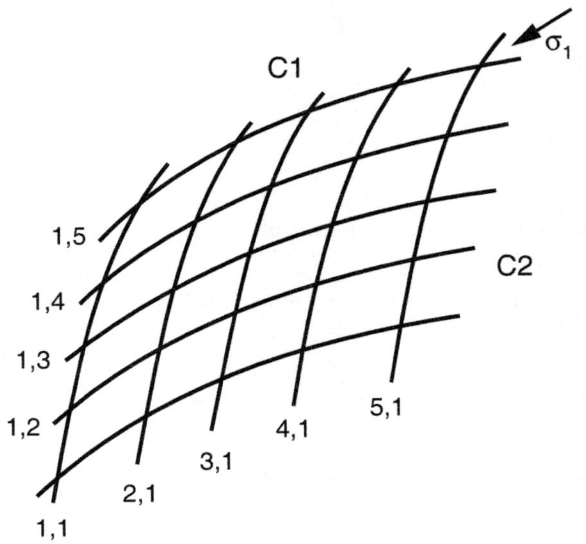

Figure 4.35 BVP II with all data given along two adjoining characteristic curves.

Formulation of a solution proceeds in much the same manner as in BVP I. With reference to Figure 4.36 that illustrates a conventional finite difference numbering scheme, translation of a double subscript notation indicating the number of the C1 and C2 curves, for example, $x(i,j)$, $I = C_1, j = C_2$, a finite difference scheme from:

$$x_p \frac{(y_2 - y_1) - [x_2 \tan(\theta_2 - \mu) - x_1 \tan(\theta_1 - \mu)]}{[\tan(\theta_1 + \mu) - \tan(\theta_2 - \mu)]} \qquad (4.151a,b)$$

$$y_p = y_1 + (x_p - x_1)\tan(\theta_1 - \mu)$$

is:

$$x(i+1, j+1) = \frac{y(i\,j+1) - y(i+1,j)}{\{\tan[(\theta(i+1,j) + \mu] - \tan[(\theta(i, j+1-\mu]\}}$$

$$-\frac{\{[x(i,j+1)\tan[\theta(i,j+1) - \mu] - x(i+1,j)\tan[(\theta(i+1,j) + \mu]\}}{\{\tan[(\theta(i+1,j) + \mu] - \tan[(\theta(i, j+1-\mu]\}} \qquad (4.151c,d)$$

$$y(i+1, j+1) = y(i+1,j) + [x(i+1,j)\tan[(\theta(i+1,j) + \mu]$$

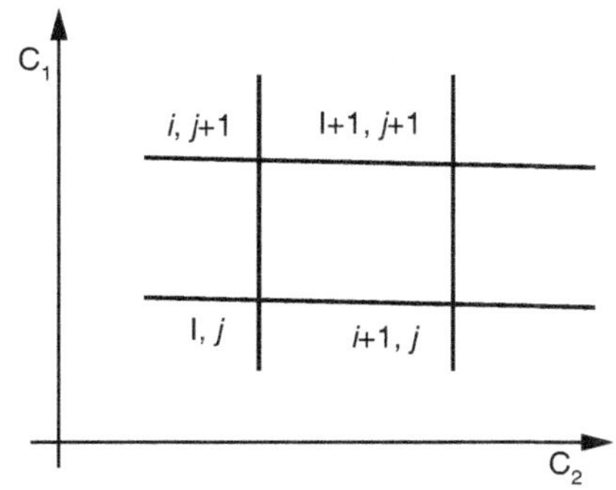

Figure 4.36 Schematic labeling of a conventional finite difference scheme with points of intersection of characteristic curves.

An analogous labeling for the unknowns $\tau_m^p = \tau_m(i+1, j+1)$ and $\theta_p = \theta(i+1, j+1)$ in Equations (149a,b) completes the system. Computer programming then follows.

BVP III Mixed data

In this third type of boundary value problem, the given data are mixed. All data are given along one characteristic curve and some data are given along a non-characteristic curve. An example is shown in Figure 4.37, where a characteristic net includes a line of symmetry. The direction of the major principal stress is known along a line of symmetry, in this example, the y-axis ($\theta = 0$) where $\theta = 0$.

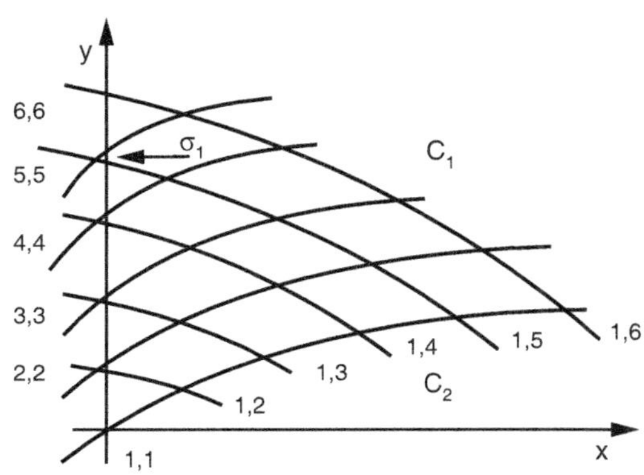

Figure 4.37 Mixed boundary value problem with data along a C_1 curve and the direction of the major principal stress known along the *y*-axis, which is a line of symmetry.

Another case of interest is a frictional boundary as illustrated in Figure 4.38. The friction condition is a restriction between normal and shear stresses acting at the wall. However, the direction of the principal stress is unknown. In fact, the principal stress direction is dependent on stress. Of course, the location of the wall is known in the figure from the x-coordinate.

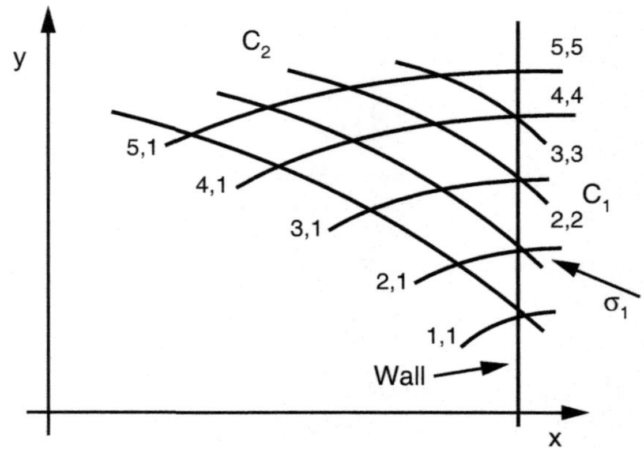

Figure 4.38 Mixed boundary value problem at a frictional wall.

Figure 4.39 shows detail of extending a solution to a frictional wall from two known points, 1 and 2, to the unknown wall point P. Reaching point P along the characteristic curve C_1 and along the wall leads to the system:

$$
\begin{aligned}
\theta_p &= \theta_p \\
y_p &= y_2 + \tan(\bar{\theta}_2 + \mu) \\
x_p &= x_o \,(\text{wall coordinate}) \\
\tau_m^p &= \tau_m^2 + 2\bar{\tau}_m^2(\theta_p - \theta_2)\tan(\phi) \\
\sigma_m^p &= [\tau_m^p - k\cos(\phi)]/\sin(\phi) \\
\bar{\theta} &= (\theta_p + \theta_2)/2,\; \bar{\tau}_m^p = (\tau_m^p + \tau_m^2)/2
\end{aligned}
\tag{4.152}
$$

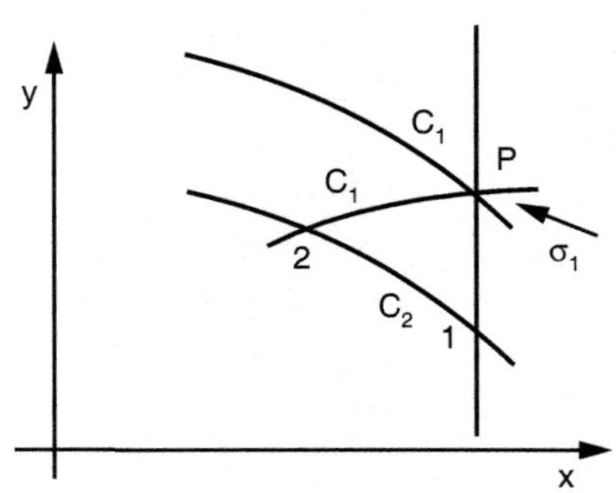

Figure 4.39 Detail of approach to a point P at a frictional wall.

The first wall point requires finding θ at the wall. Once found, the solution (Equation 4.152) is applicable. This special case uses the iteration (Equation 4.149) beginning with y_p and proceeds to σ_m^p. An estimate of θ_p is then made using:

$$\sin(2\theta_p + \phi') = \left(\frac{\sin(\phi)}{\sin(\phi)}\right)\left[\frac{K - K'}{\sigma_m + K}\right]$$
$$K = k / \tan(\phi),\ K' = k' \tan(\phi') \tag{4.153}$$

There are four solutions to Equation (4.153), so care must be taken to avoid ambiguity in using an inverse sine function that customarily returns values on the interval [0, $\pi/2$]. The desired solution is one consistent with the sign of wall shear stress. Only two of the four possible solutions to Equation (4.150) are shown in Figure 4.40.

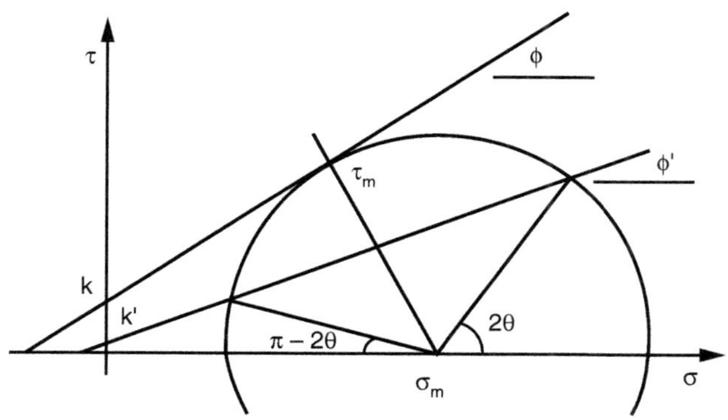

Figure 4.40 Yielding and wall friction. Negative values of θ not shown.

References

Love, A.E.H. (1944) *A Treatise on the Mathematical theory of Elasticity*. Dover, New York.

Nadai, A. (1963) *Theory of Flow and Fracture of Solids*. Vol. I. McGraw-Hill, New York.

Pariseau, W.G. (1970) Wedge indentation of anisotropic geologic media. In *Proceedings of 12th U.S. Symposium on Rock Mechanics,* Rolla, Missouri, November 16–18. Society of Mining Engineers of the American Institute of Mining, Metallurgical and Petroleum Engineers, New York.

Pariseau, W.G. (1972) Plasticity theory for anisotropic rocks and soils. In: *Proceedings of 10th U.S. Symposium on Rock Mechanics* (ed: K. Gray). Society of Mining Engineers of the American Institute of Mining, Metallurgical and Petroleum Engineers, New York, pp.267–295.

Pariseau, W.G. (2012) A symmetry requirement for failure criteria. *International Journal of Rock Mechanics and Mining Sciences*, 52, pp.42–49.

Pariseau, W.G. & Fairhurst, C. (1967) The force-penetration characteristic of a wedge penetrating rock. *International Journal of Rock Mechanics and Mining Sciences*, 4(2), 165–180.

Chapter 5

Limit theorems

Limit theorems provide bounds to collapse loads. Collapse occurs when a plastic zone encompasses an area of interest. In a finite structure such as a pipe, collapse impends when a plastic zone extends through the pipe wall. In case of a foundation, collapse impends when plastic zones extend from underneath the foundation to the sides of the structure. Tunnel collapse may occur when a plastic zone encompasses a cross-section. Generally, when the elastic zone no longer contains the plastic zone, collapse impends. Two theorems based on the stability postulate of Drucker and developed by Drucker *et al.* (1951) and Drucker and Prager (1952) provide lower and upper bounds to collapse loads. A much greater development of limit theorem applications may be found in the prodigious work by Chen (2008).

The starting point is the divergence theorem that is also a principle of virtual work and a principle of virtual power in solid mechanics application. Thus, given a "statically admissible" stress field:

$$\begin{aligned} &\sigma_{ji,j} + \gamma_i = 0 && \text{in volume V} \\ &T_i = \sigma_{ji} n_j && \text{on surface S} \\ &\sigma_{ji} = \sigma_{ij} && \text{in V+S} \end{aligned} \tag{5.1}$$

where n_j is an outward pointing unit vector on S and strain rates are related to velocities by:

$$\dot{\varepsilon}_{ij} = (1/2)(v_{i,j} + v_{j.i}) \tag{5.2}$$

the divergence theorem shows that:

$$\int_S T_i v_i dS + \int_V \gamma_i v_i dV = \int_V \sigma_{ij} \dot{\varepsilon}_{ij} dV \tag{5.3}$$

which is the principal of virtual power. Note that if tractions are not prescribed on some portion of S, then the corresponding velocities are assumed to vanish.

5.1 Lower-bound theorem

Let a *safe* state of stress σ_{ij}^* be defined as one that satisfies equilibrium and stress boundary conditions but is below yield. Let a *collapse* or unsafe state of stress σ_{ij} also satisfy equilibrium, stress boundary conditions, and yield as well. Also let velocities v_i satisfy velocity boundary conditions and the stress-strain law for stresses σ_{ij} so the strain rates $\dot{\varepsilon}_{ij}$ are the true stresses and strain rates. Then by the principle of virtual power:

$$\int_S T_i v_i dS + \int_V \gamma_i v_i dV = \int_V \sigma_{ij} \dot{\varepsilon}_{ij} dV \text{ and } \int_S T_i v_i dS + \int_V \gamma_i v_i dV = \int_V \sigma_{ij}^* \dot{\varepsilon}_{ij} dV \tag{5.4}$$

Hence $\int_V (\sigma_{ij} - \sigma_{ij}^*)\dot{\varepsilon}_{ij} dV = 0$, therefore $(\sigma_{ij} - \sigma_{ij}^*)\dot{\varepsilon}_{ij} = 0$. But, according to the stability postulate:

$$(\sigma_{ij} - \sigma_{ij}^*)\dot{\varepsilon}_{ij} = (\sigma_{ij} - \sigma_{ij}^*)d\varepsilon_{ij} \geq 0 \tag{5.5}$$

where the equality holds only if $(\sigma_{ij} - \sigma_{ij}^*) = 0$ or if $\dot{\varepsilon}_{ij} = 0$. But $\sigma_{ij} \neq \sigma_{ij}^*$ so $\dot{\varepsilon}_{ij} = 0$

This last result implies a collapse state, which requires $\dot{\varepsilon}_{ij} \neq 0$, which cannot occur under a *safe* state of stress. A collapse load must be greater than any applied load resulting in a *safe* state of stress (which nowhere violates the yield condition).

5.2 Upper-bound theorem

If an "unsafe" set of velocities can be found, then collapse must be impending or have already occurred. Let an "unsafe" motion be defined by velocities v_i^{**} that satisfy velocity boundary conditions and are at least once differentiable and thus are "kinematically" admissible. Further, let these velocities v_i^{**} be such that the actual surface tractions and body forces dissipate more energy than that computed through the stress-strain law. By definition then:

$$\int_S T_i v_i^{**} dS + \int_V \gamma_i v_i^{**} dV \geq \int_V \sigma_{ij}^{**} \dot{\varepsilon}_{ij}^{**} dV \tag{5.6}$$

The terms on the left-hand side is the work rate of the actual surface tractions and body forces on the velocities v_i^{**}. The integral on the right-hand side involves stresses σ_{ij}^{**} associated with strain rates $\dot{\varepsilon}_{ij}^{**}$ through the stress-strain law and is the rate of energy dissipation associated with the unsafe strain rates. But also:

$$\int_S T_i v_i^{**} dS + \int_V \gamma_i v_i^{**} dV = \int_V \sigma_{ij} \dot{\varepsilon}_{ij}^{**} dV \tag{5.7}$$

Here stresses σ_{ij} are the true stresses that satisfy equilibrium and traction boundary conditions and are symmetric. From these two integrals:

$$0 \geq (\sigma_{ij}^{**} - \sigma_{ij})\dot{\varepsilon}_{ij}^{**} \tag{5.8}$$

But by the stability postulate:

$$(\sigma_{ij}^{**} - \sigma_{ij})\dot{\varepsilon}_{ij}^{**} \geq 0 \tag{5.9}$$

Hence, the equality must hold and $\sigma_{ij}^{**} = \sigma_{ij}$, so the true stresses σ_{ij} are "unsafe" and collapse is impending or must have already occurred.

A useful tool for upper-bound estimates is the rate of dissipation on a discontinuity. With reference to Figure 5.1, the rate of work the normal and shear stresses do as the material yields per unit area of discontinuity is:

$$\begin{aligned} D_A &= \tau v_s - \sigma v_n \\ &= [\sigma \tan(\phi) + k][V \cos(\phi)] - \sigma V \sin(\phi) \\ D_A &= kV \cos(\phi) \end{aligned} \tag{5.10}$$

assuming a Mohr-Coulomb yield criterion.

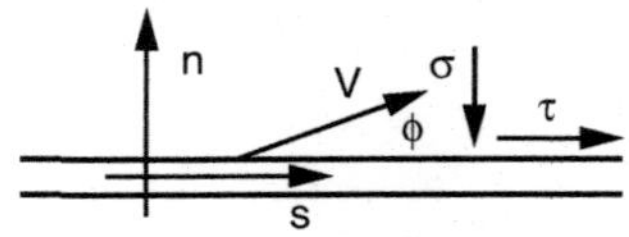

Figure 5.1 Stresses acting on a velocity discontinuity zone.

5.3 Example problems

An often-cited application relates to bearing capacity of a flexible slab illustrated in Figure 5.2a. The slab is long compared with the slab width and is thus a *strip footing*. In (a), material adjacent to the footing is not loaded. The stress field assumed below the adjacent material is just at yield. Because this material is unconfined, the horizontal stress must be near the unconfined compressive strength of the material (C_o). Continuity of normal stress across the dotted vertical line is imposed (shear stress is zero), although there is a jump in the normal stress acting parallel to the dotted lines (a stress discontinuity). The state of stress just below the applied load between the vertical lines is assumed to be very close to yielding, too, under the lateral "confining pressure." Hence, the collapse load σ_c satisfies:

$$\sigma_c \geq \sigma_o = C_p = C_o + (C_o / T_o)C_o \tag{5.11}$$

which is a lower bound to the collapse load assuming Mohr-Coulomb behavior. Sometimes this result is expressed as $\sigma_o = 2k\sqrt{N}(1+N)$ where $N = (1 + \sin(\phi))/(1-\sin(\phi)) = (C_o/T_o)$ and k is cohesion. Yet another form is the dimensionless ratio $(\sigma_o / C_o) = 1 + (C_o / T_o) = 1 + N$. If $\phi = 30°$, then $N = 3$ and $(\sigma_o / C_o) = 4$. In this case, the collapse load σ_o must be at least four times the unconfined compressive strength of the foundation material. In the case $\phi = 0$, $(\sigma_o / C_o) = 2$ or $(\sigma_o / k) = 4$. No mention is made of material weight or gravity loading in this analysis. Indeed, the stress distribution assumed is entirely fictitious, a feature that makes lower-bound estimates practical to obtain.

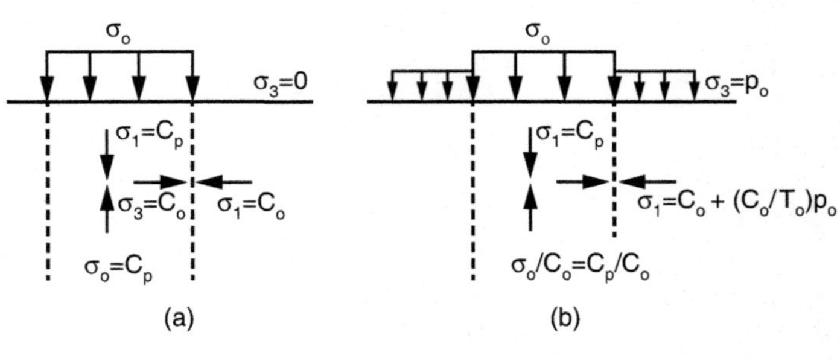

Figure 5.2 Bearing capacity example problem illustrating lower-bound theorem application. (a) No adjacent surface load. (b) Adjacent surface load p_o.

If an adjacent surface load p_o is present as in Figure 5.2b, then:

$$\sigma_c \geq (\sigma_o / C_o) = [1 + (C_o / T_o)(1 + p_o / T_o)] \tag{5.12}$$

If ϕ = 30° and $p_o = T_o$, then $(\sigma_o / C_o) = 8$. Confinement of loaded material by adjacent material provides considerable strengthening, doubling of a lower bound to bearing capacity in this example. In the case ϕ = 0, $(\sigma_o / C_o) = 2 + (p / C_o)$ or $(\sigma_o / k) = 4 + (p / k)$. If $p_o = T_o = C_o = 2k$, then $(\sigma_o / C_o) = 3$ or $(\sigma_o / k) = 6$ are lower bounds in the case $\phi = 0$.

Slipline solutions are upper-bound solutions when associated velocity fields are kinematically admissible. An example application is to the strip footing problem for which a lower bound to bearing capacity was found previously. The problem is much studied and has many variations. With reference to Figure 5.3, a solution is readily composed of special stress fields, two constant state stress regions (I and III) and one region of radial shear (II).

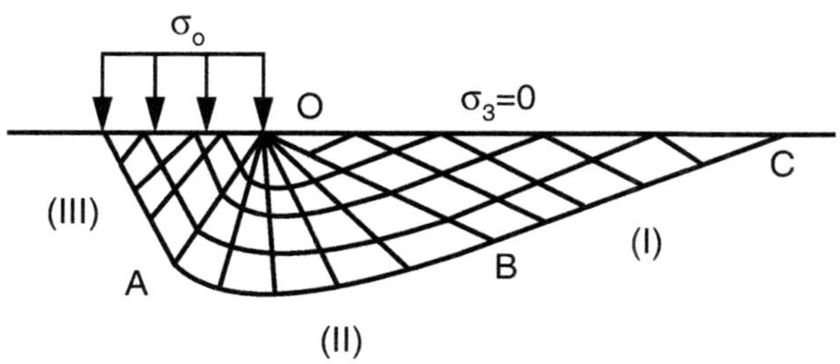

Figure 5.3 Slipline fields for estimating an upper bound to bearing capacity of a strip foundation.

Analysis begins with stress in I where the minor principal stress is 0, so the major principal stress is just the unconfined compressive strength C_o. Hence, along OB $\sigma_m = C_o / 2$ and $\tau_m = C_o / 2$ assuming a Mohr-Coulomb yield criterion. In II $\tau_m = \tau_m^0 \exp[2\beta \tan(\phi)]$ where $\beta = \pi / 2$, the fan angle. In III, $\sigma_o = \sigma_m + \tau_m$ and in consideration of yielding $\sigma_o = [\tau_m - k\cos(\phi)] / \sin(\phi) + \tau_m$ which leads to:

$$(\frac{\sigma_o}{C_o}) = [\frac{1}{(C_o / T_o) - 1}]\{(\frac{C_o}{T_o})[\exp[\pi \tan(\phi)] - 1\} \leq \sigma_c \tag{5.13}$$

as an upper bound to bearing capacity. If ϕ = 30°, $(\sigma_o / C_o) = 8.7$. In case ϕ = 0, $(\sigma_o / C_o) = (1 + \pi / 2) = 2.57$. Alternatively $(\sigma_o / k) = (2 + \pi) = 5.14$ Recall a lower bound to (σ_o / k) is 4.0. Closer bounds are certainly desirable in this example; more ingenuity in estimating the lower bound is needed.

If adjacent material is loaded as illustrated in Figure 5.4, then:

$$(\frac{\sigma_o}{C_o}) = [\frac{1}{(C_o / T_o) - 1}]\{[\frac{C_o}{T_o} + (\frac{p_o}{T_o})(\frac{C_o}{T_o} - 1)]([\exp[\pi \tan(\phi)] - 1\} \leq \sigma_c \tag{5.14}$$

If $p_o = T_o$ then an upper bound is 14.8, somewhat less than double the lower bound of 8.0. In the case of $\phi = 0$, an upper bound is $(\sigma_o / k) = (2 + \pi) + (p_o / k)$ or $(\sigma_o / C_o) = (1 + \pi / 2) + (p_o / C_o)$. When $p_o = T_o = C_o = 2k$, $(\sigma_o / k) = 7.14$ and $(\sigma_o / C_o) = 3.57$. Dimensionless lower bounds for comparison are 6.0 and 3.0, respectively.

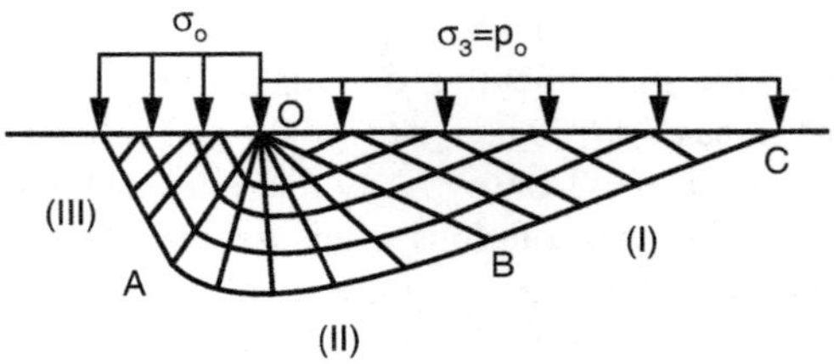

Figure 5.4 Slipline fields for estimating an upper bound to bearing capacity of a strip foundation when adjacent material is loaded.

Another problem for which bounds were obtained relates to the height of a vertical bank. Figure 5.5 illustrates a stress field for estimating a lower bound. Inspection of the figure indicates an equilibrium distribution of stress and satisfaction of stress boundary conditions. A lower bound is obtained when the stress at the bank bottom just reaches yield. Thus:

$$\sigma_c = \gamma H_c \geq C_o \tag{5.15}$$

and the critical bank height satisfies the inequality $H_c \geq C_o / \gamma$.

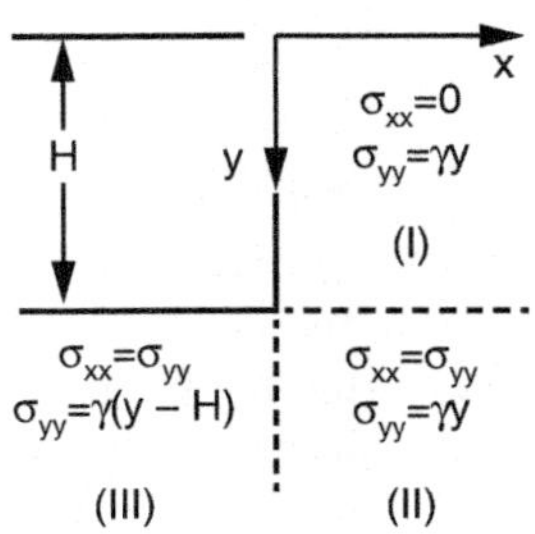

Figure 5.5 Diagram for a lower-bound estimate for height H of a vertical bank.

Figure 5.6 illustrates a velocity field for estimating an upper bound to the bank height. The collapse mechanism is simply a rigid block sliding along a velocity discontinuity.

Two work rates are required, one associated with external forces and stresses, the other associated with an unsafe motion, a dissipation rate. The external forces are just those of block weight $W = \gamma H^2 / 2\tan(\delta)$ with components normal and tangential to the shear plane $W_n = W\cos(\delta)$ and $W_s = W\sin(\delta)$. The shear plane dips at $\delta°$ as seen in the figure. The external work rate is $D = W\sin(\delta)V\cos(\phi) - W\cos(\delta)V\sin(\phi)$. Thus, $D = WV\sin(\delta - \phi)$. The dissipation rate per unit area of discontinuity from before is $D_A = kV\cos(\phi)$. Hence, the total rate is $D_T = [H/\sin(\delta)]kV\cos(\phi)$. After substituting for W and applying the upper-bound theorem, the result is:

$$H_c \leq \frac{2k\cos(\phi)}{\cos(\delta)\sin(\delta - \phi)} \tag{5.16}$$

An optimum value of δ that minimizes the bound is $\pi/4+\phi/2$. Hence:

$$H_c \leq (4k/\gamma)\tan(\pi/4+\phi/2) = (2C_o/\gamma) \quad (5.17)$$

Finally:

$$(C_o/\gamma) \leq H_c \leq (2C_o/\gamma) \quad (5.18)$$

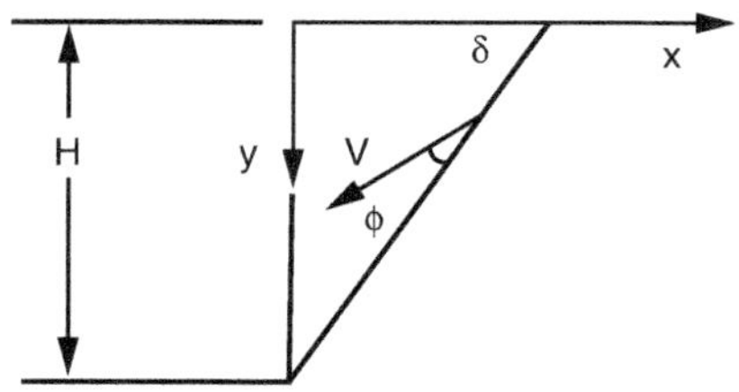

Figure 5.6 A velocity field for estimating an upper bound.

References

Chen, W.F. (2008) *Limit Analysis and Soil Plasticity*. Wai-Fah Chen, J. Ross Publishing, Fort Lauderdale, Florida.

Drucker, D.C. & Prager, W. (1952) Soil mechanics and plastic analysis or limit design. *Quarterly of Applied Mathematics*, 10, 157–164.

Drucker, D.C., Prager, W. & Greenberg, H.J. (1951) Extended limit design theorems for continuous media. *Quarterly of Applied Mathematics*, 9, 381–389.

Chapter 6

Anisotropy

Many rock formations have properties that are directional and thus are anisotropic. Examples include such obvious rock types as schist and gneiss that are anisotropic at a laboratory scale of a few centimeters. Laminated strata and rock with flow structure are also anisotropic. Rock joint sets induce anisotropy at a meter scale even when the intact rock between joints is isotropic, although random jointing is likely to result in isotropy. Some soil formations are anisotropic, varved clay for example. Alternating layers of silt and clay also show directional characteristics that need to be taken into account in geotechnical analysis.

Anisotropic rock and soil deformation begins in the elastic domain with an initial application of load. Anisotropic elasticity is well-understood as books by Lekhnitskii (1963, 1968), Green and Zerna (1968), and Ting (1996) indicate. As the limit to a purely elastic response is reached, continued loading induces an elastic-plastic response.

Anisotropic plasticity is reasonably well explored in case of metals with yielding independent of pressure, where the book by Hill (1950) is often cited. Isotropic and anisotropic plasticity in geomechanics is much less developed than in the case of metals. However, in the case of isotropy, application of plasticity theory to concrete has received considerable attention, for example, by Chen and Han (1988). In case of anisotropy, extension of the anisotropic discussion by Hill (1950) to include anisotropic, pressure-dependent yield by Pariseau (1968) is a reasonable starting point for development of anisotropic geoplasticity. Examples from engineering practice illustrate anisotropic data, including the challenge of gathering samples and property measurement.

6.1 Elasticity

Anisotropy in elasticity is accounted for in a generalized Hooke's law, which can be specialized to various forms of anisotropy that have applicability in engineering practice. Plane strain is of special interest and easily formulated in theory. Example data quantitatively illustrate anisotropy in elasticity.

Generalized Hooke's law

In matrix notation, Hooke's law for anisotropic elastic materials is:

$$\begin{Bmatrix} \varepsilon_{aa} \\ \varepsilon_{bb} \\ \varepsilon_{cc} \\ \gamma_{bc} \\ \gamma_{ca} \\ \gamma_{ab} \end{Bmatrix} = \begin{bmatrix} a_{11} & a_{12} & a_{13} & a_{14} & a_{15} & a_{16} \\ a_{21} & a_{22} & a_{23} & & & \\ a_{31} & & & & & \\ & & & \cdots & & \\ & & & & & \\ a_{61} & a_{62} & a_{63} & a_{64} & a_{65} & a_{66} \end{bmatrix} \begin{Bmatrix} \sigma_{aa} \\ \sigma_{bb} \\ \sigma_{cc} \\ \tau_{bc} \\ \tau_{ca} \\ \tau_{ab} \end{Bmatrix} \tag{6.1}$$

where ε, γ, σ and τ are normal strain, engineering shear strain, normal stress, and shear stress, respectively, the *a*s are elastic constants (compliances). For a definitive treatment of anisotropic elasticity, the oft-cited work by Lekhnitskii (1963) is an excellent source. In condensed matrix form, Equation (6.1) is:

$$\{\varepsilon\} = [S]\{\sigma\} \text{ and } \{\sigma\} = [C]\{\varepsilon\} \text{ or } \{\sigma\} = [E]\{\varepsilon\} \tag{6.2a,b}$$

where Equation (6.2b) is the inverted form of Equation (6.2a) and $[C]=[E]$ are a 6×6 matrices of elastic stiffnesses or moduli. Matrices [*C*] and [S] are mutual inverses, for example $[C]=[S]^{-1}$. These matrices are also symmetric, for example $[C]=[C]^T$ where the superscript *T* means transpose. Symmetry implies that there are at most 21 independent elastic constants. Additional symmetries reduce the number of constants.

In case of three mutually orthogonal planes of elastic symmetry, there are only nine independent elastic constants; the material is *orthotropic*. The elastic stress-strain relations then have the form:

$$\begin{Bmatrix} \varepsilon_{aa} \\ \varepsilon_{bb} \\ \varepsilon_{cc} \\ \gamma_{bc} \\ \gamma_{ca} \\ \gamma_{ab} \end{Bmatrix} = \begin{bmatrix} a_{11} & a_{12} & a_{13} & 0 & 0 & 0 \\ a_{21} & a_{22} & a_{23} & 0 & 0 & 0 \\ a_{31} & a_{32} & a_{33} & 0 & 0 & 0 \\ 0 & 0 & 0 & a_{44} & 0 & 0 \\ 0 & 0 & 0 & 0 & a_{55} & 0 \\ 0 & 0 & 0 & 0 & 0 & a_{66} \end{bmatrix} \begin{Bmatrix} \sigma_{aa} \\ \sigma_{bb} \\ \sigma_{cc} \\ \tau_{bc} \\ \tau_{ca} \\ \tau_{ab} \end{Bmatrix} \tag{6.3a}$$

where the off-diagonal terms are equal, for example $a_{12} = a_{21}$. When referred to the axes *abc* of anisotropy:

$$\begin{Bmatrix} \varepsilon_{aa} \\ \varepsilon_{bb} \\ \varepsilon_{cc} \\ \gamma_{bc} \\ \gamma_{ca} \\ \gamma_{ab} \end{Bmatrix} = \begin{bmatrix} 1/E_{aa} & -v_{ba}/E_{bb} & -v_{ca}/E_{cc} & 0 & 0 & 0 \\ -v_{ab}/E_{aa} & 1/E_{bb} & -v_{cb}/E_{cc} & 0 & 0 & 0 \\ -v_{ac}/E_{aa} & -v_{bc}/E_{bb} & 1/E_{cc} & 0 & 0 & 0 \\ 0 & 0 & 0 & 1/G_{bc} & 0 & 0 \\ 0 & 0 & 0 & 0 & 1/G_{ca} & 0 \\ 0 & 0 & 0 & 0 & 0 & 1/G_{ab} \end{bmatrix} \begin{Bmatrix} \sigma_{aa} \\ \sigma_{bb} \\ \sigma_{cc} \\ \tau_{bc} \\ \tau_{ca} \\ \tau_{ab} \end{Bmatrix} \tag{6.3b}$$

Symmetry implies $v_{ba} / E_{bb} = v_{ab} / E_{aa}$ and so on. The orthotropic model indicates there is no coupling between normal stress and shear strain and between shear stress and normal strain, although it is implied in the general anisotropic model Equation (6.1).

Some measurements of anisotropic rock properties are presented in Table 6.1. Laboratory data indicate that the ratio of Young's moduli parallel and perpendicular to bedding is seldom greater than two.

When rock cylinders are tested in uniaxial compression at an angle to bedding, the results are expected to vary according to:

$$\frac{1}{E} = \frac{\cos^2(\beta)}{E_1} + \left(\frac{1}{G_{12}} - \frac{2v_{12}}{E_1}\right)\sin^2(\beta)\cos\ (\beta) + \frac{\sin^2(\beta)}{E_2} \tag{6.4}$$

Terms in Equation (6.4) are defined in Figure 6.1. Angle $\beta = \pi / 2 - \delta$; δ = bedding plane dip. Results are plotted in Figure 6.2 with shear modulus as a parameter. Interestingly, Young's

Table 6.1 Anisotropic Young's modulus (Pariseau, 2017b).

Rock Type	*Young's Modulus GN/m² (10⁶ psi)*						
	Perpendicular		*Parallel 1*		*Parallel 2*		*Emx/Emn*
Marble	49.3	(7.15)	63.1	(9.15)	71.7	(10.4)	1.45
Limestone 1	33.4	(4.84)	41.0	(5.94)	37.2	(5.39)	1.23
Limestone 2	68.5	(9.93)	56.5	(8.19)	61.8	(8.96)	1.21
Granite 1	30.4	(4.41)	27.4	(3.97)	44.2	(6.41)	1.61
Granite 2	–	–	22.6	(3.28)	30.3	(4.39)	–
Slate	–	–	93.8	(13.6)	83.4	(12.1)	–
Granite 3	37.3	(5.41)	42.3	(6.13)	–	–	1.11
Sandstone 1	6.0	(0.87)	6.7	(0.97)	8.8	(1.28)	1.47
Sandstone 2	7.1	(1.03)	1.1	(1.54)	11.2	(1.63)	1.58
Sandstone 3	9.6	(1.39)	1.1	(1.53)	–	–	1.10
Gneiss	18.6	(2.70)	23.1	(3.35)	12.4	(1.80)	1.86
Oil Shale 1	12.4	(1.80)	21.4	(3.10)	–	–	1.72
Oil Shale 2	4.82	(3.06)	33.2	(4.82)	–	–	1.58

modulus may increase with bedding plane dip from a high parallel value to a low perpendicular value with orientation as seen in the figure before decreasing to a low value depending on the shear modulus.

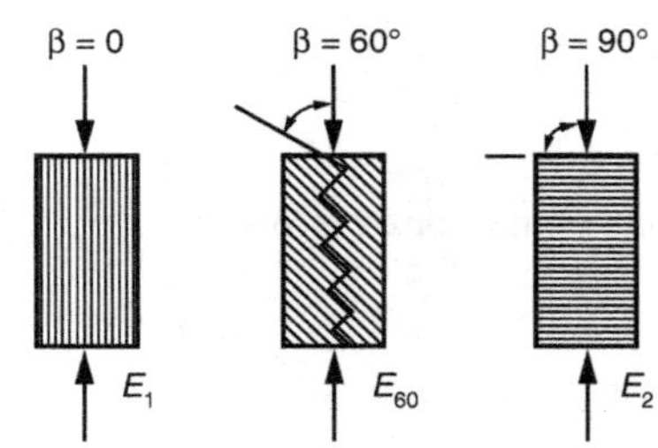

Figure 6.1 Testing at an angle to bedding.

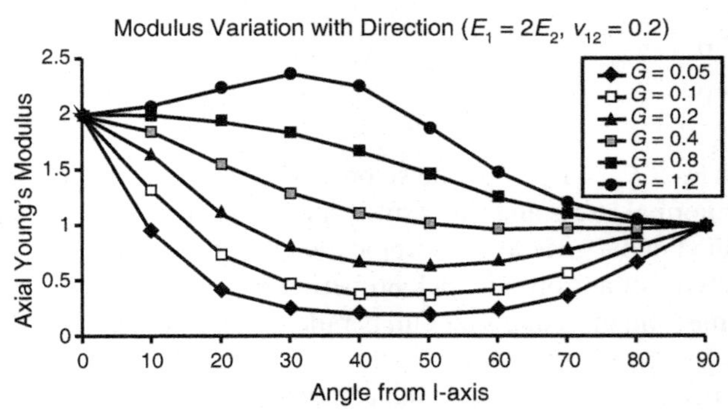

Figure 6.2 Young's modulus at an angle to bedding with a variable shear modulus (Pariseau, 2017b).

If there is an axis of rotational symmetry perpendicular to a plane of symmetry, the material is *transversely isotropic* and the number of elastic constants is reduced to five. The stress-strain law then has the form according to Lekhnitskii (1963):

$$\begin{Bmatrix} \varepsilon_{aa} \\ \varepsilon_{bb} \\ \varepsilon_{cc} \\ \gamma_{bc} \\ \gamma_{ca} \\ \gamma_{ab} \end{Bmatrix} = \begin{bmatrix} a_{11} & a_{12} & a_{13} & 0 & 0 & 0 \\ a_{12} & a_{11} & a_{13} & 0 & 0 & 0 \\ a_{13} & a_{13} & a_{33} & 0 & 0 & 0 \\ 0 & 0 & 0 & a_{44} & 0 & 0 \\ 0 & 0 & 0 & 0 & a_{44} & 0 \\ 0 & 0 & 0 & 0 & 0 & 2(a_{11}-a_{12}) \end{bmatrix} \begin{Bmatrix} \sigma_{aa} \\ \sigma_{bb} \\ \sigma_{cc} \\ \tau_{bc} \\ \tau_{ca} \\ \tau_{ab} \end{Bmatrix} \tag{6.4a}$$

When referred to the axes of anisotropy with the z-axis vertical and also the axis of rotational symmetry, so the xy plane is the plane of isotropy, the stress-strain relations may be written as:

$$\begin{Bmatrix} \varepsilon_{xx} \\ \varepsilon_{yy} \\ \varepsilon_{zz} \\ \gamma_{yz} \\ \gamma_{zx} \\ \gamma_{xy} \end{Bmatrix} = \begin{bmatrix} 1/E_h & -v_h/E_h & -v_v/E_h & 0 & 0 & 0 \\ -v_h/E_h & 1/E_h & -v_v/E_h & 0 & 0 & 0 \\ -v_v/E_h & -v_v/E_h & 1/E_v & 0 & 0 & 0 \\ 0 & 0 & 0 & 1/G_v & 0 & 0 \\ 0 & 0 & 0 & 0 & 1/G_v & 0 \\ 0 & 0 & 0 & 0 & 0 & 1/G_h \end{bmatrix} \begin{Bmatrix} \sigma_{xx} \\ \sigma_{yy} \\ \sigma_{zz} \\ \tau_{yz} \\ \tau_{zx} \\ \tau_{xy} \end{Bmatrix} \tag{6.4b}$$

where $G_h = E_h / 2(1+v_h)$. The five independent elastic constants in Equation (6.4b) are E_h, v_h, E_v, v_v, G_v. A second vertical Poisson's ratio v'_v may also be used in consideration of symmetry, that is, $v_v / E_h = v'_v / E_v$. The Poisson's ratio v_h relates a horizontal strain at right angles to a horizontal stress; v'_v relates a horizontal strain to a vertical stress (as does v_v). The transversely isotropic model is perhaps the most used anisotropic model in rock mechanics. In this case, properties parallel and perpendicular to stratification are often used. Figure 6.3 illustrates the physical meaning of the elastic constants in this case.

In case there is no directional dependency, isotropy, there are just two independent elastic constants and the form of the stress-strain relationship is the familiar Hooke's law. Thus:

$$\begin{aligned} \varepsilon_{xx} &= \frac{1}{E}\sigma_{xx} - \frac{v}{E}\sigma_{yy} - \frac{v}{E}\sigma_{zz}, & \gamma_{yz} &= \frac{1}{G}\tau_{yz} \\ \varepsilon_{yy} &= \frac{-v}{E}\sigma_{xx} + \frac{1}{E}\sigma_{yy} - \frac{v}{E}\sigma_{zz}, & \gamma_{zx} &= \frac{1}{G}\tau_{zx} \\ \varepsilon_{zz} &= \frac{-v}{E}\sigma_{xx} - \frac{v}{E}\sigma_{yy} + \frac{v}{E}\sigma_{zz}, & \gamma_{xy} &= \frac{1}{G}\tau_{xy} \end{aligned} \tag{6.4c}$$

where E, v, and G are Young's modulus, Poisson's ratio, and shear modulus, respectively, and the shear modulus is given by the usual formula: $G = E / 2(1+v)$.

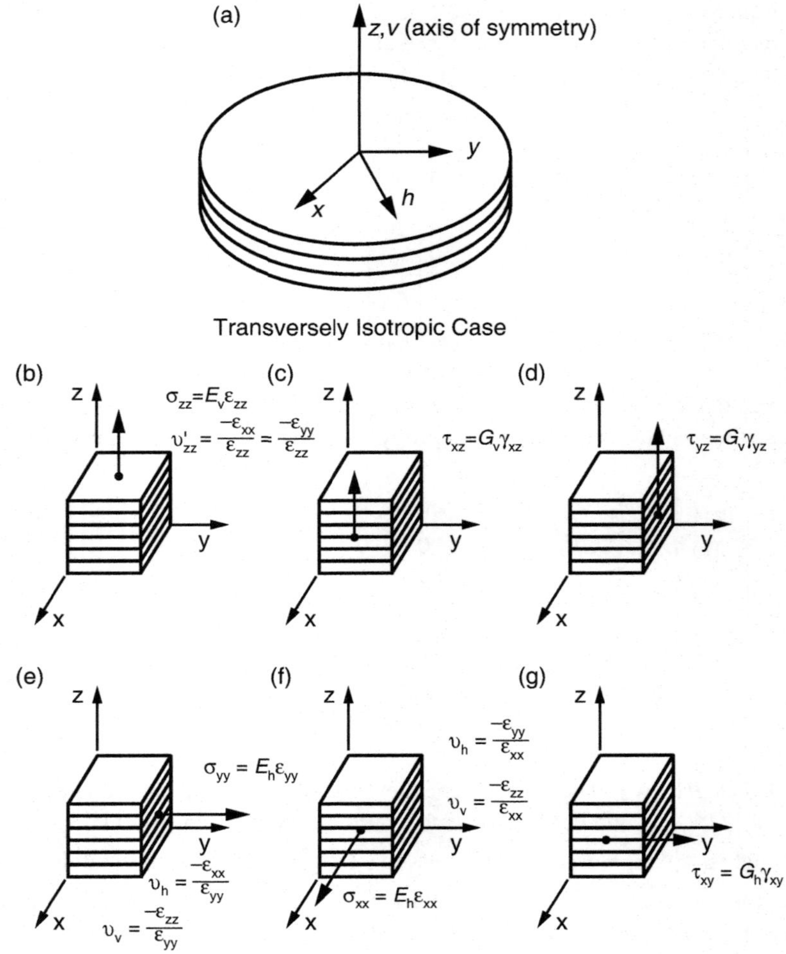

Figure 6.3 Physical interpretations of elastic constants in case of transverse isotropy (Pariseau, 2017b).

Plane strain elasticity

Plane strain is an important special case of three-dimensional elastic anisotropy. Plane stress is also a special case but of less importance in geomechanics. Degrees of plane strain are possible. Suppose the derivatives in the *z*-direction are set to zero, then with *w* as the *z*-direction displacement one has:

$$\partial w / \partial z = \varepsilon_{zz} = 0, \ \gamma_{yz} = \partial w / \partial y, \ \gamma_{zx} = \partial w / \partial x \tag{6.5a}$$

The shear strains in Equation (6.6) cause warping of the *xy*-plane, but this warping is constant along the *z*-axis. This state is sometimes referred to as generalized plane strain with simple plain strain being the case with no warping (shear strains in Equation (6.5) also zero).

Use of the first of Equation (6.5a) leads to:

$$\sigma_{zz} = (-1/a_{33})(a_{31}\sigma_{xx} + a_{32}\sigma_{yy} + a_{34}\tau_{yz} + a_{35}\tau_{zx} + a_{36}\tau_{xy}) \tag{6.5b}$$

which may be used to reduce the stress-strain law (Equation 6.1) to a 5×5 matrix of elastic properties. If warping is zero, the stress-strain relationship may be further reduced to a 3×3 matrix of elastic properties. The new elastic properties are obtained from Equation (6.1) according to Leknitskii are:

$$\beta_{ij} = a_{ij} - (\frac{a_{3i}a_{j3}}{a_{33}}), \qquad (i, j = 1,2,4,5,6) \tag{6.5c}$$

Example data

An example of extensive use of anisotropy in engineering practice at the former Homestake Mine in the Northern Black Hills of South Dakota is described in a series of reports and papers. The mine was a deep underground gold mine that extended over 2400 m (8000 ft) below ground surface and operated for over one hundred years until closure in 2002. Mitchell (2009) describes the mine history in much interesting detail and from firsthand experience at the mine. In 2007, the US National Science Foundation selected the site for conversion to a multidisciplinary underground research laboratory. The site is now the Sanford Underground Research Facility (SURF) with major emphasis on neutrino and related physics. However, physics is not new to the "Homestake." Indeed, a solar neutrino experiment was initiated in 1968 that led to a share in the Nobel Prize for physics by Dr. Ray Davis and two other physicists in 2002. The cavern that housed the experiment on the 4850 Level (4850 ft below ground surface, 1478 m), the Davis Cavern, has since been redone to house a second-generation dark matter detector. Additional excavation is planned that will require detailed geotechnical analysis based on site-specific rock and joint properties, including properties of occasional igneous dikes and joints that transect the major formations present at the site.

Major rock formations at "Homestake" are meta-sediments, mainly phyllites and schists, of the Poorman, Homestake, and Ellison formations. The Homestake formation hosted the gold ore. The Poorman formation is highly foliated and often crinkled with centimeter scale folding; it is the oldest formation of the three. The Ellison formation is also foliated; the Homestake formation less so. Folding of these formations into anticlines and synclines produces a range of strata dips, as one would expect. Effects of anisotropy on "tunnel" stability are quite noticeable. Tunnels across the foliation dip appear more stable than tunnels parallel to the foliation dip (on strike).

Elastic moduli determined from laboratory tests on drill core are given in Table 6.2 where E = Young's modulus (GPa), v = Poisson's ratio, and G = shear modulus (GPa). Subscript notation is consistent with Equation (6.3b) and an orthotropic elastic model that requires nine independent properties. Here, the *a*-axis is down the foliation dip, the *b*-axis is normal to the foliation, and the *c*-axis is on strike and thus perpendicular to the vertical *ab*-plane. Data in the table were obtained from laboratory tests on drill core taken from three holes drilled down dip, on strike, and normal to the dip. Multiple tests were done using strain gauges attached to the test cylinders in two laboratories, one at the University of Utah, Department of Mining Engineering, and the other at the South Dakota School of Mines and Technology, Department of Mining Engineering. Uniaxial compression was used to determine Young's moduli and Poisson's ratios. Shear moduli were estimated from several empirical relations and were constrained by test data from compression at an angle to the foliation.

Table 6.2 Elastic moduli of three anisotropic Precambrian meta-sedimentary formations.

Property *Formation*	E_a	E_b	E_c	v_{ab}	v_{bc}	v_{ca}	G_{ab}	G_{bc}	G_{ca}
Ellison	89.6	63.4	75.8	0.20	0.17	0.15	31.7	29.0	75.8
Homestake	88.3	64.1	62.1	0.23	0.15	0.22	33.1	26.9	29.7
Poorman	93.1	49.6	84.5	0.14	0.18	0.19	26.2	26.9	38.6

Source: US Bureau of Mines Report of Investigations 9531, Part 1. Rock Mechanics Study of Shaft Stability and Pillar Mining, Homestake Mine, Lead, SD, 1995 (Pariseau, W.G., J.C. Johnson, M.M. McDonald and M.E. Poad). E, G in GPa.

Coefficients of variation (ratio of standard deviation to mean value) of Young's moduli ranged from 5% to 35%. Ratios of mean Young's moduli parallel and perpendicular to the foliation in the table range between 1 and 2, so the elastic anisotropy is noticeable but not extreme. Indeed, statistical tests for difference in mean values of Young's moduli may show no significant difference. However, such a result is no justification for ignoring anisotropy and the evidence in the field of noticeable effects on excavation safety and stability.

The axes of anisotropy (*abc*) may be inclined to the axes of analysis (*xyz*), necessitating rotation of the stress-strain relations (Equation 6.2). An explanation of the process with details is given by Cook (1974). The process begins with a table of direction cosines that relate the *abc* and *xyz* axes as illustrated in Table 6.3.

Table 6.3 Direction cosines for axes rotation.

New *Old*	x	y	z
a	l_1	m_1	n_1
b	l_2	m_2	n_2
c	l_3	m_3	n_3

These data are then used to fill in two 6 × 6 matrices $[T_\sigma]$ and $[T_\varepsilon]$ that are used to transform 6×1 stress and strain vectors $\{\sigma(xyz)\}$ and $\{\varepsilon(xyz)\}$ to $\{\sigma(abc)\}$ and $\{\varepsilon(abc)\}$ and in the process to transform [E(abc)] to [E(xyz)]. Thus:

$$\{\sigma(abc)\} = (\sigma_{aa}, \sigma_{bb}, \sigma_{cc}, \tau_{bc}, \tau_{ca}, \tau_{ab})$$
$$\text{and} \tag{6.6a}$$
$$\{\varepsilon(abc)\} = (\varepsilon_{aa}, \varepsilon_{bb}, \varepsilon_{cc}, \gamma_{bc}, \gamma_{ca}, \gamma_{ab})$$

With the transformation matrices:

$$[T_\sigma] = \begin{bmatrix} l_1^2 & m_1^2 & n_1^2 & 2m_1n_1 & 2n_1l_1 & 2l_1m_1 \\ l_2^2 & m_2^2 & n_2^2 & 2m_2n_2 & 2n_2l_2 & 2l_1m_1 \\ l_3^2 & m_3^2 & n_3^2 & 2m_3n_3 & 2n_3l_3 & 2l_1m_1 \\ l_2l_3 & m_2m_3 & m_2n_3 & (l_2m_3 + l_3m_2) & (m_2n_3 + m_3n_2) & (n_2l_3 + n_3l_2) \\ l_3l_1 & m_3m_1 & m_3n_1 & (l_3m_1 + l_1m_3) & (m_3n_1 + m_1n_3) & (n_3l_1 + n_1l_3) \\ l_1l_2 & m_1m_2 & m_1n_2 & (l_1m_2 + l_2m_1) & (m_1n_2 + m_2n_1) & (n_1l_2 + n_2l_1) \end{bmatrix} \tag{6.6b}$$

$$[T_\varepsilon] = \begin{bmatrix} l_1^2 & m_1^2 & n_1^2 & m_1n_1 & n_1l_1 & l_1m_1 \\ l_2^2 & m_2^2 & n_2^2 & m_2n_2 & n_2l_2 & l_1m_1 \\ l_3^2 & m_3^2 & n_3^2 & m_3n_3 & n_3l_3 & l_1m_1 \\ 2l_2l_3 & 2m_2m_3 & 2m_2n_3 & (l_2m_3+l_3m_2) & (m_2n_3+m_3n_2) & (n_2l_3+n_3l_2) \\ 2l_3l_1 & 2m_3m_1 & 2m_3n_1 & (l_3m_1+l_1m_3) & (m_3n_1+m_1n_3) & (n_3l_1+n_1l_3) \\ 2l_1l_2 & 2m_1m_2 & 2m_1n_2 & (l_1m_2+l_2m_1) & (m_1n_2+m_2n_1) & (n_1l_2+n_2l_1) \end{bmatrix} \tag{6.6c}$$

one obtains:

$$\begin{aligned} \{\sigma(\mathrm{abc})\} &= [\mathrm{E(abc)}]\ \{\varepsilon(\mathrm{abc})\} \\ [\mathrm{T}_\sigma]\{\sigma(xyz)\} &= [E(abc)][T_\varepsilon]\{\varepsilon(xyz)\} \\ \{\sigma(xyz)\} &= [\mathrm{T}_\sigma]^{-1}[E(abc)][T_\varepsilon]\{\varepsilon(xyz)\} \\ \{\sigma(xyz)\} &= [\mathrm{T}_\varepsilon]^t[E(abc)][T_\varepsilon]\{\varepsilon(xyz)\} \\ \therefore\quad [E(xyz)] &= [\mathrm{T}_\varepsilon]^t[E(abc)][T_\varepsilon] \end{aligned} \tag{6.6d}$$

where the inverse $[T_\sigma]^{-1} = [\mathrm{T}_\varepsilon]^t$ and $[T_\varepsilon]^{-1} = [\mathrm{T}_\sigma]^t$; the inverse of one is the transpose of the other.

The order is important in these matrices, and one should be careful because some authors use a different order for shear stress and strain, namely $(\sigma_{aa}, \sigma_{bb}, \sigma_{cc}, \tau_{ab}, \tau_{bc}, \tau_{ca})$ and $\{\varepsilon(abc)\} = (\varepsilon_{aa}, \varepsilon_{bb}, \varepsilon_{cc}, \gamma_{ab}, \gamma_{bc}, \gamma_{ca})$, as does Cooke. The order of stresses and strains in Equation (6.6) follows the order in Lekhnitskii.

The requirement for a 6 × 6 material properties matrix may be problematic for some computer codes that are based on an assumption of isotropy and do not use a 6 × 6 material properties matrix in stress-strain relations. Such a legacy makes upgrading to handle anisotropy an expensive project. In this regard, rotation of axes often leads to a fully populated 6 × 6 material properties matrix when anisotropy is taken into account.

Joints in sets induce an anisotropic response to load in the elastic range even though intact rock between joints may be isotropic. Accounting for joint effects proceeds the same way in case of anisotropic rock between joints. Although there are numerous models of jointed rock, few are suitable for geotechnical engineering for several reasons, but mainly for the requirement of a representative volume element (RVE). The size of an RVE may be too large to accommodate closely spaced computational points where details of stress, strain, and displacement are sought, usually near excavation walls and foundations. In general, a non-RVE process for obtaining equivalent properties of jointed rock is necessary. However, joint sets may impose a periodic structure on rock at excavation scales of meters and more and thus define an RVE scale at joint spacing.

An example of the effects of joints on strata properties relates to underground mining of trona in southwest Wyoming. Trona is mined for the production of sodium carbonate and found in multiple beds in the Green River formation of Eocene age. The resource is the largest trona deposit in the world. Depth of excavation at the example site is 520 m (1700 ft). Elastic properties obtained from laboratory tests are given in Table 6.4 where E = Young's modulus (GPa), G = shear modulus (GPa), and v = Poisson's ratio. Strata are considered isotropic.

Table 6.4 Elastic properties of trona mine strata in the Green River formation.

Rock Type	*E*	*G*	*v*
Mudstone	8.62	3.61	0.20
Shale	6.01	2.45	0.22
Sandstone	13.97	5.68	0.23
Oil shale	5.86	2.13	0.33
Trona	28.16	11.26	0.25

Source: Comparison of Underground Coal and Trona Mine Seismicity, Preprint 17–027, SME Annual Meeting, Denver, CO, Feb. 19–22, 2017, Pariseau, W.G. E, G in GPa.

Joints occur in five sets as illustrated in Figure 6.4. Four sets are vertical; the fifth set is composed of horizontal bedding planes. Vertical joints are spaced 9 m; bedding plane joints are spaced 2.1 m. Joint properties were assigned values1/100th of intact strata properties.

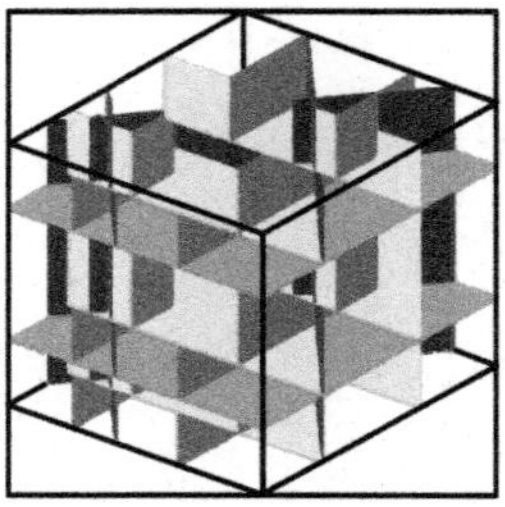

Figure 6.4 Joint set diagram related to trona mine properties (Pariseau, 2017a).

Computation of equivalent properties involves increasing a sample cube size until a steady state is reached. Results of the process are shown in Figure 6.5. The results show a sharp reduction in equivalent moduli as the sample cube size increases beyond maximum joint set spacing. This result confirms the hypothesis that a unit cell in a periodic structure is an RVE. The E1 and E2 directions are parallel to bedding and are reduced to 54% of intact moduli; E3 is perpendicular to bedding and is reduced to 44% of intact moduli. Shear moduli G1 and G2 are across bedding and are 32% of intact shear moduli; G3 is parallel to bedding and is 52% of intact moduli.

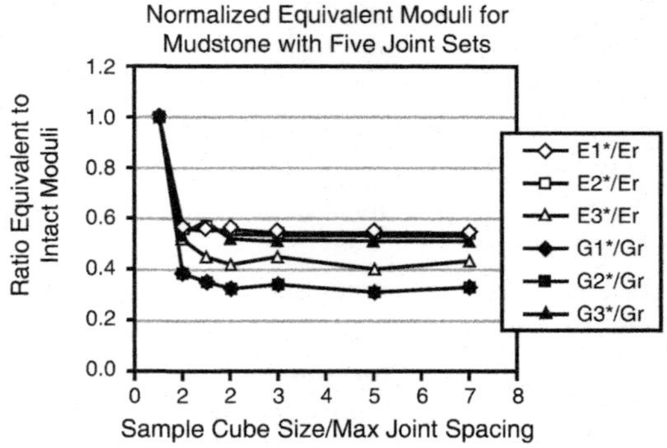

Figure 6.5 Ratio of equivalent jointed rock moduli to intact rock moduli as a function of the ratio of sample cube size to maximum joint spacing (Pariseau, 2017a).

Values of equivalent moduli are given in Table 6.5. Results in the table suggest the anisotropy is in the form of transverse isotropy. Moduli parallel to bedding (1- and 2-directions) are equal; results perpendicular to bedding (3-direction) are different and smaller. In these example data, the major effect of joints is to greatly reduce Young's moduli and shear moduli relative to intact values. Inducing anisotropy is a lesser effect. The reason is in joint set geometry, joints in vertical sets that tend to be orthogonal and horizontal joints that are orthogonal to vertical joints.

Table 6.5 Equivalent elastic properties of jointed trona mine strata in the Green River formation.

Rock Type	*E1*	*E2*	*E3*	*v12*	*v23*	*v31*	*G1*	*G2*	*G3*
Mudstone	3.76	3.76	3.17	025	0.14	0.12	1.03	1.03	1.52
Shale	4.65	4.65	3.79	0.24	0.11	0.09	1.18	1.18	1.87
Sandstone	5.88	5.88	4.55	0.27	0.10	0.08	1.33	1.33	2.32
Oil shale	3.61	3.61	3.07	0.32	0.22	0.18	0.97	0.97	1.38
Trona	7.45	7.456	5.45	0.28	0.07	0.05	1.51	1.51	2.91

E = Young's moduli (GPa), G = shear moduli (GPa), v = Poisson's ratio (Pariseau, 2017a).

6.2 Plasticity

A yield criterion is at the core of any plasticity theory. When related to the Drucker stability postulate, the yield criterion may be identified as a plastic potential. With linkage to the Prager continuity requirement, one arrives at a plasticity stress-strain relationship, that is, associated rules of flow. Distinguishing between a yield criterion or plastic potential and a failure criterion requiring continuity allows for nonassociated flow rules. The same logic applies in case of anisotropic strength. However, even within a class of yield criteria, whether dependent on the intermediate principal stress, anisotropic formulations are not unique as examples show. Moreover, orthotropic formulations of strength are as anisotropic as is practical. Indeed, transverse isotropy is perhaps the most used anisotropic model. Example strength data quantitatively illustrate anisotropic strength in geoplasticity. An important special case is plane plastic strain, which proves to be more challenging than in case of isotropy. The important problem of foundation-bearing capacity illustrates several complexities of anisotropic plane plastic strain in the case of a strip foundation.

Anisotropic strength

In metal plasticity, Hill (1950) proposed an anisotropic yield condition for a metal containing three mutually orthogonal planes of symmetry, that is, for orthotropic metals. Thus:

$$F(\sigma_b - \sigma_c)^2 + G(\sigma_c - \sigma_a)^2 + H(\sigma_a - \sigma_b)^2 + 2L\tau_{bc}^2 + 2M\tau_{ca}^2 + 2N\tau_{ab}^2 = 1 \tag{6.6a}$$

where F, G, H, L, M, N are strength parameters of the current state of anisotropy and the axes of anisotropy are abc. The use of terms $2\tau_{ab}^2$, $2\tau_{bc}^2$ $2\tau_{ca}^2$ allows for the possibility of distinguishing between τ_{xy} and τ_{yx} should the need arise. One could also simply delete the 2s in Equation (6.6). This criterion is independent of pressure, that is, the mean normal stress; only deviatoric stresses appear in Equation (6.6). The strength parameters in Equation (6.6) can be expressed in terms of tensile (or compressive) strengths in the directions of anisotropy. Thus:

$$\begin{aligned} 2F &= \frac{1}{T_b^2} + \frac{1}{T_c^2} - \frac{1}{T_a^2}, \quad 2L = \frac{1}{R_a^2} \\ 2G &= \frac{1}{T_c^2} + \frac{1}{T_a^2} - \frac{1}{T_b^2}, \quad 2M = \frac{1}{R_b^2} \\ 2H &= \frac{1}{T_a^2} + \frac{1}{T_b^2} - \frac{1}{T_c^2}, \quad 2N = \frac{1}{R_c^2} \end{aligned} \tag{6.7a}$$

where the *T*s are tensile strengths and the *R*s are yield strengths in shear about axes of anisotropy. Because yield is not pressure dependent, compressive strengths (*C*s) are equal to corresponding tensile strengths and could be used in Equation (6.7), although this is not customary in metal plasticity. In case of isotropy, $L = M = N = 3F = 3G = 3H$ and $2F = 1/T^2$ where T is tensile strength and Equation (6.6) reduces to the Mises criterion $J_2 = k^2$ with $T = k\sqrt{3}$.

Alternatively, one could postulate:

$$(F\sigma_b - G\sigma_c)^2 + (G\sigma_c - H\sigma_a)^2 + (H\sigma_a - F\sigma_b)^2 + (L\tau_{bc})^2 + (M\tau_{ca})^2 + (N\tau_{ab})^2 = 1 \tag{6.6b}$$

with properties:

$$\begin{aligned} F &= \frac{1}{T_b\sqrt{2}}, \quad G = \frac{1}{T_c}\sqrt{2}, \quad H = \frac{1}{T_a\sqrt{2}}, \\ L &= \frac{1}{R_a}, \quad M = \frac{1}{R_b}, \quad N = \frac{1}{R_c} \end{aligned} \tag{6.7b}$$

In case of isotropy, $L = M = N = \sqrt{6}F$ and $F = G = H$; $F = 1/T\sqrt{2}$, and $k = T/\sqrt{3}$ as before.

Addition of normal stress terms to Equation (6.6) creates pressure dependency and a yield criterion applicable to anisotropic geological media, rocks, and soils. Thus:

$$\begin{aligned} &[F(\sigma_b - \sigma_c)^2 + G(\sigma_c - \sigma_a)^2 + H(\sigma_a - \sigma_b)^2 + L\tau_{bc}^2 + M\tau_{ca}^2 + N\tau_{ab}^2]^{1/2} \\ &\quad - (U\sigma_a + V\sigma_b + W\sigma_c) = 1 \end{aligned} \tag{6.8a}$$

following Pariseau (1968). This criterion reduces to the Drucker-Prager criterion with vanishing anisotropy and to the Hill criterion in case of vanishing pressure dependency that in turn reduces to the Mises criterion in case of isotropy. In more compact form and with accommodation of nonlinearity, one has:

$$\bar{J}_2^{n/2} - \bar{I}_1 = 1 \tag{6.9}$$

where $\bar{J}_2$ and $\bar{I}_1$ are the anisotropic forms in Equation (6.8a). The exponent n accommodates nonlinearity and a positive square root value is implied. This yield criterion applies to orthotropic rock and soil and requires three additional strength parameters, U, V, W, for a total of nine.

The strength parameters in Equation (6.9) may be determined from unconfined compressive, tensile, and shear strengths. Thus:

$$
\begin{aligned}
2F &= \left(\frac{C_b+T_b}{C_b^nT_b+T_b^nC_b}\right)^{2/n} + \left(\frac{C_c+T_c}{C_c^nT_c+T_c^nC_c}\right)^{2/n} - \left(\frac{C_a+T_a}{C_a^nT_a+T_a^nC_a}\right)^{2/n}, \quad L=\frac{1}{R_a^2}\\
2G &= \left(\frac{C_c+T_c}{C_c^nT_c+T_c^nC_c}\right)^{2/n} + \left(\frac{C_a+T_a}{C_a^nT_a+T_a^nC_a}\right)^{2/n} - \left(\frac{C_b+T_b}{C_b^nT_b+T_b^nC_b}\right)^{2/n}, \quad M=\frac{1}{R_b^2}\\
2H &= \left(\frac{C_a+T_a}{C_a^nT_a+T_a^nC_a}\right)^{2/n} + \left(\frac{C_b+T_b}{C_b^nT_b+T_b^nC_b}\right)^{2/n} - \left(\frac{C_c+T_c}{C_c^nT_c+T_c^nC_c}\right)^{2/n}, \quad N=\frac{1}{R_c^2}\\
U &= \frac{C_a^n-T_a^n}{T_aC_a^n+C_aT_a^n}, \quad V=\frac{C_b^n-T_b^n}{T_bC_b^n+C_bT_b^n}, \quad W=\frac{C_c^n-T_c^n}{T_cC_c^n+C_cT_c^n},
\end{aligned}
\tag{6.10a}
$$

The nine independent strength parameters in the n-type yield criterion (Equation 6.9) are now the unconfined compressive, tensile, and shear strengths with respect to the axes of anisotropy: $C_a, T_a, R_a, C_b, T_b, R_b, C_c, T_c, R_c$. Laboratory measurement of compressive and tensile strength is routine, but determination of shear strength is seldom done. Rather, shear strengths are often estimated from testing at angles to foliation or planes of anisotropy.

Alternatively, one may again associate properties such that:

$$
\begin{aligned}
&[(F\sigma_b-G\sigma_c)^2+(G\sigma_c-H\sigma_a)^2+(H\sigma_a-F\sigma_b)^2+(L\tau_{bc})^2+(M\tau_{ca})^2\\
&\quad+(N\tau_{ab})^2]^{n/2}-(U\sigma_a+V\sigma_b+W\sigma_c)=1
\end{aligned}
\tag{6.8b}
$$

where an alternative n-type is postulated. Associated properties are:

$$
\begin{aligned}
\sqrt{2}F &= \left(\frac{C_b+T_b}{C_b^nT_b+T_b^nC_b}\right)^{1/n}, \quad V=\frac{C_b^n-T_b^n}{T_bC_b^n+C_bT_b^n}, \quad L=\frac{1}{R_a}\\
\sqrt{2}G &= \left(\frac{C_c+T_c}{C_c^nT_c+T_c^nC_c}\right)^{1/n}, \quad W=\frac{C_c^n-T_c^n}{T_cC_c^n+C_cT_c^n}, \quad M=\frac{1}{R_b}\\
\sqrt{2}H &= \left(\frac{C_a+T_a}{C_a^nT_a+T_a^nC_a}\right)^{1/n}, \quad U=\frac{C_a^n-T_a^n}{T_aC_a^n+C_aT_a^n}, \quad N=\frac{1}{R_c}
\end{aligned}
\tag{6.10b}
$$

In case of isotropy, $F=G=H$, $L=M=N$ and $U=V=W$, the yield criterion (Equation 6.9) may then be written as:

$$
J_2^{n/2}=AI_1+B \tag{6.11}
$$

where:

$$A = \frac{(C_o/\sqrt{3})^n - (T_o/\sqrt{3})^n}{(C_o + T_o)}, \quad B = \frac{T_o(C_o/\sqrt{3})^n + C_o(T_o/\sqrt{3})^n}{(C_o + T_o)} \tag{6.12a}$$

which reduces to the Drucker-Prager criterion when $n = 1$. The constants A and B are then:

$$A = (\frac{1}{\sqrt{3}})(\frac{C_o - T_o}{C_o + T_o}), \quad B = (\frac{2}{\sqrt{3}})(\frac{C_o T_o}{C_o + T_o}) \tag{6.12b}$$

where C_o and T_o are unconfined compressive and tensile strengths, respectively.

Laboratory tests on cylinders of anisotropic rock at angles to foliation, bedding, laminations, or schistosity requires the applied stress to be referred to the axes of anisotropy. Under uniaxial compression or tension, the transformation is given by the usual formulas:

$$\begin{aligned}
\sigma_a &= (\sigma/2)[1 - \cos(2\delta)] = \sigma\sin^2(\delta) \\
\sigma_b &= (\sigma/2)[1 + \cos(2\delta)] = \sigma\cos^2(\delta) \\
\tau_{ab} &= (\sigma/2)\sin(2\delta) \qquad = \sigma\sin(\delta)\cos(\delta) \\
\sigma_c &= \tau_{cb} = \tau_{ca} = 0
\end{aligned} \tag{6.13}$$

Figure 6.6 illustrates the orientation of axes; σ is the applied stress and δ is foliation dip. Substitution of Equation (6.13) into Equation (6.9) allows for solution of the uniaxial strength σ at an angle to the foliation. The process is lengthy but not especially complex, although a computer program for evaluation would certainly be convenient and aid in avoiding errors should the process be done very often. Uniaxial testing in compression and tension parallel to the axes of anisotropy *abc* is the first step. Evaluation of the parameters $F, G \ldots N$ using Equation (6.10) is the second step. Testing at angles to the foliation follows. Shear testing to evaluate parameters L, M, N is likely to be problematic. For this reason, N may be treated as a free parameter in a plot of σ vs δ.

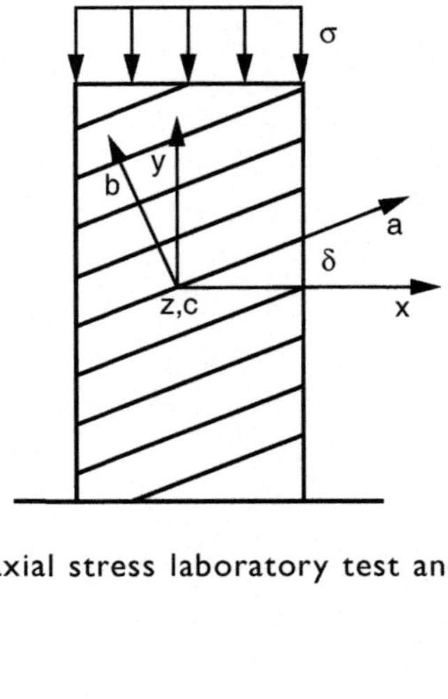

Figure 6.6 Schematic of a uniaxial stress laboratory test and an angle to foliation, a plane of anisotropy.

In case the exponent $n = 2$, solution for σ leads to a quadratic equation:

$$\begin{aligned} &A\sigma^2 - B\sigma = 1 \\ &A = Fc^4(\delta) + Gs^4(\delta) + H(s^2 - c^2)^2 + Ns^2c^2 \\ &B = Us^2 + Vc^2 \\ &c = \cos(\delta), \quad s = \sin(\delta) \end{aligned} \tag{6.14}$$

Solution of Equation (6.14) for uniaxial compressive and tensile strength, C_δ and T_δ, respectively by formula gives:

$$\begin{Bmatrix} C_\delta \\ T_\delta \end{Bmatrix} = \left(\frac{B \pm \sqrt{B^2 + 4A}}{2A}\right) \tag{6.15}$$

An example of a how unconfined compressive strength varies with foliation dip is shown in Figure 6.7. In this example, compressive strength C_1 perpendicular to the bedding is used to "normalize" the plot; all values are multiples or fractions of C_1 including shear strength. In the isotropic case, the normalized shear strength is 0.115. Compressive strength parallel to the bedding C_2 is two times C_1 as seen in the plot when the angle is 90°. Compressive strength C_3 in a direction parallel to the bedding but at 90° to C_2 is set equal to C_1. Inspection of the plot shows that unconfined compressive strength does not increase monotonically from a low to high value as bedding plane dip increases. Moreover, simply averaging strength parallel and perpendicular to bedding does not result in a useful value because of the importance of shear strength. Indeed, a relatively high shear strength leads to a maximum value of compressive strength at an angle to the bedding of approximately 55° in this example. At relatively low values of shear strength, compressive strength is minimum over much of the range of bedding plane dip, perhaps 80% as seen in the figure.

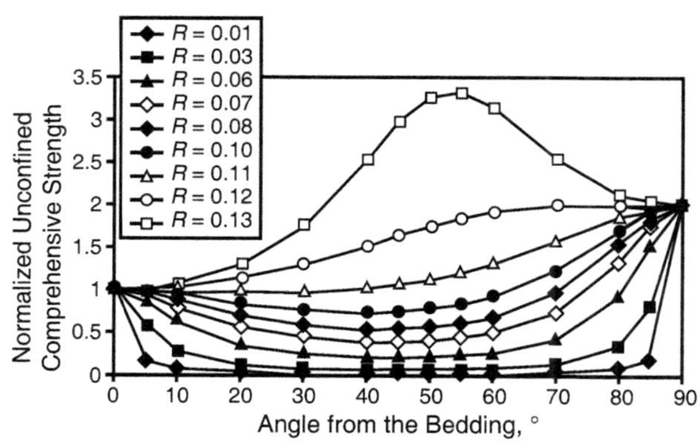

Figure 6.7 Unconfined compressive strength as a function of bedding plane dip with shear strength ***R*** as a parameter in anisotropic (orthotropic) rock (Pariseau, 2017b).

Although possible, direct pull tensile testing is not often done. The indirect Brazil test for tensile strength is more likely to be used. In case of anisotropy, one of the axes of anisotropy should be perpendicular to the test specimen. Figure 6.8 illustrates the Brazil test for tensile strength. Analytical solution of this problem is unknown. However, when the foliation is parallel to the disk axis, progress can be made.

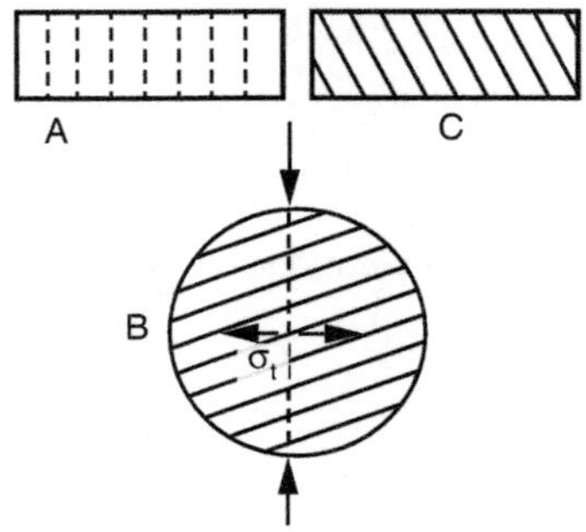

Figure 6.8 Schematic of the Brazil test for tensile strength. A – plan view of the disk showing foliation aligned with the disk axis into the page. B – tensile strength direction provided a fracture occurs along the dotted line between the load points. C – foliation not aligned.

Tensile strength varies with dip in a manner similar to compressive strength. Figure 6.9 illustrates variation of tensile strength with load angle, the angle from bedding planes to the axial load vector. This angle $\beta = \pi/2 - \delta$ where δ is bedding plane dip. At angle zero, tensile strength is parallel to the bedding; at $\pi/2$, load is perpendicular to the bedding. In this example, tensile strength perpendicular to bedding is 0.95% of the tensile strength parallel to the bedding. Shear strength is a parameter as before. Data are normalized in the plot by division by tensile strength parallel to the bedding. Thus, the first data point in the plot has coordinates (1,0); the last has coordinates (0.95,90). Inspection of the plot shows a noticeable influence of shear strength ***R*** which in the isotropic case is 0.115 as before. Relatively high shear strength gives a rise and then a fall of tensile strength with load angle. Low shear strength leads to low tensile strength over a considerable range of angles as inspection of the plot shows.

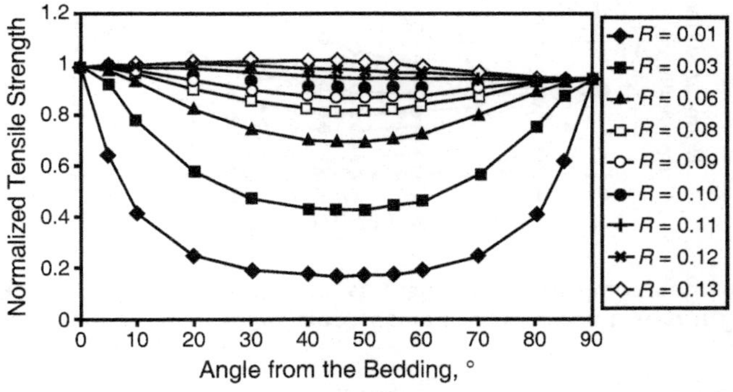

Figure 6.9 Tensile strength as a function of bedding plane dip with shear strength as a parameter in anisotropic (orthotropic) rock (Pariseau, 2017b).

Example data

Strength data pertaining to three Precambrian meta-sediments at the former Homestake Mine are given in Table 6.6. Direction a = down dip (parallel to foliation), b = perpendicular to foliation, and c = on strike. The ab plane is vertical and c is normal to this plane. C, T, and R are unconfined compressive, tensile, and shear strength, respectively. Units are MPa. Tensile strength perpendicular to foliation is the lowest of the three tensile strengths in each of the three formations, perhaps, as one would expect. Ratios of compressive strengths (parallel to perpendicular directions) are less than two. Tensile strength ratios are greater but less than four.

Table 6.6 Strengths of three anisotropic Precambrian meta-sedimentary formations.

Property *Formation*	C_a	C_b	C_c	T_a	T_b	T_c	R_{ab}	R_{bc}	R_{ca}
Ellison	78.2	78.7	56.2	16.2	4.1	11.4	7.9	14.6	8.6
Homestake	138.9	79.6	91.5	9.5	7.9	13.2	14.1	17.0	14.5
Poorman	94.0	69.0	84.6	20.6	5.7	13.2	10.3	19.3	8.8

Source: US Bureau of Mines Report of Investigations 9531, Part 1. Rock Mechanics Study of Shaft Stability and Pillar Mining, Homestake Mine, Lead, SD, 1995 (Pariseau, W.G., J.C. Johnson, M.M. McDonald and M.E. Poad). Units are MPa.

Joint sets certainly induce directional features in natural rock masses whether intact rock between joints is isotropic or not. While equivalent elastic moduli take into account the effect of joints on deformation in the purely elastic domain, joint effects on strength is problematic. The reason is physical; yield may occur along joints or in the intact rock between joints or possibly both may yield. The presence of two distinct mechanisms of plastic deformation precludes specification of a single yield criterion. Thus, unlike elastic moduli of jointed rock, no equivalent yield function exists.

Computer programs address the problem of jointed rock strength in different ways. One popular program simply assigns a yielding state to a cell if any joint segment in the cell fails or if the intact material in the cell fails. Such a procedure may seem justifiable from the viewpoint of an unconfined compressive test of a rock cylinder transected by a joint. If rock or joint fails, the test cylinder fails under application of monotonically increasing stress. Under three-dimensional stress, the logic is questionable. However, if a program recognizes individual joint segments in a cell or element, then joint and intact rock yield criteria may be used to determine whether yielding is present and compute joint and intact rock stresses accordingly. The difficulty with this approach is in the volume of data needed, a volume that may exceed the capacity of even large computers.

Scaling or calibration is often a practical solution to the problem of equivalent strength. A calibration process is simply a process of changing input properties (moduli, strengths) until reasonable agreement with excavation scale measurements is obtained. Elastic moduli dominate displacements in computer models; strengths dominate extent of yield zones.

Guidance for estimating scale factors for moduli and strengths may be obtained from a simple dimensionless argument that uniaxial compressive strain to failure is the same at the laboratory and excavation scales. Thus $\varepsilon_f = (C_o / E)$. If Young's modulus E is reduced by a factor of 0.36, then so is the unconfined compressive strength C_o. Alternatively, if the energy to failure $U = (C_o^2 / E)$ is the same at the laboratory and excavation scales, then strength scales as a square root. A modulus reduction of 0.36 leads to a strength reduction of 0.6 by the energy guideline.

Table 6.7 presents equivalent strengths of strata associated with underground trona mining in southwestern Wyoming. A uniaxial strain scaling guideline was used to obtain the table values in conjunction with a ratio of unconfined compressive to tensile strength ratio. Thus, this ratio is the same for the scaled properties as for laboratory properties.

Table 6.7 Scaled strength properties of jointed trona mine strata in the Green River formation (Pariseau, 2017a).

Rock Type	*C1*	*C2*	*C3*	*T1*	*T2*	*T3*	*R1*	*R2*	*R3*
Shale	21.2	21.2	17.9	2.24	2.24	1.90	3.99	3.99	3.37
Mudstone	13.2	13.2	10.8	1.48	1.48	1.50	5.31	5.31	5.31
Sandstone	18.4	18.4	14.2	1.51	1.51	1.17	3.04	3.04	2.35
Oil shale	23.3	23.3	19.8	2.03	2.03	1.72	3.98	3.98	3.34
Trona	12.4	12.4	9.10	0.75	0.75	0.55	1.77	1.77	1.30

Plane strain

Anisotropic plane plastic strain follows much the same pattern of development as in the isotropic case described in Section 4.2. Plane plastic strain implies the elastic part of a strain increment in the direction normal to the plane of deformation (z-axis) is negligible and the plastic part is zero. This material model is sometimes referred to as a rigid-plastic model. In case of a $J_2^{1/2} - I_1$ yield criterion where the intermediate principal stress affects yield, one has:

$$(d\varepsilon_z^p / \lambda) = 0 = nJ_2^{(n/2-1)}[-F(\sigma_y - \sigma_z) + G(\sigma_z - \sigma_x)] - W \tag{6.14}$$

Solution of Equations (6.8) and (6.14) for σ_z in the special case $n = 1$ leads to:

$$\sigma_z = \frac{\sigma_x(UW + G) + \sigma_y(VW + F) + W}{(F + G) - W^2} \tag{6.15a}$$

If $n = 2$, then:

$$\sigma_z = \frac{G\sigma_x + F\sigma_y + W / 2}{(F + G)} \tag{6.15b}$$

Values of $n > 2$ lead to a lengthy quadratic equation in σ_z.

Use of Equation (6.15a, $n = 1$) to eliminate σ_z from the yield condition results in a yield condition for anisotropic plane plastic strain:

$$[(\frac{\sigma_{xx} - \sigma_{yy}}{2})^2(\frac{1}{1-c}) + \tau_{xy}^2]^{1/2} = (A_1 / 2)\sigma_{xx} + (A_2 / 2)\sigma_{yy} + B_3 \tag{6.16a}$$

where:

$$
\begin{aligned}
A_1 &= 2(FU + UG + GW)/[N(F+G)(F+G-W^2)]^{1/2} \\
A_2 &= 2(GV + VF + FW)/[N(F+G)(F+G-W^2)]^{1/2} \\
B_3 &= (F+G)/[N(F+G)(F+G-W^2)]^{1/2} \\
c &= 1 - N(F+G)/4(FG+GH+HF)
\end{aligned}
\tag{6.16b}
$$

The Cartesian stresses may be expressed in terms of principal stresses that in turn may be expressed in terms of mean normal stress and maximum shear stress $\sigma_m = (\sigma_1 + \sigma_3)/2$, $\tau_m = (\sigma_1 - \sigma_3)/2$ in the xy plane. Thus:

$$
\begin{aligned}
\sigma_{xx} &= \sigma_m + \tau_m \cos(2\theta) \\
\sigma_{yy} &= \sigma_m - \tau_m \cos(2\theta) \\
\tau_{xy} &= \tau_m \sin(2\theta)
\end{aligned}
\tag{6.17}
$$

where θ is the clockwise (negative) angle from the 1-axis to the x-axis.

After substituting Equation (6.17) into Equation (6.16a), one obtains:

$$
\begin{aligned}
\sigma_m &= \tau_m[(\frac{1-c\sin^2(2\theta)}{1-c})^{1/2} - (\frac{A_1 - A_2}{2})\cos(2\theta)]/(\frac{A_1 + A_2}{2}) - B_3/(\frac{A_1 + A_2}{2}) \\
&= F(\tau_m, \theta) \\
\sigma_m &= \tau_m f(\theta) - B_3/(\frac{A_1 + A_2}{2})
\end{aligned}
\tag{6.18}
$$

Derivatives of Equation (6.18) are:

$$
\begin{aligned}
&\frac{\partial \sigma_m}{\partial \tau_m} = f(\theta) = [h - (\frac{A_1 - A_2}{2})\cos(2\theta)]/(\frac{A_1 + A_2}{2}) \\
&\frac{\partial \sigma_m}{\partial \theta} = \frac{df(\theta)}{d\theta} = \tau_m f' \\
&\text{with } h = [\frac{1-c\sin^2(2\theta)}{1-c}]^{1/2}
\end{aligned}
\tag{6.19}
$$

Derivatives needed in the equations of equilibrium are:

$$
\begin{aligned}
\frac{\partial \sigma_{xx}}{\partial x} &= \frac{\partial \tau_m}{\partial x}[f(\theta) + \cos(2\theta)] + \tau_m[f'(\theta) - 2\sin(2\theta)]\frac{\partial \theta}{\partial x} \\
\frac{\partial \sigma_{yy}}{\partial y} &= \frac{\partial \tau_m}{\partial y}[f(\theta) - \cos(2\theta)] + \tau_m[f'(\theta) + 2\sin(2\theta)]\frac{\partial \theta}{\partial y} \\
\frac{\partial \tau_{xy}}{\partial x} &= \frac{\partial \tau_m}{\partial x}\sin(2\theta) + 2\tau_m \cos(2\theta)\frac{\partial \theta}{\partial x} \\
\frac{\partial \tau_{xy}}{\partial y} &= \frac{\partial \tau_m}{\partial y}\sin(2\theta) + 2\tau_m \cos(2\theta)\frac{\partial \theta}{\partial y}
\end{aligned}
\tag{6.20}
$$

The formalism of **4.2** applies. Thus, the characteristic directions are given by:

$$(dy/dx)^2[AC]+(dy/dx)(-[BC]-[AD])+[BD]=0 \tag{4.17}$$

where:

$$\begin{aligned}
[AC] &= \tau_m[(2f\cos(2\theta)+2-f'\sin(2\theta)] \\
[BC] &= \tau_m[2f\sin(2\theta)+f'\cos(2\theta)-ff'] \\
[AD] &= \tau_m[2f\sin(2\theta)+f'\cos(2\theta)+ff'] \\
[BD] &= \tau_m[2-2f\cos(2\theta)+f'\sin(2\theta)]
\end{aligned} \tag{6.21}$$

The derivative of f is given by:

$$f' = \tau_m[h'+(A_1-A_2)\sin(2\theta)]/(A_1+A_2) \tag{6.22}$$

where:

$$h' = \frac{-2\sin(2\theta)\cos(2\theta)}{h(1-c)} \tag{6.23}$$

Solution of Equation (4.17) gives expressions for the characteristic directions of C_1 and C_2. Thus:

$$(dy/dx) = \frac{(1-c)\sin(2\theta)\pm\sqrt{(1-c)^2\sin^2(2\theta)-[A_2h(1-c)+\cos(2\theta)][A_1h(1-c)-\cos(2\theta)]})}{(1-c)hA_2+\cos(2\theta)} \tag{6.24a}$$

that may be shortened to $(dy/dx)=\tan(\theta+\mu_1)$, $dy/dx=\tan(\theta-\mu_2)$ with μ_1 and μ_2 being angles from the 1-axis to the C_1 and C_2 characteristics, respectively (negative is clockwise).

Alternatively:

$$(dx/dy) = \frac{(1-c)\sin(2\theta)\pm\sqrt{(1-c)^2\sin^2(2\theta)-[A_2h(1-c)+\cos(2\theta)][A_1h(1-c)-\cos(2\theta)]})}{(1-c)hA_1-\cos(2\theta)} \tag{6.24b}$$

In case of isotropy $h=1$, $c=0$, $A_1=A_2=A$, and Equation (6.24a) reduces to:

$$\begin{aligned}
(dy/dx) &= \frac{\sin(2\theta)\pm\sqrt{\sin^2(2\theta)-[A+\cos(2\theta)][A-\cos(2\theta)]})}{A+\cos(2\theta)} \\
&= \frac{\sin(2\theta)\pm\sqrt{1-A^2})}{A+\cos(2\theta)} \\
&= \frac{\sin(2\theta)\pm\cos(\phi)}{\sin(\phi)+\cos(2\theta)} \\
(dy/dx) &= \tan(\theta\pm\mu)
\end{aligned} \tag{6.24c}$$

As a reminder, $2\mu = \pi / 2 - \phi$ and ϕ is the inclination of the failure envelope in the $\sigma - \tau$ plane.

Stress variables τ_m and θ vary along the characteristics according to the same formalism in **4.2**. Thus:

$$d\tau_m\{[BC]-[BD\}(dx/dy)\} - d\theta[CD] + [CE]dy - [DE]dx = 0 \tag{4.23}$$

where dx and dy are increments along the characteristic curves (Equation 6.24a). The new terms in square brackets are:

$$\begin{aligned}
[CD] &= (\tau_m)^2[(f')^2 - 4] \\
[CE] &= \tau_m[f' - E_2\sin(2\theta)] - 2\tau_m E_1\cos(2\theta) \\
[DE] &= 2\tau_m E_2\cos(2\theta) - \tau_m E_1[f' + 2\sin(2\theta)]
\end{aligned} \tag{6.25}$$

Reduction of Equation (4.23) leads to differential relationships along the characteristic curves:

$$\begin{aligned}
& d\tau_m f_1(\theta)/2\tau_m + d\theta + \{E^*{}_1\sin(\alpha) - E^*{}_2\cos(\alpha)\}ds = 0 \quad :C_1 \\
& d\tau_m f_2(\theta)/2\tau_m + d\theta + \{E^*{}_1\sin(\beta) - E^*{}_2\cos(\beta)\}ds = 0 \quad :C_2
\end{aligned} \tag{6.26}$$

where ds is an element of arc length along a characteristic curve with:

$$\begin{aligned}
& dx = \cos(\alpha)ds,\ dy = \sin(\alpha)ds \quad :C_1 \\
& dx = \cos(\beta)ds,\ dy = \sin(\beta)ds \quad :C_2
\end{aligned} \tag{6.27}$$

Angles α and β are inclinations of C_1 and C_2 to the x = axis, respectively. Expressions E_1^* and E_2^* are modifications of E_1 and E_2. Functions f_1 and f_2 are:

$$\begin{aligned}
\left.\begin{matrix} f_1 \\ f_2 \end{matrix}\right) &= h(1-c)SC\left(\frac{A_1+A_2}{2}\right)^2 + S[(h-A_4C)][cC + hA_2(1-c)] \\
&-\left(\frac{A_1+A_2}{2}\right)\{(1-c)S \pm [(1-c)^2S^2 + (C-(1-c)hA_1)(C+(1-c)hA_5)]^{1/2}\}/ \\
&\{\left(\frac{S^2}{h(1-c)}\right)[-cC - A_2(1-c)h][-C + A_1(1-c)h] - [C^2\left(\frac{A_1+A_2}{2}\right)^2(1-c)h]\}
\end{aligned} \tag{6.28}$$

where an abbreviated notation is used: $S = \sin(2\theta)$ and $C = \cos(2\theta)$. In the isotropic case, Equation (6.26) reduces to:

$$\pm d\tau_m\cot(\phi)/2\tau_m + d\theta + \{E_1\sin(\theta \mp \mu) - E_2\cos(\mp\mu)\}ds = 0 \quad :C_1, C_2 \tag{4.25}$$

as required.

In case of anisotropic metal plasticity ($n = 1$), $U = V = W = 0$ and the preceding equations reduce to those in Hill (1950) with characteristic directions given by:

$$dy/dx = \tan(\theta' \pm \pi/4) = (1-c)\tan(\theta \pm \pi/4) \tag{6.29}$$

which shows that the characteristic directions are not directions of maximum and minimum shear stress. In consideration of the directions of maximum plastic strain rate, one has:

$$\tan(2\theta_e) = \frac{d\varepsilon^p_{xy}}{(d\varepsilon^p_{xy} - d\varepsilon^p_{xy})/2} = \frac{(1-c)\tau_{xy}}{(\sigma_{xx} - \sigma_{yy})/2} = (1-c)\tan(2\theta) \tag{6.30}$$

So the directions of maximum plastic strain and principal stress do not coincide as they do in the isotropic case ($c = 0$) of metal plasticity.

An example illustrates several important features of anisotropic plane plastic strain. Suppose tensile strengths T_a, T_b T_c are 2, 1, and 3, respectively, in arbitrary units and the compressive strengths C_a, C_b C_c are 24, 18, and 30, so strength ratios are 12, 18, and 10, respectively. Further let shear strengths R_a, R_b R_c be given by the formula in Equation (6.34) that has the interesting effect: $H = N$, $c = 0$. Evaluation of the characteristic equations Equation (6.24a) leads to a plot of the angles $\theta + \mu_1$ and $\theta + \mu_2$ shown in Figure 6.10. The gaps in the plots of the characteristic angles occur when the value under the square root sign in Equation (6.24a) becomes imaginary. In this regard, the characteristic directions are determined by a solution to a quadratic equation, so the discriminant determines whether the characteristics are real or imaginary (and whether the system of equations is hyperbolic, parabolic, or elliptic). Also evident in the figure is the lack of symmetry of the characteristic angles with respect to the direction of major principal stress. As in isotropic theory, imaginary characteristics imply envelopes of Mohr stress circles do not exist, and there are plastic states not given by envelop failure criteria but rather by Equation (6.35).

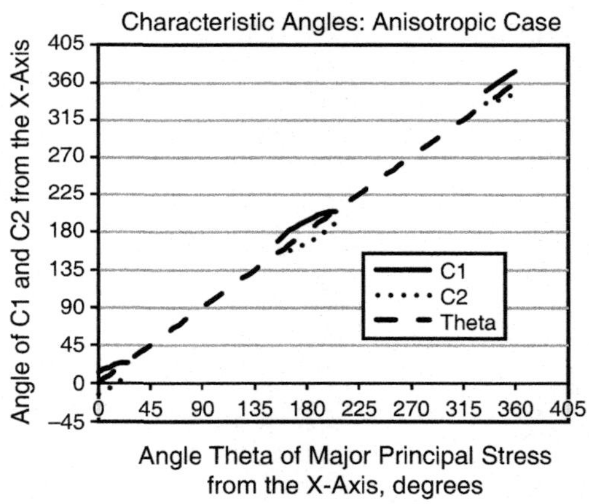

Figure 6.10 Characteristic angles as functions of the angle θ of major principal stress from the *x*-axis in the case of isotropy.

The isotropic case is shown in Figure 6.11 where the characteristic directions are always real and symmetrically disposed with respect to the direction of major principal stress. The isotropic case is obtained by setting $c = 0$ and $A_1 = A_2 = A < 1$. Surprisingly, the isotropic case becomes problematic, too, when the ratio of unconfined compressive to tensile strength exceeds 2.25 or so. In the figure, the strength ratio is just 2.25, which is quite low for rocks and soils.

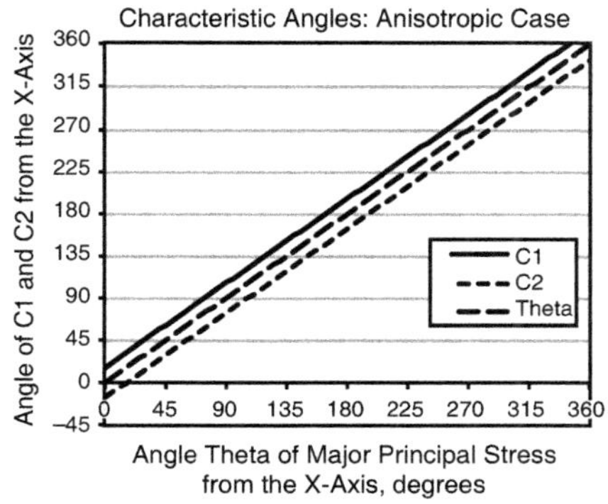

Figure 6.11 Characteristic angles as functions of the angle θ of major principal stress from the *x*-axis in the case of isotropy.

Application of the anisotropic characteristic system, Equations (6.24a) and (6.26), to a drilling problem where a bit tooth was idealized as a wedge penetrating a flat surface was made by Pariseau (1970). Figure 6.12 illustrates the geometry of the problem, the applied forces, and the pressure applied adjacent to the bit tooth. The face of the wedge is considered to be well lubricated and therefore smooth and friction-free. Numerical results for forces in several cases of anisotropy are shown in Figure 6.13. Details may be found in the reference.

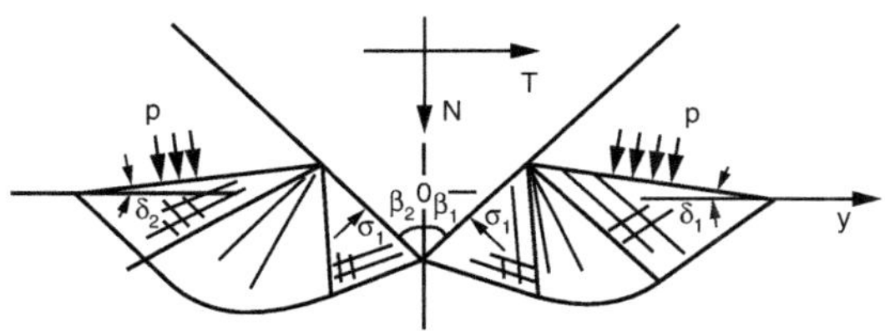

Figure 6.12 Wedge indentation geometry during penetration of an anisotropic rock (Pariseau, 1970).

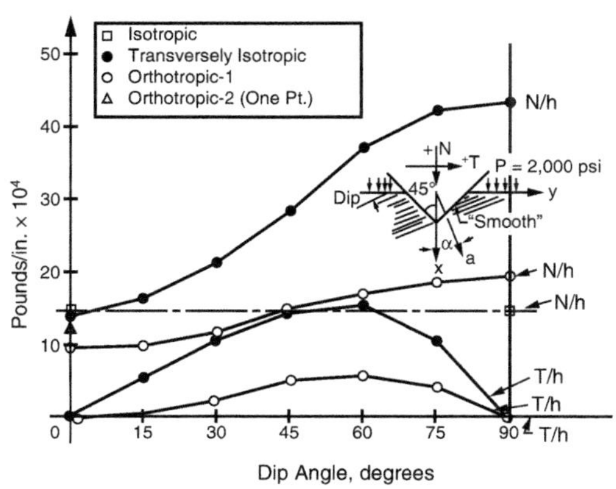

Figure 6.13 Normal and shear forces required for penetration a unit distance as a function of strata dip (Pariseau, 1970).

Just as the isotropic Mises criterion led to the anisotropic Hill criterion where both include a dependency on the intermediate principal stress σ_2 and are applicable to metal plasticity, the Tresca criterion $|\tau_m| = \text{constant}$ could also be developed into an anisotropic form such as $|F\sigma_b - H\sigma_a| = 1$, although the absence of pressure dependency negates any general application in geoplasticity.

However, the Mohr-Coulomb criterion does invite an anisotropic (orthotropic) form, as does the Drucker-Prager criterion. Much the same analysis applies. One possible formulation is:

$$\begin{aligned}
&\left|[F(\frac{\sigma_{yy}-\sigma_{zz}}{2})^2 + L\tau_{yz}^2]^{1/2}\right| = V\sigma_{yy} + W\sigma_{zz} + 1 \\
&\left|[G(\frac{\sigma_{zz}-\sigma_{xx}}{2})^2 + M\tau_{zx}^2]^{1/2}\right| = W\sigma_{zz} + U\sigma_{xx} + 1 \\
&\left|[H(\frac{\sigma_{xx}-\sigma_{yy}}{2})^2 + N\tau_{xy}^2]^{1/2}\right| = U\sigma_{xx} + V\sigma_{yy} + 1
\end{aligned} \tag{6.31}$$

where the absolute value allows for positive and negative shear stress. Equations (6.31) form a six-sided pyramid in principal stress space. The nine strength parameters *F, G, . . ., W* are not the same as before, of course. Thought experiments for unconfined compressive (C_a, C_b C_c), tensile (T_a, T_b T_c), and shear (R_a, R_b R_c) strengths lead to expressions for the nine strength parameters in terms of laboratory tests. For example, from the last of Equation (6.31):

$$H = (\frac{C_a + T_a}{C_a T_a})^2, \quad U = (\frac{C_a - T_a}{2C_a T_a}), \quad N = \frac{1}{R_a^2} \tag{6.32}$$

But also from the second of Equation (6.31), $H = (\frac{C_b + T_b}{C_b T_b})^2$, which contradicts the assumption of anisotropy. Further examination of the strength parameters leads to additional contradictions and the conclusion that Equation (6.31) is *not* a useful anisotropic form of Mohr-Coulomb yield.

Exploration of other potential formulations for anisotropic Mohr-Coulomb yield and extension to *n*-type criteria that are free of contradictions suggests the form:

$$\begin{aligned}
&\left|[(F\sigma_{bb} - G\sigma_{cc})^2 + (L\tau_{bc})^2]^{n/2}\right| = V\sigma_{bb} + W\sigma_{cc} + 1 \\
&\left|[(G\sigma_{cc} - H\sigma_{aa})^2 + (M\tau_{ca})^2]^{n/2}\right| = W\sigma_{cc} + U\sigma_{aa} + 1 \\
&\left|[(H\sigma_{aa} - F\sigma_{bb})^2 + (N\tau_{ab})^2]^{n/2}\right| = U\sigma_{aa} + V\sigma_{bb} + 1
\end{aligned} \tag{6.33}$$

relative to the *abc* axes of anisotropy. The nine anisotropic strength parameters may be expressed in terms of unconfined compressive, tensile, and shear strengths. Thus:

$$H = (\frac{C_a + T_a}{T_a C_a^n + C_a T_a^n})^{1/n}, \quad U = (\frac{C_a^n - T_a^n}{T_a C_a^n + C_a T_a^n}), \quad N = \frac{1}{R_{ab}} \tag{6.34}$$

$$F = (\frac{C_b + T_b}{T_b C_b^n + C_b T_b^n})^{1/n},\ V = (\frac{C_b^n - T_b^n}{T_b C_b^n + C_b T_b^n}),\ L = \frac{1}{R_{bc}}$$

$$G = (\frac{C_c + T_c}{T_c C_c^n + C_c T_c^n})^{1/n},\ W = (\frac{C_c^n - T_c^n}{T_c C_c^n + C_c T_c^n}),\ N = \frac{1}{R_{ca}}$$

which are similar to the properties in Equation (6.10b).

In case of isotropy, $U = V = W,\ F = G = H,\ L = M = N$. The last of Equation (6.33) may be brought into the familiar form:

$$\left| [(\frac{\sigma_{aa} - \sigma_{bb}}{2})^2 + (\tau_{ab})^2]^{n/2} \right| = A(\frac{\sigma_{aa} + \sigma_{bb}}{2}) + B \text{ with} \tag{6.35a}$$

$$A = 2(\frac{C_o^n - T_o^n}{C_o + T_o})(\frac{1}{2})^n,\ B = (\frac{C_o^n T_o + T_o^n C_o}{C_o + T_o})(\frac{1}{2})^n \tag{6.35b}$$

This case (isotropic) with $n = 2$ is plotted in Figure 6.14 and illustrates the parabolic form of the yield criterion.

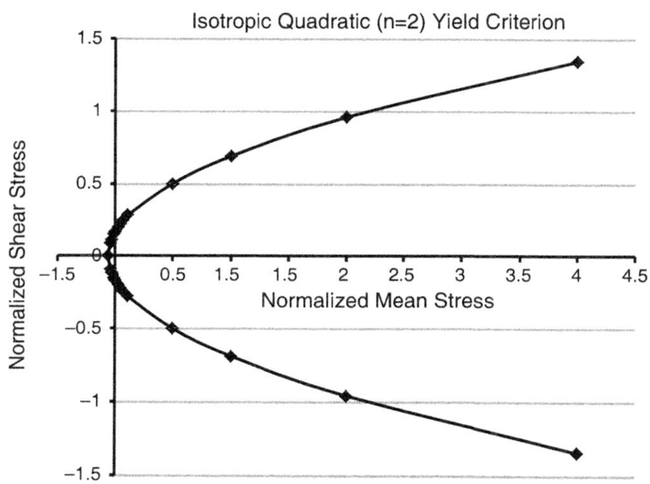

Figure 6.14 Graph of an isotropic quadratic yield criterion independent of the intermediate principal stress. Mean normal stress and maximum shear stress are normalized with division by unconfined compressive strength. Normalized stresses are dimensionless.

An expectation when testing at an angle to foliation as illustrated in Figure 6.4 is a decrease in strength with dip of the foliation, although an increase is possible. A monotonic change would seem unrealistic and thus bring any anisotropic formulation into question. Figure 6.15 shows the results of unconfined compressive strength variation with foliation dip in the quadratic case, $n = 2$. The normalized strength data are actual strength divided by the unconfined compressive strength parallel to the foliation. Unconfined compressive and tensile strengths parallel and perpendicular to foliation C_a, T_a and C_b, T_b are 9, 1 and 18, 2, respectively. The ratio of unconfined compressive to tensile strength is nine in both directions. The normalized shear strength in the isotropic case is 0.167. As the parameter R is increased, shear strength is increased. Higher and lower values are given as a parameter in the figure. Data in the figure show that the anisotropic formulation (Equation 6.33) meets the expectation of decreasing or possibly increasing strength with foliation angle relative to uniaxial compressive load direction.

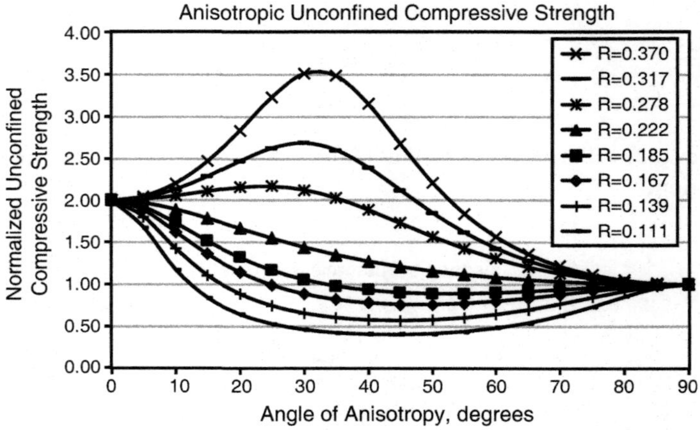

Figure 6.15 Normalized unconfined compressive strength as a function of "foliation dip" or angle of anisotropy according to the formulation Equations (6.33) with shear strength as a parameter and n=2, quadratic case.

Yet another anisotropic formulation that is reminiscent of the well-known Mohr-Coulomb yield criterion is the n-type:

$$[(\frac{\sigma_a - \sigma_b}{2})^2 + \tau_{ab}^2]^{n/2} = A(\eta)(\frac{\sigma_a + \sigma_b}{2}) + B(\eta) \tag{6.36a}$$

where $\eta = \theta - \delta$ is a difference in angles that is illustrated in Figure 6.16 and the positive square root is implied. In consideration of the invariance of the terms involving stress, Equation (6.36a) may be written with reference to axes x,y and with respect to the principal directions 1,3. Thus:

$$[(\frac{\sigma_{xx} - \sigma_{xx}}{2})^2 + \tau_{xy}^2]^{n/2} = A(\eta)(\frac{\sigma_{xx} + \sigma_{yy}}{2}) + B(\eta) \tag{6.36b}$$

$$\left|(\frac{\sigma_1 - \sigma_3}{2})\right|^n = A(\eta)(\frac{\sigma_1 + \sigma_3}{2}) + B(\eta) \tag{6.36c}$$

$$|\tau_m|^n = A(\eta)\sigma_m + B(\eta) \tag{6.36d}$$

The form Equation (6.36) allows the reference axes to be specified relative to the axes of anisotropy and is independent of the direction of principal stress given by the angle θ. Noteworthy also is the symmetry of Equation (6.36) with respect to the normal stress axis in the normal stress shear stress plane where Mohr circles representing stress at yield may be plotted. As a reminder, such symmetry is essential to any acceptable yield criterion. This last n-type form is similar to the Mohr-Coulomb criterion when $n = 1$.

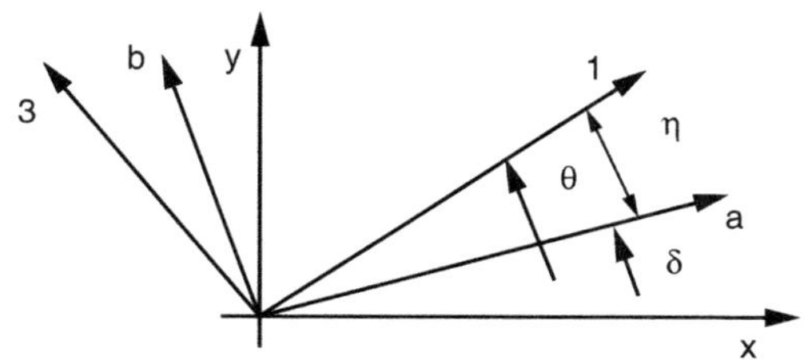

Figure 6.16 Angle definitions relative to the formulation Equations (6.36).

Specification of the functions A and B is constrained by δ values of 0° and 90°. Simple candidate functions that meet these conditions are $A(\delta) = A_a \cos^2(\eta) + A_b \sin^2(\eta)$ and $B(\eta) = B_a \cos^2(\eta) + B_b \sin^2(\eta)$. Constants A_a, A_b, B_a and B_b in terms of unconfined compressive and tensile strengths are:

$$A_a = \frac{(C_a/2)^n - (T_a/2)^n}{(C_a/2) + (T_a/2)}, \quad B_a = \frac{(C_a/2)^n (T_a/2) + (T_a/2)^n (C_a/2)}{(C_a/2) + (T_a/2)} \tag{6.37}$$

and similarly for A_b and B_b. However, the functions A and B do not allow for decreasing or possibly increasing shear strength with the anisotropy angle as physical evidence indicates. Introduction of an additional parameter meets this requirement for both functions. Thus:

$$\begin{aligned} A(\eta) &= A_a \cos^2(\eta) + A_3 \cos(\eta)\sin(\eta) + A_b \sin^2(\eta) \\ B(\eta) &= B_a \cos^2(\eta) + B_3 \cos(\eta)\sin(\eta) + B_b \sin^2(\eta) \end{aligned} \tag{6.38a,b}$$

Alternatively:

$$\begin{aligned} A(\eta) &= \frac{(A_a + A_b)}{2} + \frac{(A_a - A_b)}{2}\cos(2\eta) + \frac{A_3}{2}\sin(2\eta) \\ B(\eta) &= \frac{(B_a + B_b)}{2} + \frac{(B_a - B_b)}{2}\cos(2\eta) + \frac{B_3}{2}\sin(2\eta) \end{aligned} \tag{6.38c,d}$$

where angular variation occurs about mean values. Figure 6.17 illustrates the form of Equations (6.38a,c) and the effect of the parameter A_3 and B_3 on $A(\delta)$. The function B in the figure is normalized by division by C_a. The results in Figure 6.17 indicate all constraints are met by the formulation Equations (6.36) using the functions Equation (6.38).

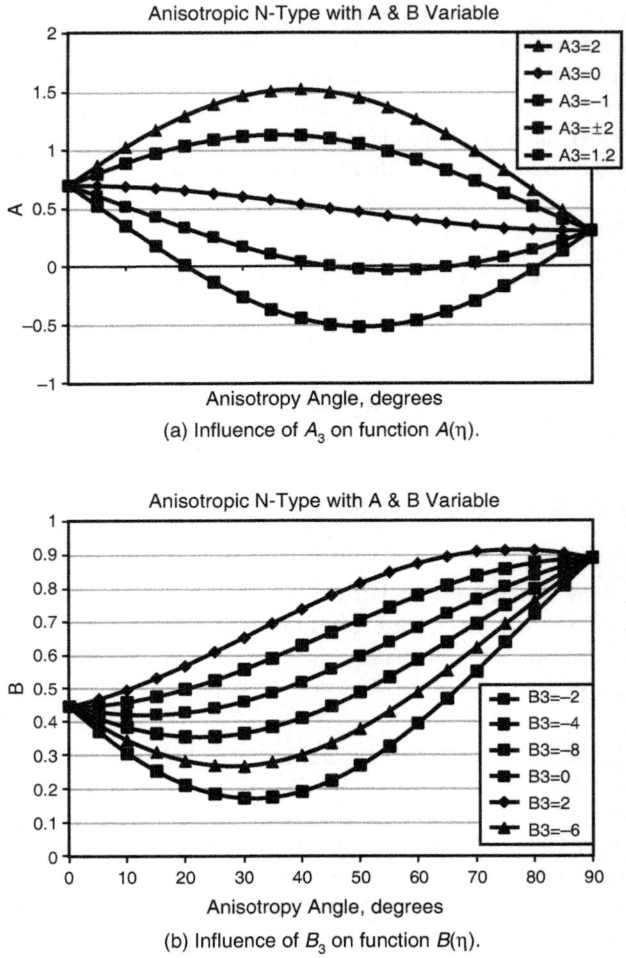

Figure 6.17 Influence of parameters on functions A_3 and B_3 on anisotropic functions A and B:

Input data for the results in Figure 6.13 are $C_a = 9$, $T_a = 1$, $C_b = 18$, $T_b = 2$.

An example of the criterion Equations (6.36) with $n = 1$ is plotted in Figure 6.18. In this example, slopes are $A = \sin(\phi)$ and shear axis intercepts are $B = k\cos(\phi)$, k is cohesion.

The anisotropic formulation Equations (6.36) leads to a characteristic system that is complicated by dependency on η, that is, on θ in the functions $A(\eta)$ and $B(\eta)$ in Equation (6.36) for a fixed foliation dip (δ). Elements of the determinants that lead to the characteristic system are:

$$\Delta = \begin{vmatrix} (n\tau_m^{n-1}/A + C) & S & (A'\sigma_m + B') - 2\tau_m S & 2\tau_m C \\ S & (n\tau_m^{n-1}/A - C) & 2\tau_m C & (A'\sigma_m + B') + 2\tau_m S \\ dx & dy & 0 & 0 \\ 0 & 0 & dx & dy \end{vmatrix} \quad (6.39a)$$

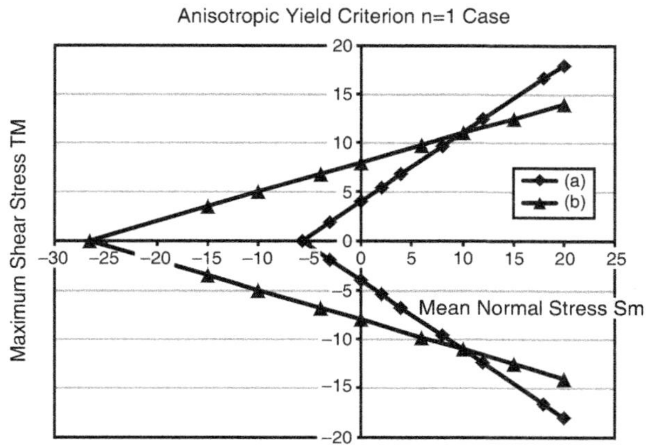

Figure 6.18 Anisotropic yield criteria.

and:

$$N_1 = \begin{vmatrix} E_1 & S & (A'\sigma_m + B) - 2\tau_m S & 2\tau_m C \\ E_2 & (n\tau_m^{n-1}/A - C) & 2\tau_m C & (A'\sigma_m + B') + 2\tau_m S \\ d\tau_m & dy & 0 & 0 \\ d\theta & 0 & dx & dy \end{vmatrix} \quad (6.39b)$$

where $C = \cos(2\theta)$, $S = \sin(2\theta)$, and $\sigma_m = (\tau_m^n - B)/A$. In the isotropic case the derivatives indicated with a prime (′) vanish.

Following the usual path to the characteristic system, one obtains the system:

$$\begin{aligned} & dy = f_1 dx, \ dy = f_2 dx, \\ & f_3 d\tau_m + f_5 d\theta + f_7 dx + f_9 dy = 0 \\ & f_4 d\tau_m + f_6 d\theta + f_8 dx + f_{10} dy = 0 \end{aligned} \quad (6.39)$$

where the even number functions apply to the first C_1 family of characteristics and the odd number functions apply to the second family of C_2 family of characteristics. This system is decidedly nonlinear because all functions depend on τ_m and θ despite the seeming simplicity of the anisotropic formulation Equations (6.36). In this regard, the anisotropic formulation Equations (6.8a) leads to a system in plane plastic strain where the functional coefficients in a system like Equation (6.39) depend only on θ, which thus offers a computational advantage.

The anisotropic Mohr-Coulomb *n*-type formulation Equations (6.36), (6.37), and (6.38) relative to the *ab* axes of anisotropy is also applicable to the *bc* and *ca* axes of anisotropy. Each of the three axes of anisotropy requires at least two parameters (A_a, B_a, A_b, B_b, A_c, B_c) that may be computed in terms of unconfined compressive and tensile strengths C_a, T_a, C_b, T_b, C_c, T_c. Each pair of the three axes may also involve two additional parameters, for example, axes *ab* may involve A_3, B_3. Values of these parameters are guided by consideration of shear strengths R_a, R_b, R_c. In effect, this extended formulation describes an orthotropic material.

Bearing capacity

Bearing capacity of a strip footing is a problem of some practical importance. A strip footing is one that is long compared with width and is therefore amenable to a plane strain analysis. As a practical matter, strip footings would be used under walls or rows of closely spaced columns. The "footing" could also be the bottom of a rock pillar that is several times longer than wide in a mine, an underground storage cavern, or some other underground excavation that relies on rock pillars for primary support and therefore where floor bearing capacity is under consideration. Several anisotropic formulations are available, but the *n*-type Mohr-Coulomb approach (Equations 6.36, 6.37, and 6.38) is most appealing because the widespread use of the linear Mohr-Coulomb criteria is addressing bearing capacity of strip footings.

The anisotropic yield criterion extremes are shown in Figure 6.19 where the exponent $n = 2$ in Equation (6.36d). The *a* and *b* axes of anisotropy are horizontal and vertical, respectively; the *c*-axis is horizontal and normal to the page (parallel to the footing). Unconfined compressive, tensile, and shear strengths are $C_a = 28, T_a = 4, R_a = 28, C_b = 14, T_b = 1, R_b = 3.5$ in arbitrary units. The strength anisotropy is 2:1 relative to unconfined compressive strength and 4:1 relative to tensile strength. Slopes of the yield criteria, $\tan(\psi)$, are variable and are greater than 45° for Mohr circles centered to the left of the dotted vertical lines, for example, stress circles representing tensile failures. Stress circles centered on the origin (pure shear) are also in the region where $\tan(\psi)>1$. Recall, the significance of $\tan(\psi)>1$ is the occurrence of imaginary characteristics that necessitates a different solution procedure than the method of characteristics. Fortunately, Mohr circles representing unconfined compressive strength are centered to the right of the vertical lines and thus the characteristics are real in the case of the bearing capacity problem.

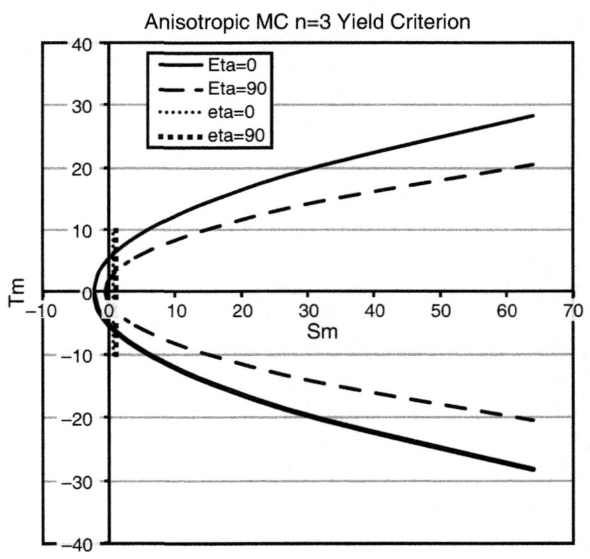

Figure 6.19 Anisotropic Mohr-Coulomb yield criteria parallel (Eta=0) and perpendicular (Eta=90) to "foliation."

Figure 6.20 shows the characteristics for the bearing capacity problem. Because the material adjacent to the footing is unconfined, yield occurs under uniaxial compression in Region I, a region of constant state indicated by straight characteristics. Confining pressure, so to speak, increases below the footing, so $d\tau_m / d\sigma_m = tan(\psi) < 1$ and $\sin(\phi) = \tan(\psi)$ everywhere. Region II is a region of radial shear with one family of straight characteristics. Region III is another region of constant state.

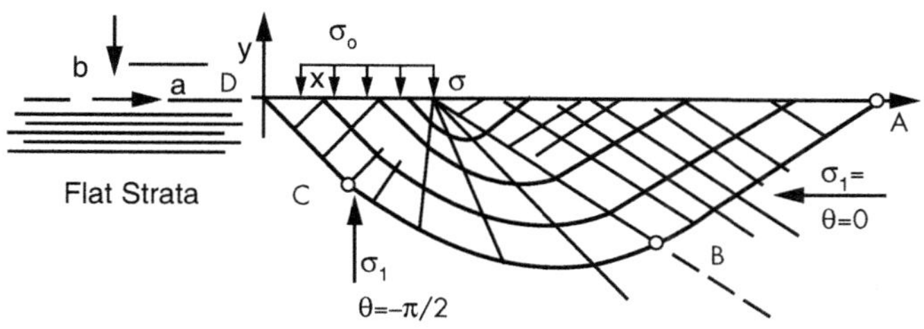

Figure 6.20 Bearing capacity geometry and characteristic regions for an anisotropic, Mohr-Coulomb n=2 type of yield criterion.

Boundary conditions for Region I are $\sigma_3 = p = 0$ and $\theta = 0$. Hence $\sigma_1 = C_a$, thus $\tau_m = C_a / 2$. Hence, on OB, $\tau_m = C_a / 2$ and the requisite boundary conditions for Region II are known. Integration of the stress field along a C_1 characteristic, say, along BC, gives $\tau_m = \tau_m^o$ in Region III. The footing is assumed to be smooth, so $\theta = \pi / 2$. The yield criterion then leads to $\sigma_m = (\tau_m^2 - B_b) / A_b$. Finally, the bearing capacity is obtained: $\sigma_o = \sigma_m + \tau_m$ in Region III supporting the strip footing where a uniform distribution of load is assumed.

Solution of the characteristic system in *radial shear* in this example requires consideration of a the highly nonlinear system (Equation 6.39) in polar-cylindrical coordinates (r, ω, z) and in consideration of the characteristic equations in the form:

$$dr / r = d\omega \cot[(\theta - \omega \pm \mu(\theta - \delta)] \quad : C_1, C_2 \tag{6.40}$$

where $\theta - \delta = \eta$ and $d\omega = d\theta$ for fixed "foliation" dip angle δ. In Region II, the characteristics of the C_2 family are straight lines and $\omega = \theta - \mu$. The C_1 characteristics are given by:

$$dr / r = d\omega \cot(2\mu) = d\theta \tan[\phi(\theta - \delta)] \tag{6.41}$$

The relations of the stress variables τ_m, θ along the characteristics are complex as reduction of the determinants (Equation 6.39) indicate. However, some simplification can be obtained with neglect of the properties derivatives A' and B'. The result is:

$$\pm d\tau_m \cot(2\mu) + 2\tau_m d\theta = 0 \tag{6.42}$$

with neglect of weight. Neglect of weight may not be justified in case of weak, anisotropic soils, but is reasonable for rock and even closely jointed rock. The system (Equation 6.42) appears deceptively similar to the same system in the isotropic, linear Mohr-Coulomb case. However, the dependency of μ on τ_m and θ makes solution problematic.

One solution approach begins with rewriting Equation (6.42) using Equation (6.41). Thus, $d\tau_m / \tau_m = -2\tan(\phi)d\theta$ that integrates to:

$$\tau_m / \tau_m^o = \exp\{\int_0^{\pi/2} 2\theta \tan[\phi(\theta)]d\theta\} \tag{6.43}$$

The change in θ between OB and OC is a known $\pi/2$, so decreasing θ, say, in intervals of 5° from 0° to −90° while also incrementing σ_m and thus τ_m via the yield criterion $\tau_m^2 = A(\eta)\sigma_m + B(\eta)$ generates and estimate of τ_m. To ensure satisfaction of equilibrium, an estimate of the integral in Equation (6.43) is necessary. An estimate begins with an expression for $\sin(\phi) = A(\eta)/2\tau_m$ using the value of τ_m obtained previously through the anisotropic yield criterion. Summation of increments associated with increments in θ gives a second estimate of τ_m that should match the estimate obtained through the yield criterion. If not, then another estimate of τ_m using the yield criterion is made. The iteration continues until satisfactory agreement is reached. Usually, just a few iterations are needed in a spreadsheet to reach agreement within a fraction of 1%. The process may be repeated for any desired dip angle δ.

Figure 6.21 shows bearing capacity of a strip footing for two cases of anisotropy and two cases of isotropy. Units are arbitrary but serve the purpose of comparison. In the first anisotropic case, $A_3 = 0$, $B_3 = 0$ and no dip or rise in bearing capacity occurs between dip angle extremes of 0° and 90°. In the second case of anisotropy, $A_3 = 10$, $B_3 = 10$ and minimum in bearing capacity occurs near 55° as seen in the figure. In this regard, values of A and B as functions of dip angle are given in Figure 6.22 where A_3 and B_3 are parameters. The two isotropic cases are based on two yield criteria that correspond to the extremes of the anisotropic yield criterion in Figure 6.19.

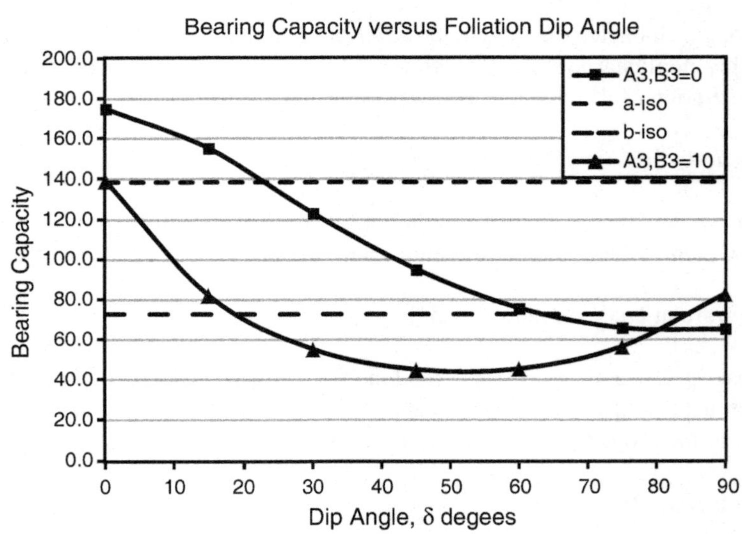

Figure 6.21 Bearing capacity of a strip footing: n=2 Mohr-Coulomb yield criterion with control parameters A3 = 0, 10, B3 = 0, 10 anisotropic cases and isotropic cases a-iso and b-iso related to properties C_a, T_a, R_a, C_b, T_b, R_b,.

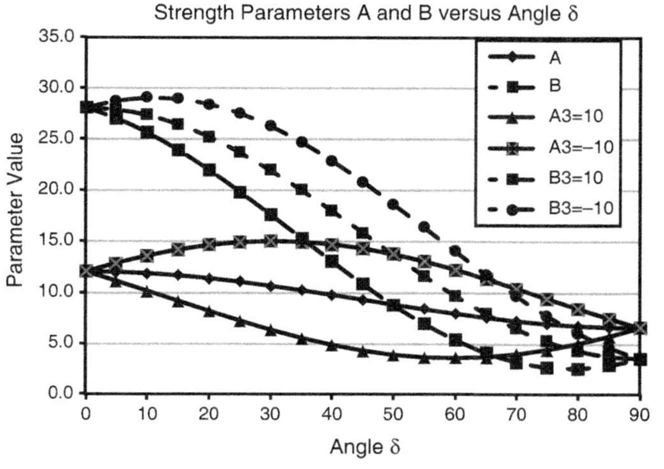

Figure 6.22 Strength parameters variation with "foliation" dip. Arbitrary units.

Unconfined compressive strength as a function of dip angle is shown in Figure 6.23. Interestingly, the minimum of bearing capacity does not coincide with the minimum of unconfined compressive strength, which occurs near 65°.

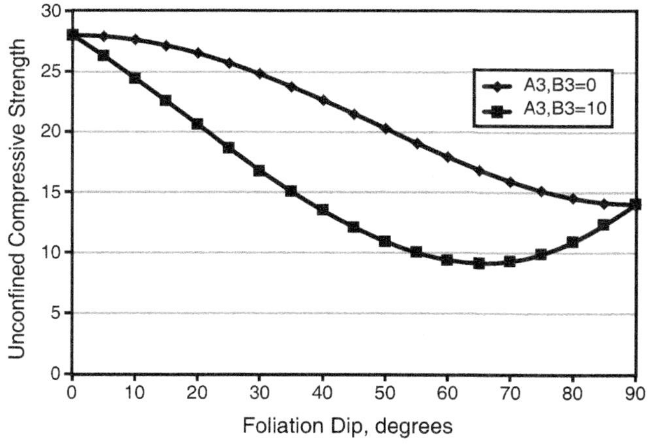

Figure 6.23 Unconfined compressive strength as a function of "foliation" dip with and without shear strength effects on A and B.

If the isotropic cases were linear ($n = 1$), the associated bearing capacities would be much greater than in the nonlinear examples. The reason is in the rate of increase of strength with confining pressure. Although this rate is initially high in the nonlinear case, a rapid decrease occurs as the confining pressure increases. In the linear case, the initially high rate remains constant and leads to bearing capacities more than twice the greatest nonlinear estimates of bearing capacity (dip angle of zero).

References

Chen, W.F. & Han, D.J. (1988) *Plasticity of Structural Engineers*, Springer-Verlag, New York.

Cook, R.D. (1974) *Concepts and Applications of Finite Element Analysis*. John Wiley & Sons, New York.

Green, A.E. & Zerna, W. (1968) *Theoretical Elasticity*. Oxford University Press, Oxford.

Hill, R. (1950) *The Mathematical Theory of Plasticity*. Clarendon Press, Oxford.

Lekhnitskii, S.G. (1963) *Theory of Elasticity of an Anisotropic Elastic Body* (Translated by P. Fern). Holden-Day, San Francisco.

Lekhnitskii, S.G. (1968) *Anisotropic Plates*. (Translated by S.W. Tsai & T. Cheron). Gordon and Breach Science Publishers, New York.

Mitchell, S. (2009) *Nuggets to Neutrinos: The Homestake Story*. Steve Mitchell, Xlibris Corporation, Orders@Xlibris.com.

Pariseau, W.G. (1968) Plasticity theory for anisotropic rocks and soils. *Proceedings 10th U.S. Symposium on Rock Mechanics*. University of Texas at Austin, May 20–22, Society of Mining Engineers of the American Institute of Mining, Metallurgical and Petroleum Engineers, New York, 1972.

Pariseau, W.G. (1970) Wedge indentation of anisotropic geologic media. *Proceedings 12th U.S. Symposium on Rock Mechanics*. Rolla, MO, November 16–18, Society of Mining Engineers of the American Institute of Mining, Metallurgical and Petroleum Engineers, New York, 1971.

Pariseau, W.G. (2017a) Comparison of underground coal and trona mine seismicity, Preprint 17–027. *SME Annual Meeting*. Denver, CO, February 19–22.

Pariseau, W.G. (2017b) *Design Analysis in Rock Mechanics* (3rd ed.). CRC Press/Balkema, Taylor & Francis, London.

Pariseau, W.G., Johnson, J.C., McDonald, M.M. & Poad, M.E. (1995) Rock Mechanics Study of Shaft Stability and Pillar Mining, Homestake Mine, Lead, SD, U.S. Part 1. Bureau of Mines Report of Investigations 9531.

Ting, T.C.T. (1996) *Anisotropic Elasticity: Theory and Applications*. Oxford University Press, Oxford.

Chapter 7

Viscoplasticity[1]

Viscoplasticity introduces a rate dependency into plasticity theory. An initial elastic response to load is still assumed before the onset of plastic strain, but the plastic strain is no longer time-independent. In this regard, conventional wisdom considers intact rock to be a brittle material at the laboratory scale. However, stiff testing machine experiments have long shown that microcracking allows for inelasticity and some time-dependent behavior. In field-scale rock masses, slip and separation of joints allow for inelastic deformation. Here, "joints" refer to a variety of structural discontinuities including bedding planes, faults, and so forth. However, a casual inspection of symposia and conferences the past several decades that were dedicated especially to joints shows little regard for time dependency, despite much observational evidence and experience. A viscoplastic material responds reversibly within the elastic limit, but viscoplastically above. Viscoplastic effects are synonymous with strain-rate effects and time dependency. Such effects range over orders of magnitude in time from fractions of a second in blast wave dynamics (e.g., DiMaggio and Sandler, 1971; Isenberg and Wong, 1971) to hundreds of years in creep of geologic media, especially salt (e.g., Munson *et al.*, 1989; Arguello *et al.*, 1989). Strictly speaking, the latter references involve fluid-like viscoelastic media that have no well-defined elastic limit. However, concern here is with extension of what may be called the standard model in rock mechanics, an elastic material limited by strength or stability criteria. The objective of an elastic-viscoplastic model (e.g., Prager, 1961; Perzyna, 1963, 1966; Naghdi and Murch, 1963; Malvern, 1969; Kachanov, 1971; Cristescu and Hunsche, 1998) is to achieve greater realism in engineering design by accounting for time-dependent rock mass displacement when evidence indicates a need to do so. Evidence may be found in field observations at an engineering scale and in laboratory measurements as well.

A snapshot of some field data from underground and surface mines is examined for time-dependent behavior first in this chapter. Laboratory test data are examined next. After formulations of theory, these data are qualitatively compared with a conventional elastic-viscoplastic model. A critique of the advantages and limitations of the model is then presented, followed by several example problems. These example problems are approached numerically with the popular finite element method, which is representative of current geoplasticity analysis capability. The necessity of distinguishing between pseudo-time or load increments used in elastic-plastic models and real time used in elastic-viscoplastic models is made evident.

1 This chapter is based on a study done by the author for the Spokane Mining Research Laboratory/National Institute of Occupational Health and Safety.

7.1 Some field data

Although the main applications of viscoplasticity to date appear to be at extreme time scales, there is much observational evidence for time-dependent behavior in rock mechanics that occurs at intermediate time scales, ranging from hours to days to a few years. These time intervals are important to engineering safe, stable excavations for civil and mining purposes. For example, the well-known concept of standup time in tunneling is based on observations of ground falls that occur in time when the face is not being advanced and before support is installed. The stair-step shape of plots of mine roof sag as a function of time often reveal a sharp rise followed by a rounded run as excavation proceeds. The first part of a step is readily associated with an instantaneous elastic response, while the second part takes time. An example is given in Figure 7.1 (Merrill, 1957), which shows roof sag measured during a field study in an underground limestone mine. According to Merrill,

> In general, the sag rate – that is, the sag per unit time – was high immediately after a widening blast. This sag rate fell considerably in a few days and eventually reached a "zero" rate, usually in about 4 months or less, depending upon the width of the room and the anchor depth of the roof pin. Elastic theory accounts only for the sag immediately after the blast and the zero sag rate a few months after the blasts.

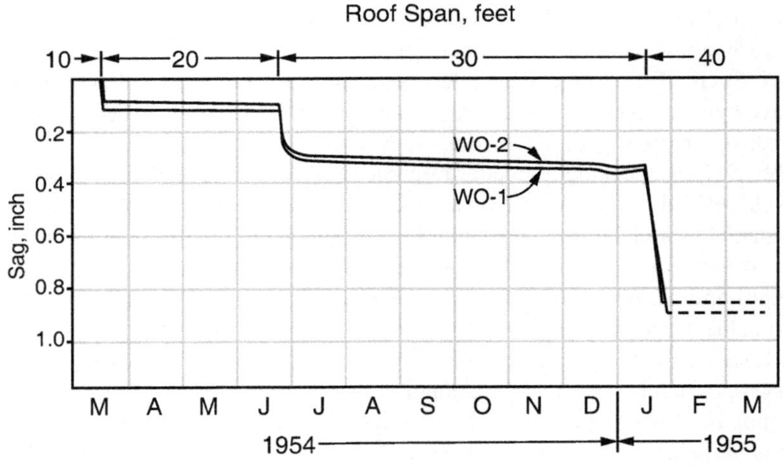

Figure 7.1 Roof sag versus time in an underground limestone mine (Merrill, 1957).

Figure 7.2 shows sag as a function of roof span as measured and calculated from elastic theory under clamped and simply supported ends. The simply supported ends case is added to the original plot to show that the measured data fall between the two theoretical cases. These two cases bound the set of beam solutions that allow for some rotation of ends and confirm the intuitive notion that the actual end condition is in between. In this experiment, the time-dependent sag was about 10% of the instantaneous elastic sag. Bed separation occurred during the test and roof failure followed after the span reached 12.2 m (40 ft). The test panel in this study was located away from other active sections of the mine, so the post-blast displacements were certainly time-dependent. Because of bed separation, there is a possible time-dependent advance of the bedding plane fracture that defines the separation horizon rather than time-dependent flexure of the roof strata proper. Both mechanisms could be present, of course.

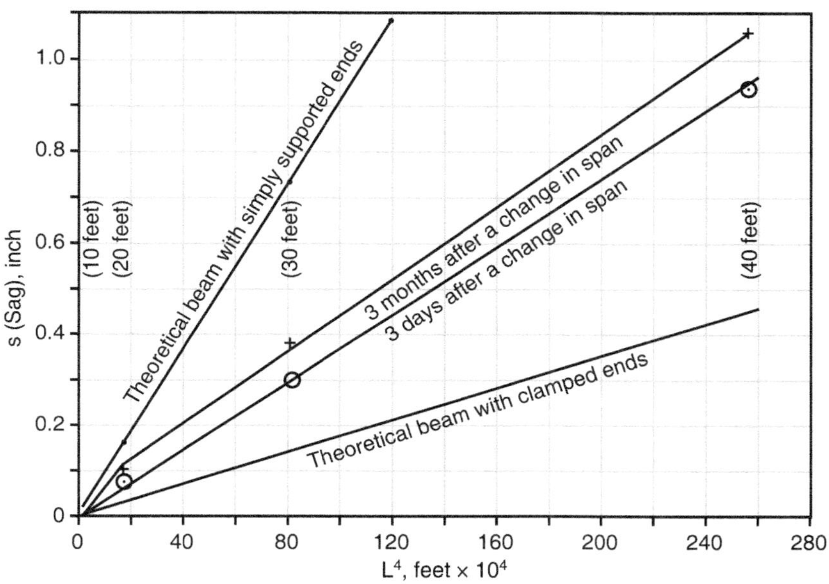

Figure 7.2 Roof sag versus span in an underground limestone mine (Merrill, 1957). Simply supported ends case added.

Another example of time-dependent displacement is presented in Figure 7.3, which shows borehole extensometer readings obtained in an underground copper mine (Pariseau *et al.*, 1984). According to the authors,

> The maximum displacement indicated by the extensometers immediately after the last blast is of the order of 2.5 cm (1.0 in.). Two weeks after the last blast the readings more than doubled. Higher readings were obtained on occasion, but these are attributed to anchors lost during overbreak. . . . a continued increase in readings [occurred] between blasts but at a decreasing rate with time.

In this study, the test stope was remote from other stopes and no other faces were being advanced. Thus, post-blast displacements were unambiguously time-dependent. Despite the variety of rock types present, no dominant joint sets were identified in the complex metamorphic geology of the study site.

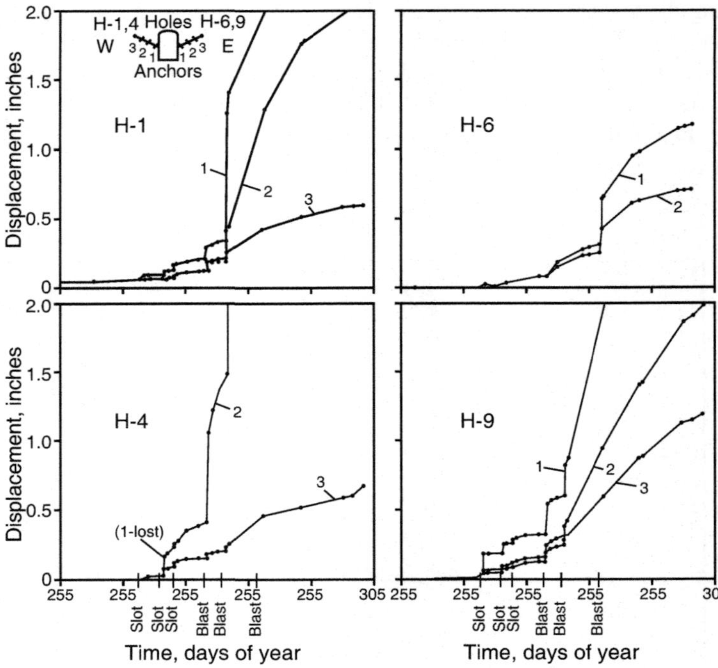

Figure 7.3 Borehole extensometer displacement readings versus time in an underground copper mine test stope (Pariseau *et al.*, 1984).

Similar observations of displacement versus time are reported in open pit copper mining. An example is shown in Figure 7.4, taken from Zavodni and Broadbent (1980) who indicate that the "cycles" or steps are generally associated with an external event such as a blast, bench removal, or drainage process. They examined 13 major open pit mine slope failures ranging from 120,000 to 30,000,000 tons and noted that sharp breaks in velocity occurred at the onset of instability. The slide types and initiating events were varied, but in all cases were controlled by geologic structure, joints, and faults.

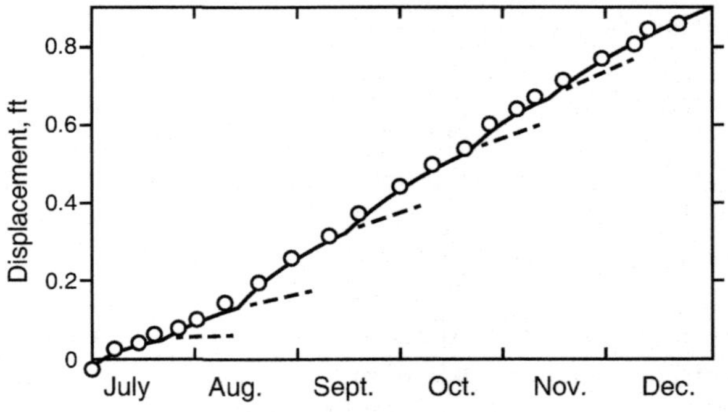

Figure 7.4 Slide movement illustrating long cycle periods in the regressive stage, Liberty Pit, Slide No. 1 (Zavodni and Broadbent, 1980).

7.2 Some laboratory data

Although the focus here is on rock mass engineering, some mention of rate-dependent laboratory data is in order. Figure 7.5 from Serdengecti and Boozer (1961) shows the combined effects of confining pressure and strain rate on Berea sandstone in compression under confining pressure. The strain rate varied over more than three orders of magnitude. The data show increased stiffness and strength with increased strain rate and also a noticeable departure from linearity at about one-half the peak stress obtained under confining pressure. Bieniawski (1970) and Rummel and Fairhurst (1970) obtained similar results in laboratory tests on sandstone and Tennessee marble, respectively.

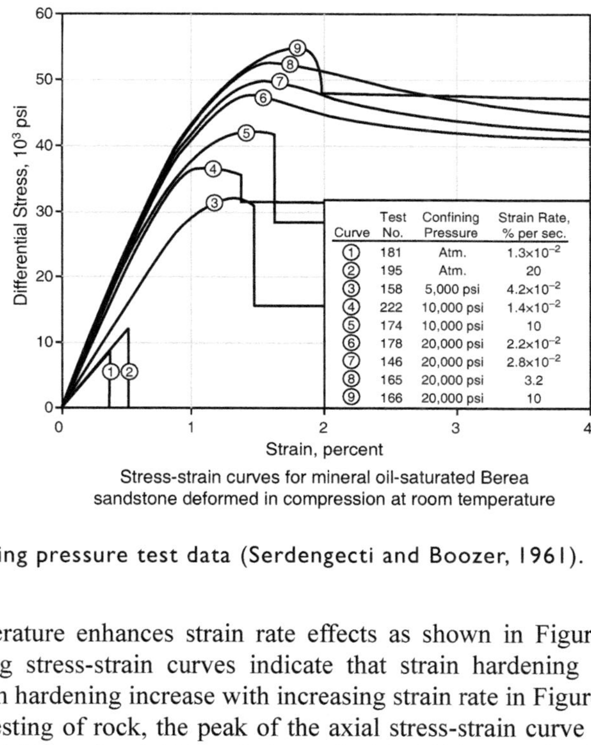

Figure 7.5 Confining pressure test data (Serdengecti and Boozer, 1961).

Elevated temperature enhances strain rate effects as shown in Figure 7.6 from Heard (1963). The rising stress-strain curves indicate that strain hardening is rate-dependent. Strength and strain hardening increase with increasing strain rate in Figure 7.6.

In laboratory testing of rock, the peak of the axial stress-strain curve is usually reached at about 0.1% to 1.5% strain. A 1% strain is reached in 1000 seconds, about17 minutes, at an applied strain rate of 10^{-5}/sec. An applied stress rate of 100 psi/sec will reach 20,000 psi in 200 sec, about 3 minutes. One may therefore identify a strain rate of 10^{-5}/sec as corresponding to a laboratory test strain rate. Slow creep rates would be orders of magnitude less. A strain rate of about 3×10^{-10}/sec would reach a 1% strain in one year. Such small strain rates amount to about 30×10^{-6} or 30 micro-inches/inch strain in one day. A temperature drift of a few degrees would account for strain of a similar magnitude, as may other environmental changes. Because of these experimental difficulties, time-dependent behavior is often studied at stress near the peak of the stress-strain curve and at elevated temperature. Extrapolation of test data to lower stress, temperature, and strain rate is problematic, especially on the rising portion of the stress-strain curve, and raises the question of whether rock has a fundamental strength below which time is unimportant.

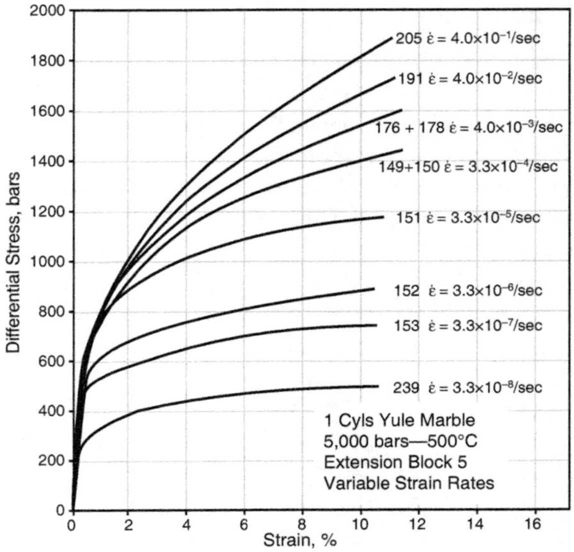

Figure 7.6 Stress-strain curves for cylinders of Yule Marble at 500 °C and 75,000 psi confining pressure (Heard, 1963).

The question is a persistent one and cannot be answered with great certainty, but as Price (1966) argues, it may be of no great consequence. With reference to Figure 7.7, F is the peak of the stress curve and marks the "instantaneous" strength determined during an ordinary laboratory test. If the load is held at D, say, 80% of the instantaneous strength, then creep to failure at E requires some time *t*. At a lower load B, more time will be required before the failure strain is reached. Price points out that if the time to failure at the lower load is millions of years, then as a practical matter, the material responds as if it had a fundamental strength. Another argument is that given the viscosity of solids, the maintenance of regions of high topographic relief through geologic time is impossible if rock masses did not have a fundamental strength but instead were simply fluids with very high viscosity. Salt seems to be an exception to the tentative rule that rock has a fundamental strength.

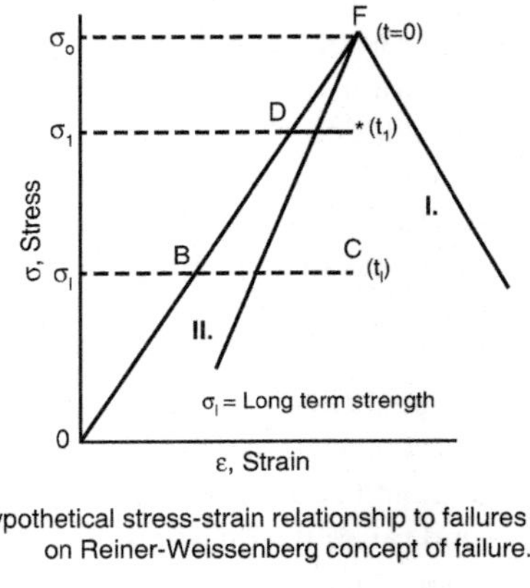

Figure 7.7 Conceptual stress–strength and time relationship (Price, 1966).

One may also argue that the question of whether rock has a fundamental strength is actually quite important, because the answer dictates the basic type of material model to be used to insure safety and stability over the service life of excavations in rock. The decision is whether rock masses are fluid-like or solid-like in response to load. Elastic-viscoplastic models imply a solid-like response where the rate dependency occurs above an elastic limit.

An extension to the concept of long-term strength by Price (1966) was explored by Wawersik and Brown (1971), who tested the hypothesis that strain to failure is governed by the falling portion of the complete stress-strain curve. A basic distinction is made between rocks that have negative post-peak slopes and those that have positive post-peak slopes (Types I and II). These lines have been added to Figure 7.7. The latter are unstable in the sense that failure occurs even at fixed displacement; unloading would be required to arrest the failure process. The former is stable under controlled closure (displacement) of testing machine platens during a laboratory test for compressive strength. Because test data also show less negative slopes (flatter slope) at slower strain rates (Type I), the strain and time to failure increase with decreasing strain rate.

Fabric analyses revealed development of extensive microcracking in the three rock types tested – Westerly granite, Nugget sandstone, and Tennessee marble. The first two were unstable in the post-peak range (Type II); the last was Type I. Strains rates, at room temperature, ranged from about 10^{-5}/sec to 10^{-3}/sec (at room temperature) and were thus relatively fast or "laboratory test scale." Evolution of microcracking accelerated at about 50% of the peak strength. Microcrack density was postulated to be in direct relationship with creep strain. In this regard, Scholz (1968) much earlier tested six rock types (granite, quartzite, marble, rhyolite tuff, sandstone, and gabbro) in uniaxial compression and under confining pressure and found quite similar behavior, in that microcracking increased noticeably at about 50% of peak strength as did acoustic emissions with a noticeable departure from linearity. Figure 7.8 shows results for two rock types and the association of acoustic emissions with microcracking. The strain rate was 10^{-5}/sec in all tests.

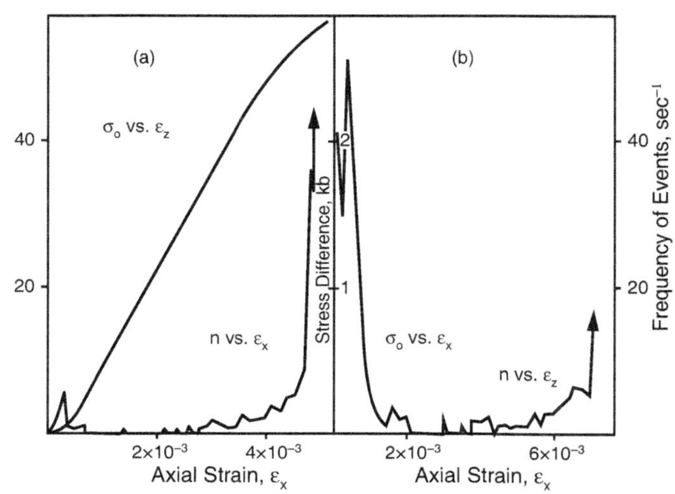

Stress versus strain and microfracturing frequency versus strain for two rocks in uniaxial compression. (a) Westerly granite; (b) Colorado rhyolite tuff.

Figure 7.8 Stress – strain – acoustic emission plot (Scholz, 1968).

With the advent of servo-controlled stiff testing machines during the 1970s, many other investigators (e.g., Peng and Podnieks, 1972; Hudson and Brown, 1972, Peng, 1973) noted similar results: an increasing microcrack density above about 50% of the peak stress accompanied by an increase in acoustic emissions with the formation of macroscopic fractures and rapid test specimen collapse. Interestingly, Hudson and Brown noted a decrease in post-peak slope with decreasing strain rate. In this regard, Brady *et al.* (1973) pointed out that the falling portion of the complete stress-strain curve in Figure 7.9 was not a material property because of the reduction in intact rock area during progressive failure of the test specimen in the post-peak region *at constant strain rate* [emphasis added].

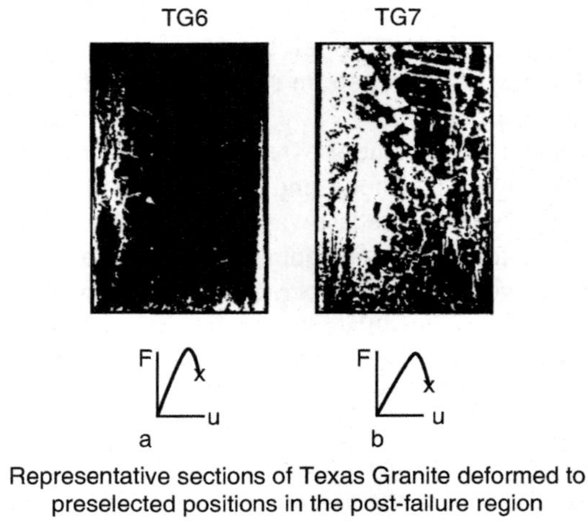

Figure 7.9 Damage (white) during post-peak compression (Brady *et al.*, 1973).

Whether a limit to peak stress, an instantaneous strength, actually exists may also be open to question, because very high strain rates associated with impact loading have not been considered. In fact, the peak strength, referred to as instantaneous strength, observed in the test data reviewed here is always reached quasi-statically. At very high strain rates, microcracks may not have time to grow, so the dynamic response is purely elastic. Figure 7.10 shows quasi-dynamic laboratory test results from Logan and Handin (1971) that indicate an increase in strength up to a strain rate of 1/sec in granite occurs over a considerable range of confining pressure and that the effect is greater at higher confining pressure. A 1% strain is reached in 1/100th second at the highest rate used. (The data in Figure 7.5 show strain rate effects an order of magnitude higher, to 10/sec.) The almost straight line plots in Figure 7.10 of the square root of the second invariant of deviatoric stress versus mean normal stress are indicative of the well-known Drucker-Prager yield condition ($J_2^{1/2} = AI_1 + B$, where J_2 is the second invariant of deviatoric stress, I_1 is the first invariant of stress ($3P_m$), A and B are material properties). The curved plots indicate a n-type criterion, $(J_2^{1/2})^n = AI_1 + B$, may be appropriate (Pariseau, 1972), perhaps with provision for strain hardening.

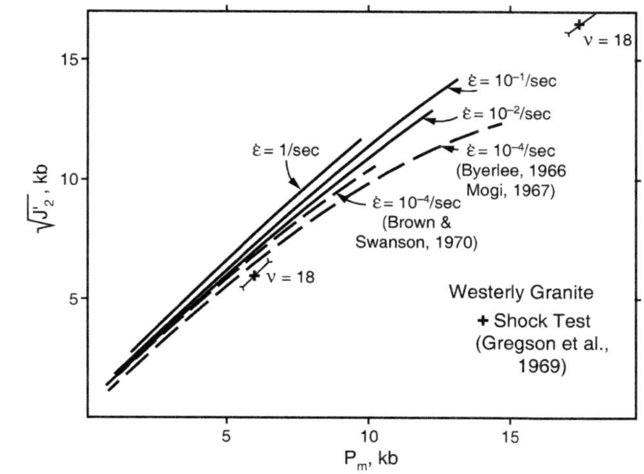

The square root of the second deviatoric stress invariant versus mean pressure for Westerly granite for strain rates of 10^{-2} to 1/sec. Data at a strain rate of 10^{-4}/sec and shock data are from the sources noted.

Figure 7.10 Quasi-dynamic strain rate effect on strength versus confining pressure (Logan and Handin, 1971).

Figure 7.11 shows dynamic test results for limestone and granite. Strength increases (non-linearly) with strain rates to 1000/s, which reaches 1% strain in 10 microseconds. A straight line fit of limestone data to the semi-log scale suggests an exponential relationship between shear stress and strain rate. A similar fit to the granite data could also be done, although the correlation is not as high. Other fits to the data such as a power law could also be done.

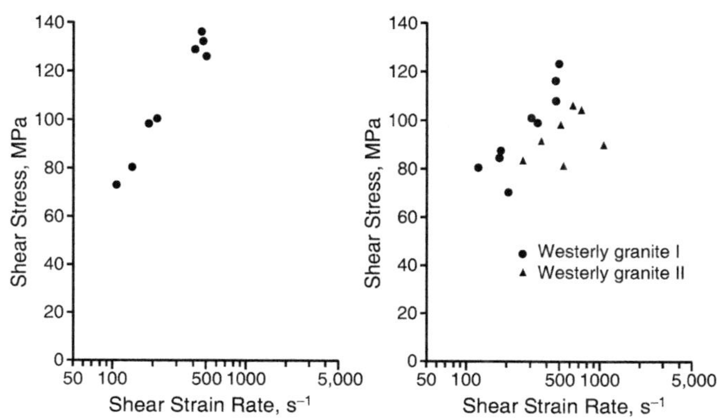

Strain rate dependence of failure strength in Solenhofen limestone

Strain rate dependence of failure strength in Westerly granite

Figure 7.11 Dynamic strain rate effect on strength (Lipkin *et al.*, 1977).

At the laboratory scale, this data snapshot indicates that many rock types show:

1. an elastic range of deformation
2. time-dependent, strain-rate dependent behavior above the elastic limit
3. the higher the strain rate, the greater is the observed stress rise above the elastic limit before peak stress is reached
4. extensive microcracking evolves in advance of macroscopic fracturing in many cases
5. at high confining pressure, a transition between brittle fracture and ductile deformation may occur in some rock types.

While the data indicate an increase in peak stress with strain rate from quasi-static through quasi-dynamic to well into the dynamic range, a caution is needed in arriving at these observations because of the practical limit to very slow or small laboratory strain rates, say, less than 10^{-7}/sec. An elastic-viscoplastic material model should be consistent with these experimental observations including a forecast of a rounded stair-step response.

7.3 Elastic-viscoplastic models

The original viscoplastic model that is much referenced in the literature is due to Bingham (1922), who made the distinction between viscous fluids and solids by drawing attention to the presence of an elastic range of shear deformation in the case of solids. In fluids, flow always occurs under shear stress; there is no well-defined yield stress as in the case of solids. However, any excess stress above the yield point resulted in viscous flow. Bingham cites a variety of test data from different materials flowing in tubes, deformed in torsion and beams sagging under self-weight in support of viscoplastic deformation. The concept was subsequently extended to three dimensions.

One-Dimensional Model: According to the Bingham plastic model, a material deforms elastically below a yield stress τ_Y but viscously above. With neglect of the elastic strain (rigid-plastic model), in one dimension, $\gamma = 0$ when $\tau < \tau_Y$, but $d\gamma/dt = (\tau - \tau_Y)/\eta$ when $\tau \geq \tau_Y$ and where γ, τ, and η are shear strain, shear stress, and viscosity, respectively. By contrast, a fluid has no yield stress; fluid flow occurs under the slightest shear stress. The rigid Bingham plastic model is represented rheologically by a parallel connection of a dashpot and slider block. Addition of a spring in series with this parallel element results in a simple elastic-viscoplastic model. The stress-strain law for the Bingham solid then has two parts:

$$\dot{\gamma} = \frac{\dot{\tau}}{G} \qquad (\tau < \tau_Y)$$
$$\dot{\gamma} = \frac{\dot{\tau}}{G} + \left(\frac{1}{\eta}\right)(\tau - \tau_Y) \qquad (\tau < \tau_Y) \qquad (7.1, 7.2)$$

where G is the elastic shear modulus. In special cases of constant stress (creep) or constant displacement (relaxation), integration of this one-dimensional model is possible. If a stress τ exceeding the yield stress τ_Y is applied at time zero, then an instantaneous elastic strain occurs in magnitude $(1/G)\tau$. This strain is followed by a linearly increasing (with time) viscoplastic strain $(1/\eta)(\tau - \tau_Y)\Delta t$ that accumulates at constant velocity. The total strain rate is the sum of the two in Equation (7.2), which shows the interesting feature that the viscoplastic strain rate does not depend on the stress rate, but only on the stress above the elastic limit.

If the applied shear stress τ is suddenly released at the end of the interval Δt, then an instantaneous recovery of elastic strain $(1/G)\tau$ occurs as the velocity drops to zero. The viscoplastic strain $(1/\eta)(\tau - \tau_Y)\Delta t$ is not recovered, nor is the displacement $(1/\eta)(\tau - \tau_Y)h\Delta t$ where h is the thickness of the considered region. The shear velocity is just $(1/\eta)(\tau - \tau_Y)h$. Repetition of the process ratchets upwards the viscoplastic strain. At the end of a second cycle of load application and removal, the viscoplastic strain is doubled. A permanent set in the form of unrecovered displacement is also doubled. In theory, the process can be continued indefinitely. In practice, accumulation of an indefinitely large permanent strain is unrealistic, so the process must be terminated by some physical event not taken into account by the model, for example rupture or some phenomenon such as strain softening that leads to instability.

The yield stress τ_Y may be given by the well-known Mohr-Coulomb criterion with angle of internal friction ϕ and cohesion c. If cohesion is made displacement dependent, so that increasing displacement decreases c, then this form of "strain softening" may create a "trigger" instability where an additional repetition of a load-unload cycle causes an acceleration of the region considered. A mechanism for progressive destruction of cohesion is the shearing of intact rock bridges between joints in a field-scale rock mass. The concept is based on the Terzaghi (1962) estimate of jointed rock mass strength. Such a mechanism is consistent with the often-observed phenomenon of peak-residual shear strength. Figure 7.12 illustrates the concept of peak-residual shear strength associated with an initial elastic range of deformation to peak strength followed by progressive loss of cohesion to residual shear strength.

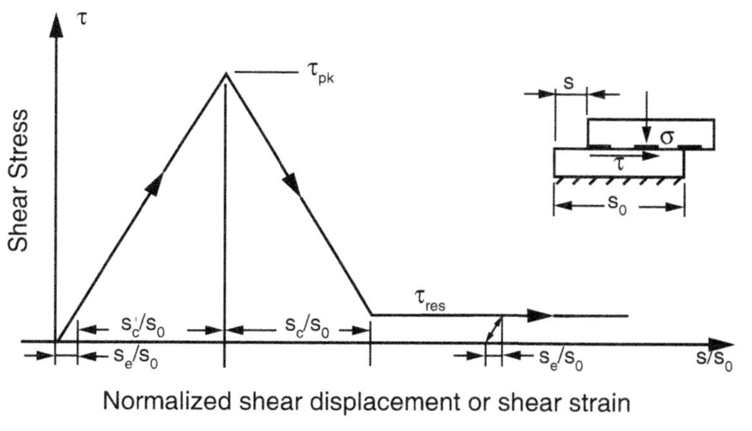

Figure 7.12 Idealized shear stress versus normalized displacement ("strain") showing peak-residual shear strength as a function of displacement along a shear plane of intact rock bridges and joint surface (Pariseau and Voight, 1979).

Figure 7.13 shows an approximate graphical analysis of a series of hypothetical load-unload cycles associated with blasting in an open pit mine. Each blast momentarily raises the shear stress above the yield stress, which shears intact rock bridges and lowers cohesion and therefore reduces the yield stress in advance of the next blast. The last blast triggers an exponential increase in velocity according to the Bingham plastic model. An event of this type represents the "break" in velocity from slow to fast that marks slope instability according to Zavodni and Broadbent (1980).

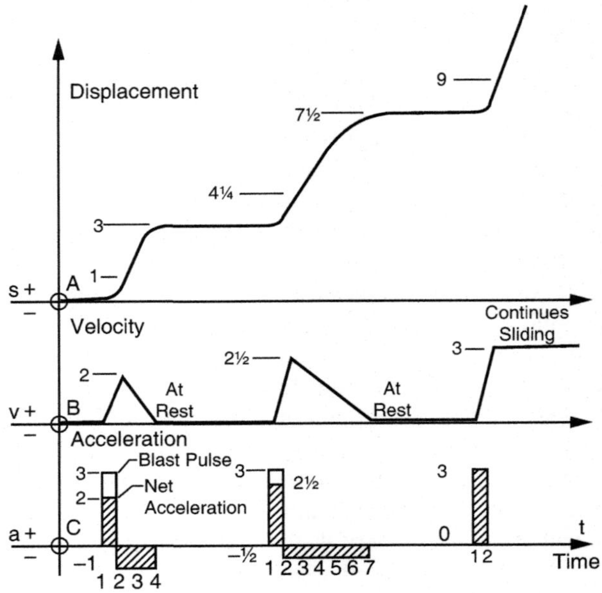

Figure 7.13 Graphical analysis of repeated blasting and cohesion loss with displacement that leads to triggering and exponential increase in slide velocity of a Bingham plastic slope (Pariseau and Voight, 1979).

Strain softening can be associated with a reduction in the frictional component of strength as well as cohesion. On fresh joints, frictional resistance to shearing is mobilized through the product of normal stress and $\tan(\varphi+i)$ where i is the angle an asperity forms with the joint plane proper (Patton, 1966). As asperities are sheared, i is reduced and joint strength is decreased. Where the failure surface is fixed by geologic structure, as is the case in most planar block slides and wedge failures in surface excavations, shearing of intact rock bridges between joint segments and asperities along the failure surface are simple, realistic physical mechanisms that account for strain softening and allow for instability in time as motion continues under the weight of the rock.

The Bingham plastic model has also been used to estimate creep of a surface layer on a long natural slope (Chen, 1969). Equilibrium, kinematics, and the Bingham plastic material law enter the solution. The region of interest is a slab of indefinite extent parallel to the slope and h units into the slope. Velocity is zero at the bottom of the slab that is being driven by gravity alone. Slope angle from the horizontal is θ, viscosity is η, angle of internal friction is ϕ, and cohesion is c (Mohr-Coulomb yield condition assumed). The slab moves parallel to the slope with velocity distribution v given by:

$$v = (\frac{\gamma \sin(\theta - \phi)}{2\eta\cos(\Phi)}(h^2 - x^2) - (\frac{c}{\eta})(h - x) \qquad (7.3)$$

where x is distance into the slope. The requirement implied in Equation (7.2) that the slope angle be greater than the friction angle is simply a criterion for sliding, so $\sin(\theta\text{-}\phi)$ is always positive. When cohesion is zero, the maximum velocity occurs at the surface and is given by the first term on the right of Equation (7.3). The velocity profile is a parabola. With cohesion, the maximum occurs below the surface.

An alternative velocity distribution is possible, although such a possibility was not discussed by Chen (1969). At depths x less than $h_o = c\cos(\varphi)/\gamma\sin(\theta - \varphi)$, the shear stress is less than the yield stress and plastic flow does not occur. Hence, cohesive slopes present the possibility of an elastically deformed slab riding atop a viscoplastic shear zone of thickness $h - h_o$. The velocity distribution within the shear zone is given by Equation (7.3), while the overall velocity distribution is somewhat similar to plug flow in tubes as described by Bingham (1922). Interestingly enough, slope stratigraphy is often preserved even when fast landsliding occurs, so there is qualitative evidence for the Bingham plastic block slide model outlined here.

The simple one-dimensional elastic-viscoplastic model Equation (7.1 7.2), originally proposed by Bingham (1922), meets the three macroscopic requirements of:

1. an elastic limit
2. time-dependent strain occurs above the elastic limit
3. the greater the stress above the elastic limit, the greater the strain rate.

As long as the applied shear stress exceeds the elastic limit, a non-zero strain rate is sustained; strain and displacement increase with time. Instability in the form of a runaway increase in velocity is not indicated with strength τ_Y held constant because shear velocity is only directly proportional to excess shear stress $(\tau-\tau_y)$ and shear zone thickness. The model could easily be made nonlinear by introducing an exponent n for the excess shear stress term, that is, by making the viscoplastic strain proportional to $(\tau-\tau_y)^n$. In case of fluid flow, the constant of proportionality $(1/\eta)$ is referred to as the coefficient of "fluidity" by Bingham (1922). In the case of viscoplastic behavior, Bingham (1922) refers to $(1/\eta)$, used in Equation (7.2b), as the coefficient of mobility. In any nonlinear extension, the conventional viscosity associated with fluid-like behavior and found in tables of material properties will be obscured and the units of η (viscosity) will change, say, to $(\text{psi})^n$-sec from psi-sec. A similar situation can be expected in any three-dimensional formulation of viscoplasticity.

Introduction of strain softening in the form of strain-dependent cohesion and concomitant peak-residual strength creates a potential for instability, as illustrated in the example analysis involving repetitive blasting in surface mines where planar block slides and wedge failures are common. However, as a macroscopic model, nothing is said about microscopic (grain scale) behavior, such as acoustic emissions associated with microcracking. Generally, microstructural considerations guide the development of macroscopic models. In this regard, laboratory test data from intact rock study suggest that some form of strain softening associated with microcrack damage and loss of cohesion may be needed for instability. At the field scale of rock masses, shear of joint asperities and intact rock bridges provide "microscale" damage mechanisms for viscoplasticity that lead to strain softening.

The study of laboratory test behavior and the response of rock masses require a three-dimensional analysis. Even uniaxial compression testing of intact rock cylinders in the laboratory requires consideration of the axially symmetric test geometry. In some instances, analytical solution is possible, but generally numerical methods and computer codes are needed for three-dimensional problem solution.

Three-Dimensional Model: Generalization of the one-dimensional model to three dimensions requires specification of the viscoplastic strain contribution to the total strain. Two functions are needed for this purpose: a yield function F and a viscoplastic potential function Y. The yield function is introduced to define a limit to a purely elastic response. Curly brackets {} indicate a 6×1 column matrix in the following; square brackets [ ] indicate 6 × 6 square matrices Thus, if:

$$F(\{\sigma(t)\},\{\varepsilon(t)\}) = f(\{\sigma(t)\},\{\varepsilon(t)\}) - \kappa \le 0 \tag{7.4}$$

where κ is a material constant, then viscoplastic deformation is not possible. If $F>0$, so that $f>\kappa$, then the elastic limit is exceeded and the response is elastic-viscoplastic. Under a uniaxial compressive stress, the function $f=\sigma$ and $\kappa = C_o$ where C_o is the unconfined compressive strength. Inspection of Equation (7.2b) shows that the stress is then equal to the *static* yield stress or elastic limit σ_Y where $\sigma_Y = C_o$. Under simple shear τ, $f=\tau$ and $\kappa=\tau_Y$, as before. In three dimensions, the yield criterion f then plays the role of the yield stress in one dimension.

A viscoplastic potential function $Y(\{\sigma(t)\},\{\varepsilon^{vp}(t)\})$ when differentiated leads to the viscoplastic strain rate. The existence of such a potential is inferred from thermodynamic considerations (Rice, 1975). Thus:

$$\{\dot{\varepsilon}^{vp}\} = \Lambda\{\frac{\partial Y}{\partial \sigma}\} \tag{7.5}$$

where Λ is a function that does not depend on strain or stress rates and $\{\partial Y/\partial\sigma\} = (\partial Y/\partial\sigma_{xx}, \partial Y/\partial\sigma_{yy}, \ldots, \partial Y/\partial\sigma_{xy})$ is a vector of partial derivatives of the yield function with respect to the six components of symmetric stress. Equation (7.5) shows that the viscoplastic strain rate is parallel to the gradient of Y. When the viscoplastic potential and yield criterion are the same, that is, "associated," then the viscoplastic flow rule, Equation (7.5), is associated. Graphically, the strain rate is perpendicular to the instantaneous yield surface and the "principle of normality" holds, as in conventional plasticity with associated flow rules but with a static yield function independent of time.

While Equation (7.5) provides the direction of viscoplastic strain, magnitude is obtained through prescription of the function Λ. Squaring both sides of Equation (7.5) and solving for Λ leads to:

$$\{\dot{\varepsilon}^{vp}\} = \sqrt{\{\dot{\varepsilon}^{vp}\}^T\{\dot{\varepsilon}^{vp}\}}\left(\frac{\{\partial Y/\partial\sigma\}}{\sqrt{\{\partial Y/\partial\sigma\}^T\{\partial Y/\partial\sigma\}}}\right) \tag{7.6}$$

The square root term on the right in Equation (7.6) is just the magnitude of the strain rate; the term in parentheses is an outward unit normal vector to Y. The product defines the direction and magnitude of the viscoplastic strain rates on the left of Equation (7.6). However, the computation is merely symbolic because the magnitude of the strain rate is still unknown.

If one follows the one-dimensional model using the three-dimensional generalization of the yield stress or elastic limit concept, then:

$$\{\dot{\varepsilon}^{vp}\} = (1/\eta)(f-\kappa)\{\partial Y/\partial\sigma\} \tag{7.7}$$

which is essentially the one discussed by Prager (1961) and amplified in some detail by Perzyna (1966). In this form, the viscosity is the conventional one for fluid-like behavior, provided the function Y has the dimensions of stress, which it does when the rules of flow are associated. The actual form is normalized and made dimensionless (Perzyna, 1966):

$$\{\dot{\varepsilon}^{vp}\} = (\gamma)[(f / \kappa) - 1]\{\partial Y / \partial \sigma\} \tag{7.8}$$

where $\gamma^{o} = 1/\eta$ and $\gamma = \gamma^{o}/\kappa$ is called a "viscosity constant of the material," which must have dimensions of 1/sec (strain rate). Equation (7.8) is generalized further to allow for nonlinearity and other forms of the yield condition. Thus:

$$\{\dot{\varepsilon}^{vp}\} = \gamma \Phi(F)\{\partial f / \partial \sigma\} \tag{7.9}$$

where F is the dimensionless yield function in parentheses in Equation (7.8) and Φ is a function of F. A common choice is a power law form $(f/\kappa - 1)^n$ that is quite analogous to the one-dimensional possibility for nonlinearity mentioned previously.

Inversion of Equation (7.9) indicates the time-dependent nature of viscoplastic yielding. Thus:

$$[(f / \kappa) - 1] = \Phi^{-1}[(1/\gamma)(\sqrt{\{\dot{\varepsilon}^{vp}\}^{T}\{\dot{\varepsilon}^{vp}\}} / \sqrt{\{\partial f / \partial \sigma\}^{T}\{\partial f / \partial \sigma\}} \tag{7.10}$$

where the right side indicates the increase in dimensionless yield stress over the elastic limit κ. An instantaneous application of load that would generate an indefinitely large strain rate would also increase strength by an indefinitely large amount. Thus, in theory, instantaneous load application results in a purely elastic response. This logical extreme is reasonable because failure such as microcracking, macroscopic fracturing, or slip on joints cannot occur instantaneously. Of course, all real processes including the application of load take time; instantaneous loading can only be achieved analytically or numerically with the aid of the computer. If the load is subsequently held constant, then plastic deformation continues unabated. If the load is removed, then deformation continues until the stresses decay to the static yield condition (elastic limit). Although the form Equation (7.10) indicates something of the nature of viscoplasticity, the crux of the problem is calculating the viscoplastic strains that occur on the right side of Equation (7.10) or the left side of Equation (7.9), for example.

Work hardening and softening are expressed symbolically by including the inelastic strain in the yield function F. The inelastic strain is the viscoplastic strain when viscous effects are present. If the material neither strain hardens or softens, then it is elastic "perfectly" plastic. In rate-independent plasticity, under uniaxial load, a rising stress-strain curve beyond the yield point indicates strain hardening, while a falling curve indicates softening. In viscoplasticity, a rising or falling stress-strain curve may simply be transitory rate effects that leave the material unchanged, for example the dynamic strengthening or stress relaxation mentioned previously. This point seems to be overlooked by Nawrocki and Mroz (1998), who outline a viscoplastic model with provision for "degradation." Others have proposed similar models that seek to combine viscoplasticity and "damage" in recognition of the physical nature of time-dependent microcracking in laboratory test cylinders of intact rock. A great variety of one-dimensional rheologic models are available for generalization to three dimensions as pointed out by Singh (1989). For example, Fakhimi and Fairhurst (1994) form a series linkage of a spring, slider, and three-element assemblage of dashpots and a spring to produce a complicated elastic-plastic-viscous model with provision for plastic and viscous softening. The motivation is to estimate tunnel standup time. Whether such complexity is needed or logically consistent are open questions. Most nonlinear models, whether rate-dependent or not, can be made to follow trends observed in laboratory stress-strain curves. Time-independent damage concepts have led to a large class of models that more or less mimic the complete stress-strain curve (e.g., Singh *et al.*, 1987; Costin and Stone, 1987; Kemeny and Cook, 1987; Klisinski and Mroz, 1987, which are papers that appeared in a symposium on constitutive equations). Some form of macroscopic softening seems justified in consideration of microcrack damage and is probably needed to allow for instability. Moreover, the growth of microcracks is observed to take time, hence a time-dependent model is also needed for laboratory analysis, while field observations show that rock mass behavior is also noticeably time-dependent above the elastic limit. Again, the simplest consistent extension of the elastic model that accommodates time-dependent deformation is the elastic-viscoplastic (linear) model.

The total strain is the sum of the elastic and viscoplastic strains. Hence:

$$\{\dot{\varepsilon}\} = \{\dot{\varepsilon}^e\} + \{\dot{\varepsilon}^{vp}\} = [E]^{-1}\{\dot{\sigma}\} + \gamma\Phi(F)\{\partial f / \partial\sigma\}) \tag{7.11}$$

where $[]^{-1}$ means inverse. Equation (7.11) may be solved for the stress rate. Thus:

$$\{\dot{\sigma}\} = [E](\{\dot{\varepsilon}\} - \{\dot{\varepsilon}^{vp}\}) \tag{7.12}$$

which can be integrated immediately with respect to time and is the form most convenient for finite element analysis where displacements are the basic unknowns. The total strains can be found immediately from displacement derivatives, for example, $\varepsilon_{xx} = \partial u/\partial x$, $2\varepsilon_{xy} = \partial u/\partial y + \partial v/\partial x$ and so forth. Again, the crux of the problem is calculating the viscoplastic strains $\varepsilon_{ij}^{vp} = \int (d\varepsilon_{ij}^{vp}/dt)dt$ over some time interval Δt. By comparison, in conventional elastic-plastic (rate-independent) analysis, the stress-strain law has the simple form:

$$\{\dot{\sigma}\} = ([E] - [E^p])\{\dot{\varepsilon}\} \tag{7.13}$$

where the second term is a plastic correction to the elastic moduli. Thus in rate-independent elastic-plastic analysis, one can proceed directly from displacements to strains and then to stresses. Although rates are indicated in Equation (7.13), time does not enter the analysis, so the rates can be interpreted as small increments after multiplication by *dt*.

7.4 Finite element formulation

An incremental form of the finite element statement for equilibrium is:

$$\int_S [N]^T \{\Delta T\} dS + \int_V [N]^T \{\Delta\gamma\} dV = \int_V [B]^T \{\Delta\sigma\} dV \tag{7.14}$$

where $[N]$ is a node displacement to element displacement transformation matrix, $[B]$ is a node displacement to element strain transformation matrix, that is:

$$\{\Delta u\} = [N]\{\delta\} \text{ and } \{\Delta\varepsilon\} = [B]\{\delta\} \tag{7.15}$$

in a notation that is in widespread use. The superscript $()^T$ means transpose. With the understanding that:

$$\{\Delta\sigma\} = \int_{\Delta t}\{\dot{\sigma}\}dt \text{ and } \{\Delta\varepsilon\} = \int_{\Delta t}\{\dot{\varepsilon}\}dt \tag{7.16}$$

the equilibrium requirement becomes:

$$\{f_m\} = \int_V ([B]^T[E^e]\{\Delta\varepsilon\} - [B]^t[E]\{\varepsilon^{vp}\})dV = [K]\{\Delta\} - \{f_V\} \tag{7.17}$$

where $\{f_m\}$, $\{f_v\}$, and $\{\Delta\}$ are vectors of incremental mechanical force, viscous force, and displacement, respectively; $[K]$ is the elastic stiffness matrix. This simple formulation tacitly assumes small time steps and negligible change in the elastic and viscoplastic properties matrices during the short time step assumed ("piecewise constant"). This same assumption is often made in time-independent plasticity.

The two- and three-dimensional finite element codes UTAH2 and UTAH3 that were originally elastic-plastic codes were modified according to the preceding development to include an elastic-viscoplastic analysis option. The original UTAH2 code is in the public domain, while the modified version is not. UTAH3 is the property of the writer and is not in the public domain either in the original or modified forms. In all cases, elastic anisotropy and strength anisotropy can be independently specified up to orthotropic media that have three axes of material symmetry. This capability is useful in dealing with the realities of anisotropic rock masses formed by schists, gneisses, slates, shales, rocks that have flow structures, and so forth.

7.5 Example problems

A comparison of shapes of stress-strain curves obtained during compression tests of intact rock cylinders under confining pressure with analytic solution to the elastic-viscoplastic problem provides a qualitative test of the model. A close link to time-dependent microcracking may be anticipated at the macroscopic level in the appearance of volumetric expansion beyond the elastic limit. This expansion under generally compressive stress states is *dilatancy* and has been observed by many investigators as a departure from linearity with increasing axial load. With stiff testing machine equipment, the axial stress volumetric strain curve may be followed to a point of reversal of slope and even into the region where the total volumetric strain is expansive.

Figure 7.14 shows the general nature of such test data and the interpretation of processes during each stage of a laboratory compression test. A feature of particular interest is the departure from linearity in the volumetric stress-strain curve at about one-half of the peak axial load eventually reached. A second feature of importance is the turning point of the volumetric stress strain from positive to negative at about 85% of the peak axial stress. A final point is the loss of data very soon after the peak axial stress is reached. This loss occurs because of loss of testing machine control. With control, the post-peak behavior can be followed, as all strains increase rapidly as the axial stress falls.

For comparison, Figure 7.15 summarizes analytical and finite element simulations of a laboratory test of an elastic-viscoplastic cylinder under a constant confining pressure (Pariseau, 1998). The finite element results are in excellent agreement with the analytical solutions; each serves to check the other. The simulations proceeded in several stages. During the

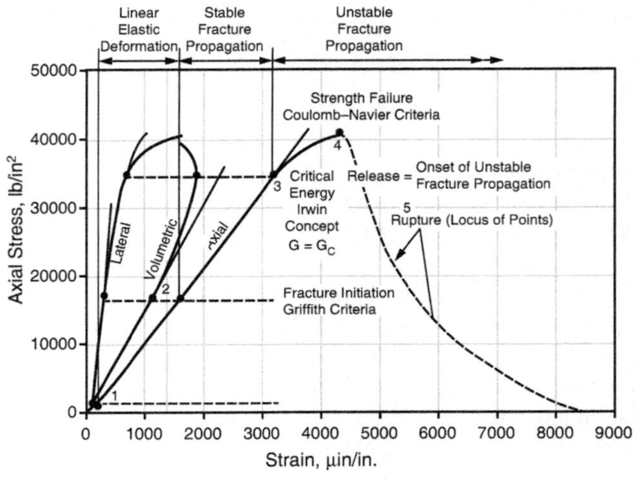

Figure 7.14 Axial stress versus axial strain, lateral strain, and volumetric strain during a laboratory test of a quartzite (Bieniawski, 1967).

first stage, axial stress and confining pressure were increased equally at the same rate until the test pressure of 10,000 psi was reached. The confining pressure was then held constant while the axial stress continued to be increased at the original rate to 60,000 psi, the peak stress shown in Figure 7.15. Comparison of Figure 7.15 with Figure 7.14 shows that the model produces the same shapes of stress-strain curves, including a turning point and reversal in the volumetric stress-strain curve and a rapid decrease in slope of the axial stress-strain curve. However, there is no early departure from linearity prior to the elastic limit at 50,000 psi. In Figure 7.15, ANL = analytical model, VOL = volumetric response, AXI = axial response, TRA = transverse response, FE2 = two-dimensional finite element analysis, and FE3 = three-dimensional finite element analysis.

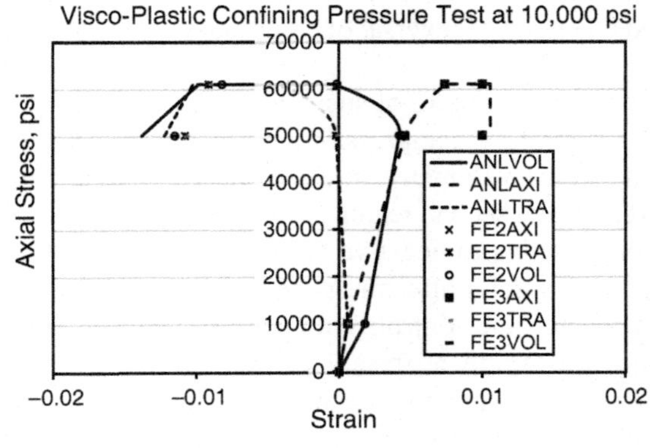

Figure 7.15 Analytical and finite element results of an elastic-viscoplastic simulation of a laboratory test of rock cylinder under axial compression and confining pressure (Pariseau, 1998).

Once the peak load is reached, the third stage of the simulation is entered by holding the load constant to observe creep strains. All strains will accumulate indefinitely as long as the axial load is held above the elastic limit. At present, there is no restriction in the model to prevent physically impossible strains from occurring, although some numerical instability would likely occur during any finite element analysis. In reality, as indicated in Figure 7.14, instability ensues after peak stress is reached, probably because of macroscopic fracturing as the microcrack network forms localized bands of connected crack segments.

The last stage of the simulations occurs when the relative displacement of the ends is fixed, so the axial strain rate instantaneously drops to zero. The axial stress then "relaxes" at fixed confining pressure. Growth in transverse and volumetric strains continues. As with the previous creep stage, this relaxation stage of the simulation has no counterpart for comparison with the experimental data in Figure 7.14.

The apparent agreement between the simple Bingham elastic-viscoplastic model extension to three dimensions and laboratory test data is encouraging. Some rather complicated models based on microstructural considerations show rather poor agreement. However, the objective here is to evaluate the viscoplastic model only. In this regard, while the action that occurs near peak axial stress appears well modeled, there is no time-dependent data available and only the shape of the stress-strain curves can be considered. This is perhaps just as well at this stage of investigation, because any attempt to simulate a specific set of test data would require the same data as input for either analytical or finite element solution. Thus, the attempt to reproduce laboratory test data would be somewhat circular. If such an attempt were unsuccessful, then a conclusion would be that the test specimen did not follow the model. A test specimen from a different source may be modeled quite well. The question that arises is whether there is a broad range of rock types and conditions in the field where the viscoplastic model is useful. Recall of the fundamental assumptions of an elastic range, an elastic limit, and time-dependent behavior above the elastic limit indicates an affirmative answer at least to the extent of a working hypothesis worth investigating with the practical objective of improved safety and stability.

The stresses, strains, and strain rates are uniform throughout the test specimen during the laboratory test under confining pressure. Although important features of the elastic-viscoplastic model are revealed in simulation of the test in comparison with actual test data, what happens under nonuniform stress is of considerable interest, too. The obvious reason is that nonuniform stress is the rule rather than the exception in reality. Lack of familiarity with the model suggests that a simple, well-known, but technologically important problem be explored first. Redistribution of stress about a circular excavation in an initially stressed but materially homogeneous rock mass is just such a problem. The excavation could be a mine shaft, bored raise, or tunnel driven by a boring machine. Figures 7.16 and 7.17 show the stresses and displacements obtained for five comparison runs.

Theory and finite element elastic analysis are in close agreement, so there is some reason to suppose that the other results for which there are no analytical solutions are also reliable. The elastic-viscoplastic results are of particular interest. Figure 7.16 shows the final stress distribution in all cases but indicates that the elastic-viscoplastic stress peak is less than the companion elastic-plastic peak. This result suggests that viscoplasticity is not merely delayed conventional plasticity. However, more study is needed before arriving at a definite conclusion. Figure 7.17 shows the displacements, which are radial in all cases because the basic problem is axially symmetric. The elastic-plastic analysis shows greater hole wall displacement ($r/a = 1$) than in the purely elastic case. The low strength elastic-plastic case shows

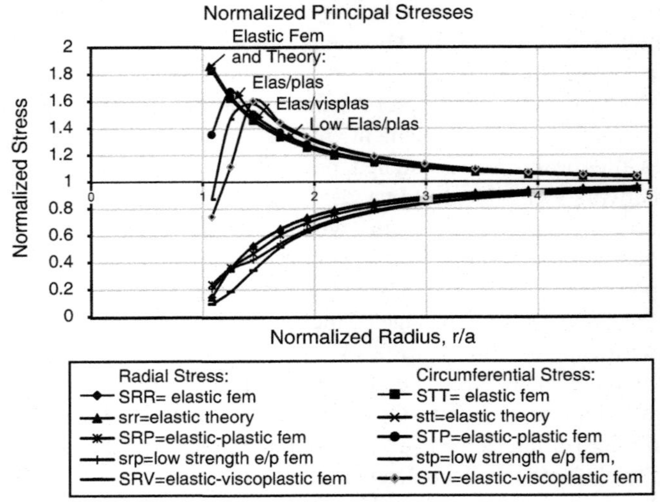

Figure 7.16 Stress analysis results for five cases (Pariseau, 1999).

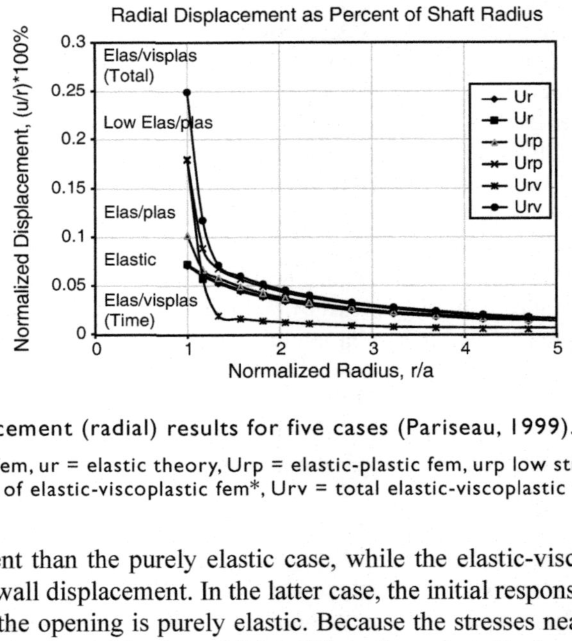

Figure 7.17 Displacement (radial) results for five cases (Pariseau, 1999).

Source: Ur = elastic fem, ur = elastic theory, Urp = elastic-plastic fem, urp low strength e/p fem, Urv = time-dependent part of elastic-viscoplastic fem*, Urv = total elastic-viscoplastic fem•.

greater displacement than the purely elastic case, while the elastic-viscoplastic shows the greatest hole total wall displacement. In the latter case, the initial response to the instantaneous excavation of the opening is purely elastic. Because the stresses near the hole wall are instantaneously above the yield point, additional time-dependent displacement follows excavation. In this case, the time-dependent displacement hole wall displacement is about twice the purely elastic displacement. The total displacement is the sum, as shown in Figure 7.17.

The axially symmetric circular hole problem (shaft/tunnel) is a step up in complexity from the axially symmetric solid cylinder problem (laboratory test) problem, which is more complicated than the one-dimensional block shear example problem and reveals additional features of the elastic-viscoplastic material model in comparison with the standard model of elastic behavior limited by strength.

The next step in complexity is to consider a more complicated excavation shape such as a rectangle that represents some mine shafts and most entries (and cross cuts) in room and pillar mines. An additional complexity is to consider pre-excavation stress states where the principal stresses are distinct. No analytical solutions are available for such problems, so one must necessarily rely on numerical methods and computer programs. Modifications to the UTAH2/3 finite element codes were made for this reason and for considering time-dependent deformation about mine openings where the elastic limit may be exceeded.

An example problem of practical importance concerns time-dependent deformation of the Ross Shaft at the Homestake Mine in Lead, SD. Ross Shaft stability was the subject of a cooperative research study by the Spokane Research Laboratory, the Homestake Mining Company, and the University of Utah. The influence of pillar mining on shaft stability is described in a number of publications and is summarized in Bureau of Mines publications (Johnson *et al.*, 1993; Pariseau *et al.*, 1995a, 1995b, 1996). The Ross Shaft is a rectangle about 15 ft by 21 ft (14 ft by 19-1/4 ft outside the buntons) and was placed in operation in 1934 (Bjorge *et al.*, 1935). A two-dimensional elastic-viscoplastic analysis of the shaft considered in isolation from other mining is a first step to addressing time-dependent aspects of shaft wall stability. The experience obtained in the case study at the Homestake Mine is important to mine safety throughout the industry where shafts are the primary accessways to and from a mine. Figure 7.18 shows the finite element mesh used in the example problem. The dimensions of the mesh are 10 times the dimensions of the shaft; the outer boundaries are thus sufficiently remote to have negligible effect on the response of the shaft walls.

Details of the region inside the first five "rings" of elements marked by the heavy line are shown in Figure 7.19. The preshaft stress field was calculated from equations fit to measurements made at the mine (Johnson *et al.*, 1993). The 3500 Level (3500 ft below surface) is in the vicinity of shaft pillar mining and was selected for this analysis. Orientation of the shaft with respect to compass coordinates and principal stress directions is also shown in Figure 7.19.

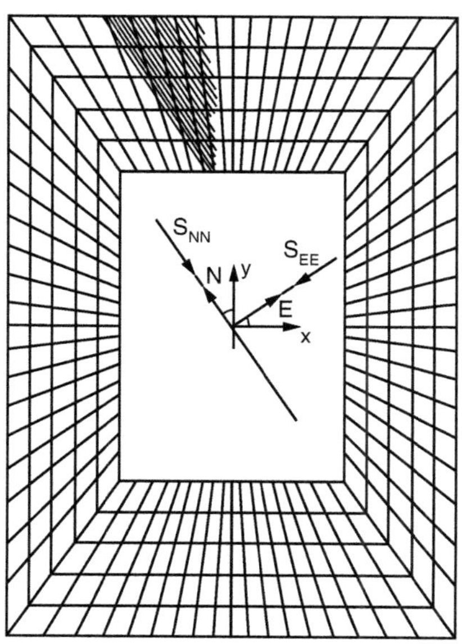

Figure 7.18 Finite element mesh used for a preliminary analysis of Ross Shaft time-dependent deformation.

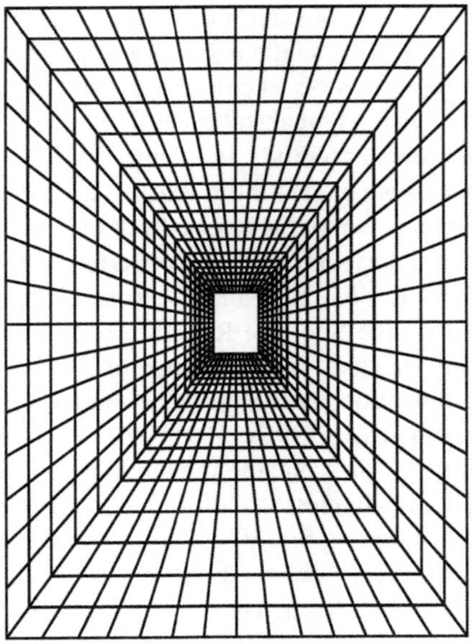

Figure 7.19 Mesh refinement near shaft walls (15 ft by 21 ft), principal stress directions (S_{EE}, S_{NN}) and foliation shown in hatched area.

Because the pre-excavation principal stresses are skewed relative to the shaft walls, the region is loaded in shear as well as compression. Symmetry with respect to the *x*- and *y*-axes shown in Figure 7.19 was thus not available to reduce the size of the mesh; the entire mesh shown in Figure 7.18 was used in the analysis. However, there is a rough, fortuitous symmetry of about the shaft diagonals.

The shaft was considered to be entirely in the Poorman Formation, which shows a pronounced anisotropy associated with foliation. Orientation of the foliation is shown by the hatched area in Figure 7.19. Material axes are parallel to the foliation, normal to the foliation, and parallel to vertical (into the page). The three material axes define an orthotropic rock mass with respect to elasticity and independently with respect to strengths. Nine independent elastic and nine independent strengths are required. Determination of these constants is described in the references cited. A reduction factor of 0.36 was applied to the laboratory rock moduli for application to the field-scale rock mass. The same reduction was applied to laboratory strengths. The strength anisotropy causes anisotropic viscoplastic behavior in regions that are stressed above the elastic limit.

Under these simplified conditions, instantaneous excavation of the shaft causes 28 elements to be stressed above the elastic limit. These elements are in opposite shaft corners that experience the greatest shear deformation. The location of these elements is shown in Figure 7.20.

After excavation, time-dependent deformation occurs as excess stress in the corner elements is relieved with additional post-excavation displacement about the shaft. The point of greatest time-dependent displacement is also shown in Figure 7.11. The rate of displacement (velocity) diminishes with time as shown in Figure 7.21 for the point of maximum (post-excavation) displacement. As the rate of displacement diminishes to a very low value, the time-dependent displacement amounts to about 15% of the displacement attendant excavation. This additional displacement relieves the initial excess stress. The combination of stress relief with additional strain is neither creep nor relaxation, but rather a combination associated with the nonuniform distribution of stress about the shaft.

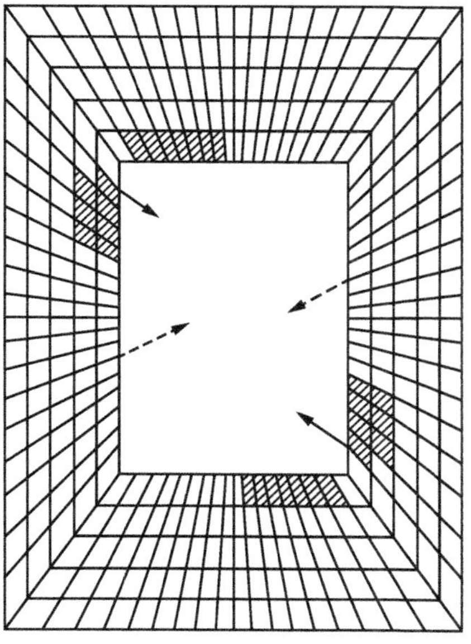

Figure 7.20 Shaft wall viscoplastic elements and maximum displacement. Dotted arrow indicates point and direction of maximum excavation displacement (instantaneous). Solid arrow indicates maximum post-excavation time-dependent displacement (Pariseau, 1999).

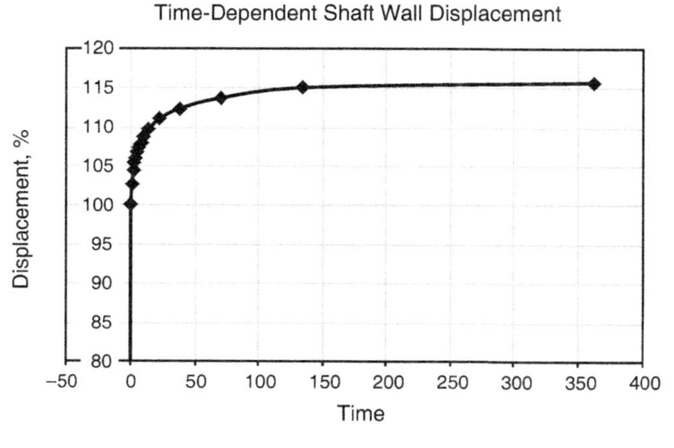

Figure 7.21 Time-dependent shaft wall displacement as a percent of initial displacement (Pariseau, 1999).

Another problem of practical importance where time-dependent deformation is evident concerned stability of a test stope at the Carr Fork Mine. The geological environment is that of contact metamorphic rock, especially in copper-bearing garnetite. Figure 7.3 shows clear evidence of time-dependent behavior.

The test stope was 12 m (36 ft) wide by 17.3 m (51.9 ft) in length and 47 m (141 ft) high from a flat bottom to an arched crown top. Some readings more than doubled between blasts, a period of several weeks. Retrospective finite element analyses were two- and three-dimensional as indicated in Figure 7.22. Center of the study stope was at a depth of 4228 ft (1288 m). *In situ* stress measurements were made to quantify the pre-stope stress state. Mining proceeded in six steps, beginning with excavation of development drifts, then proceeding to overcut and undercut drifts that were slashed to design width. Steps 3 through 6 were production blasts. No other mining was being done, so extensometer readings were entirely associated with test stope excavation. Instrumentation was installed after access was obtained with completion of development drifts. Some yielding of stope walls was inferred from loss of extensometer anchors as mining progressed.

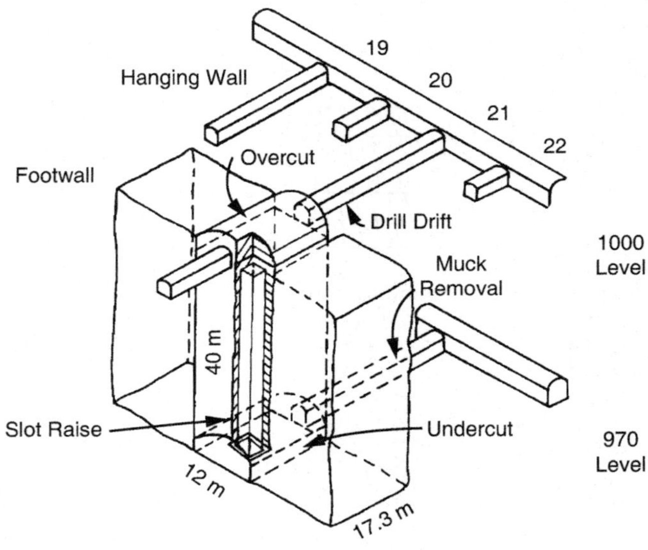

Figure 7.22 Geometry of the Carr Fork Mine study stope (Pariseau *et al.*, 1984).

Figure 7.23 shows correlations of short-term elastic extensometer readings and regression analyses of calculated on measured displacements. The regression points marked as Elab were obtained using rock properties obtained from laboratory test measurements. Scaling of readings to obtain a regression line slope (Escaled) leads to a slope of 1.0 and an elastic modulus scale factor. Such a scale factor is used to multiply elastic moduli, Young's modulus and shear modulus, obtained from laboratory tests on intact rock cylinders to obtain elastic moduli values representative of rock masses at an engineering scale in the mine where joints and similar structural features are present. The same applies to strength. Elastic and strength scale factors were 0.22 and 0.15, respectively. Scaling does not change the correlation coefficient, of course.

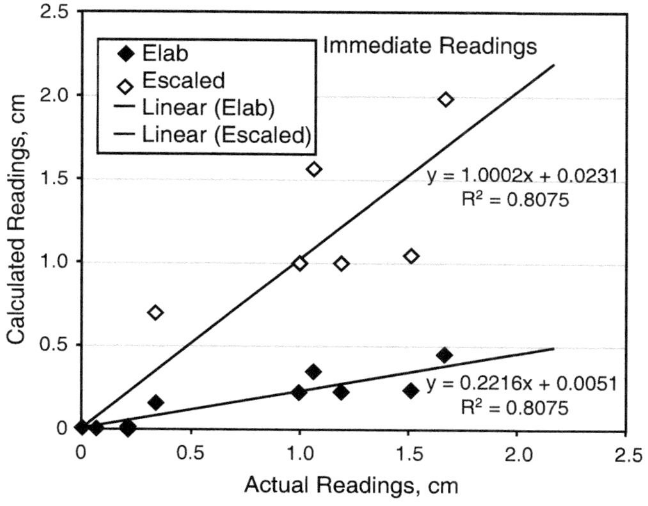

Figure 7.23 Regression of calculated two-dimensional finite element readings on actual extensometer readings in the elastic domain (Pariseau, 2007).

Later readings after noticeable time-dependent displacement occurred were modeled as elastic-viscoplastic in two- and three-dimensional finite element analyses. Figure 7.24 shows results from three-dimensional elastic-viscoplastic analyses. Scaled results lead to a regression line slope of 1.0 and an $R^2 = 0.6$ immediate (elastic). Later (viscoplastic) readings that were also synthesized from finite element results show a reduction in R^2 to 0.5. Elastic and strength scale factors were the same as before.

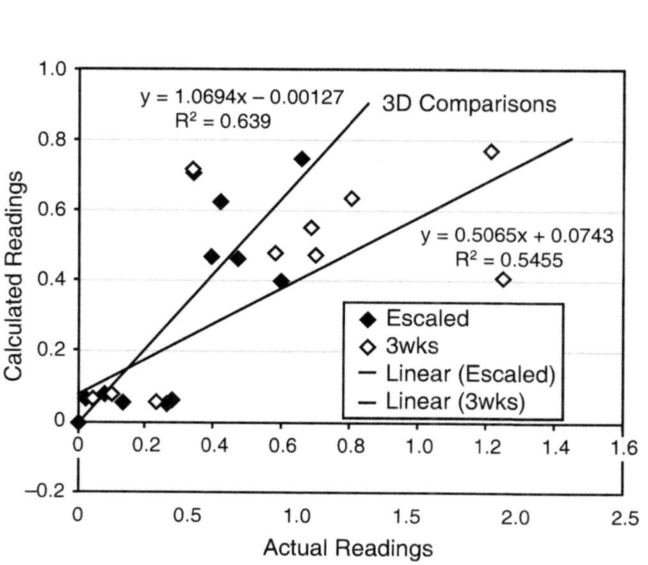

Figure 7.24 Actual readings compared with computed three-dimensional elastic-viscoplastic finite element readings (Pariseau, 2007).

The limited elastic-viscoplastic numerical experiments suggest that the viscosity parameter η and power law exponent n influence strain rate through the viscoplastic stress-strain relationship, but have little effect on total time-dependent strain and displacement. The main driving force of the time-dependent, viscoplastic strain is the excess stress above the elastic limit that occurs immediately after a blast but which diminishes in time as the associated stress returns to the static yield surface. These inferences are apparent in the viscoplastic stress-strain rate relationships. Equations (7.7), (7.8), and (7.9) and are confirmed by measurements made during the Carr Fork Mine study.

7.6 Discussion

What may be called the standard model in rock mechanics considers rock to respond elastically to an initial application of load, while the elastic range is limited by strength. A critical review of basic viscoplasticity was undertaken for the purpose of extending this *de facto* standard model to allow for time-dependent deformation that is so often observed in practice. The review began with an examination of physical evidence obtained during field studies of underground and surface mines and proceeded to laboratory test data. Although by no means exhaustive, the physical evidence clearly shows significant time-dependent deformation in rock masses and in laboratory study of intact rock test cylinders. There is a dearth of laboratory and field evidence for time-dependent behavior of rock joints, although the importance of such measurements is recognized in the laboratory (e.g., Howing and Kutter, 1985; Olofsson, 1985) and in application to mine stability (Napier and Malan, 1997). Development of theory followed first according to the historical one-dimensional Bingham plastic model and subsequently in three dimensions, where a distinction between the viscoplastic potential and yield condition is possible. A finite element formulation was outlined that had been implemented earlier in the UTAH2 and UTAH3 computer codes.

A problem of surface slope creep amenable to analytical solution illustrated some of the consequences of viscoplasticity in one-dimensional, homogenous stress states. One important consequence was the possibility of accumulating indefinitely large strains under a constant strain rate and thus the conclusion that some physical event not included in the model must eventually intervene to terminate the process.

Several additional problems of increasing complexity were also examined. These example problems began by considering uniform stress in the laboratory test for compressive strength under confining pressure. The next problem considered the circular hole (shaft/bored raise/tunnel) under nonuniform but axially symmetric stress and then a problem under nonuniform stress in anisotropic rock. The laboratory test problem was solved analytically and numerically using the UTAH2 and UTAH3 finite element codes. The results show creep and relaxation phenomena mimicked the shapes of laboratory stress-strain curves. In particular, the turning point on the axial stress volumetric strain curve that is associated with dilatancy was reproduced by the elastic-viscoplastic model. The analytical and finite element solutions served as mutual checks on code validity and analytical integration. Several solutions to the circular hole problem were obtained for comparison. These solutions included elastic, elastic-plastic, and elastic-viscoplastic analyses. Hole wall displacement was greatest in the viscoplastic analysis and, in part, time-dependent, as is often the case in practice.

The next test problem concerned the Ross Shaft at the Homestake Mine. A simplified two-dimensional analysis was done. Although simplified with respect to actual three-dimensional mine geometry and geology, the analysis accommodated the rectangular shaft geometry, the rock mass elastic and strength anisotropy, and the combined compression and shear loading of the shaft caused by skewness of the pre-mining principal stresses with respect to the shaft walls. The elastic-viscoplastic model showed that maximum displacement attendant on excavation occurred near the mid-side of the long dimension of the shaft cross-section. Maximum time-dependent deformation following excavation occurred in opposite corners of the shaft where shear loading is most intense. The time-dependent shaft wall displacement was about 15% of the initial excavation displacement. The shape of the displacement-time curve was characterized by a sharp rise followed by a rounded run that seems typical of field data, for example, that obtained at the underground Carr Fork Mine and from the open pit Bingham Canyon Mine.

The last test problem was a retrospective analysis of the test stope at the Carr Fork Mine. Elastic and viscoplastic finite element analyses in two and three dimensions allowed for synthesis of multiple position borehole extensometer readings. These computed readings were directly compared with readings taken immediately after a blast and readings taken several weeks later. The immediate readings indicated an elastic response. Finite element results compared favorably in regression analyses where R^2 values of 0.8 were obtained in the elastic domain. Synthesized readings in the viscoplastic domain were less favorable with R^2 values of 0.6.

The simple elastic-viscoplastic formulation examined thus has the capability to reproduce shapes of laboratory stress-strain curves and displacement-time curves obtained in the field. However, a field-scale study at the underground Carr Fork Mine indicated elastic strains are useful for estimating moduli scale factor, as were strength scale factors. Viscoplastic strains were not as favorably estimated using regression analyses of calculated on measured data. The indication is other time-dependent strain models are worthy of study. One candidate is a delayed elasticity model, which does not admit an indefinitely large creep strain but does produce rounded stair steps (with finite strength to limit the range of elasticity), as so often observed in practice.

References

Arguello, J.G., Molecke, M.A. & Beraun, R. (1989) 3D Thermal stress analysis of WIPP room TRH TRU experiments. *Proceedings 30th U.S. Symposium on Rock Mechanics*. Balkema, Rotterdam, pp 681–688.

Bingham, E.C. (1922) *Fluidity and Plasticity*. McGraw-Hill, New York, p.217.

Bieniawski, Z.T. (1967) Mechanism of brittle fracture of rock part I-theory of the fracture process. *International Journal Rock Mechanics and Mining Science*, 4(4), pp 395–406.

Bieniawski, Z.T. (1970) Time-dependent behavior of fractured Rock. *Rock Mechanics*, 2, 123–147.

Bjorge, G.N., Ross, A.J.M., Johnson, J.D., Staple, S.J. & Wiggert, J.F. (1935) *Construction and Equipment of the Ross Shaft, Homestake Mining Company*. Technical Publication No. 621. AIME, New York, p.43.

Brady, B.T., Duvall, W.I. & Horino, F.G. (1973) An experimental determination of the true uniaxial stress-strain behavior of brittle rock. *Rock Mechanics*, 5(2), 107–120.

Chen, B.C. (1969) Stability of slopes undergoing creep deformation. *Journal of Soil Mechanics Foundations Div*. ASCE (SM4), 1075–1096.

Costin, L.S. & C.M. Stone (1987) Implementation of a finite element damage model for rock. In: *Constitutive Laws for Engineering Materials: Theory and Applications*. Elsevier, New York, pp 829–840.

Cristescu, N.D. & Hunsche, U. (1998) *Time Effects in Rock Mechanics*. Wiley, New York, pp 342.

DiMaggio, R.L. & Sandler, I. (1971) *The Effect of Strain Rate on the Constitutive Equations of Rocks*. Report by Paul Weidlinger, Consulting Engineer, to Headquarters – Defense Nuclear Agency, Washington, DC, Contract No. DNA001–72-C-003.

Fakhimi, A.A. & Fairhurst, C. (1994) A model for the time-dependent behavior of Rock. *International Journal Rock Mechanics Mining Science & Geomechanics Abstracts*, 31(2), 177–126.

Heard, H.C. (1963) Effect of large changes in strain rate in the experimental deformation of yule marble. *Journal of Geology*, 71(2), 162–196.

Howing, K-D. & Kutter, H.K. (1985) Time-dependent shear deformation of filled Rock joints. In: *Proceedings International Symposium Fundamentals of Rock Joints*. Centek Publishers, Lulea, Sweden, pp.113–122.

Hudson, J.A. & Brown, E.T. (1972) Studying time-dependent effects in failed Rock. *Proceedings 14th U.S. Symposium Rock Mech*. ASCE, New York, pp.25–34.

Isenberg, J. & Wong, F.S. (1971) *A Viscoplastic Model with a Discussion of its Stress/Strain and Wave Propagation Properties*. Report by Agababian-Jacobsen Associates to the Defense Atomic Support Agency, Washington, DC, Contract No. DASA 01-70-6-0065.

Johnson, J.C., Pariseau, W.G., Scott, D.F. & Jenkins, F.M. (1993) In Situ Stress Measurements Near the Ross Shaft Pillar, Homestake Mine, South Dakota. U.S. Bureau of Mines Report of Investigations 9446, p.17.

Kachanov, L.M. (1971) *Foundations of the Theory of Plasticity*. American Elsevier, Inc., New York.

Kemeny, J.M. & Cook, N.G.W. (1987) Crack models for the failure of rocks in compression. In: *Constitutive Laws for Engineering Materials: Theory and Applications*. Elsevier, New York, pp.879–887.

Klisinski, M. & Mroz, Z. (1987) Description of inelastic deformation and degradation of concrete. In: *Constitutive Laws for Engineering Materials: Theory and Applications*. Elsevier, New York, pp.889–896.

Lipkin, J., Grady, D.E. & Campbell, J.D. (1977) Dynamic flow and fracture of Rock in pure shear. *Proceedings 18th U.S. Symposium Rock Mechanics*. Johnson Pub. Co., Boulder, pp.3B2-1 3B2-7.

Logan, J.M. & Handin, J. (1971) Triaxial compression testing at intermediate strain rates. *Proceedings 12th U.S. Symposium Rock Mechanics*. SME/AIME, New York, pp.167–194.

Malvern, L.E. (1969) *Introduction to the Mechanics of a Continuous Medium*. Prentice-Hall, Englewood Cliffs, NJ.

Merrill, R.H. (1957) Roof span studies in limestone. *U.S. Bureau of Mines Report of Investigations 5346*, p.38.

Munson, D.E., Fossum, A.F. & Senseny, P.E. (1989) Approach to first principles model prediction of measured WIPP in situ room closure in salt. *Proceedings 30th U.S. Symposium Rock Mechanics*. Balkema, Rotterdam, pp.673–680.

Naghdi, P.M. & Murch, S.A. (1963) On the mechanical behavior of viscoelastic/plastic solids. *Journal of Applied Mechanics*, 30, 321–328.

Napier, J.A.L. & Malan, D.F. (1997) A viscoplastic discontinuum model of time-dependent fracture and seismicity effects in brittle Rock. *International Journal Rock Mechanics and Mining Science & Geomechanics Abstracts*, 34(7), 1075–1089.

Nawrocki, P.A. & Mroz, Z. (1998) A viscoplastic degradation model for Rocks. *International Journal Rock Mechanics and Mining Science & Geomechanics Abstracts*, 35(7), 991–1000.

Olofsson, T. (1985) A Non-linear model for the mechanical behavior of continuous Rock joints. In: *Proceedings International Symposium Fundamentals of Rock Joints*. Centek Publishers, Lulea, Sweden, pp.395–404.

Pariseau, W.G. (1972) Plasticity theory for anisotropic rocks and soils. *Proceedings 10th U.S. Symposium Rock Mechanics*. SME/AIME, New York, pp.267–295.

Pariseau, W.G. (1998) Modification of UTAH2 and UTAH3 finite element codes for viscoplastic failure criteria. Final Report. Spokane Research Laboratory/NIOSH.

Pariseau, W.G. (1999) Some consequences of viscoplastic approach to time-dependent behavior. *Proceedings 37th U.S. Symposium Rock Mechanics*, Volume 2. Balkema, Rotterdam, pp.927–934.

Pariseau, W.G. (2007) Time-dependent Rock mass behavior: Car Fork mine test Stope. *Proceedings 1st Canada-U.S. Symposium on Rock Mechanics*. Taylor & Francis, London, pp.1475–1481.

Pariseau, W.G., Fowler, M.E., Johnson, J.C., Poad, M. & Corp, E.L. (1984) Geomechanics of the car fork mine test stope. In: *Geomechanics Applications in Underground Hardrock Mining*. SME/AIME, New York, pp.3–38.

Pariseau, W.G., Johnson, J.C., McDonald, M.M. & Poad, M.E. (1995a) Rock Mechanics Study of Shaft Stability and Pillar Mining, Homestake Mine, Lead, SD. Part 1. Premining Geomechanical Modeling Using UTAH2. U.S. Bureau of Mines Report of Investigations 9531, p.20.

Pariseau, W.G., Johnson, J.C., McDonald, M.M. & Poad, M.E. (1995b) Rock Mechanics Study of Shaft Stability and Pillar Mining, Homestake Mine, Lead, SD. Part 2. Mine Measurements and Confirmation of Premining Results. U.S. Bureau of Mines Report of Investigations 9576, p.13.

Pariseau, W.G., Johnson, J.C., McDonald, M.M. & Poad, M.E. (1996) Rock Mechanics Study of Shaft Stability and Pillar Mining, Homestake Mine, Lead, SD. Part 3. Geomechanical Monitoring and Modeling Using UTAH3. U.S. Bureau of Mines Report of Investigations 9618, p 29.

Pariseau, W.G. & Voight, B. (1979) Rockslides and avalanches: Basic principles and perspectives in the realm of civil and mining operations. In: *Rockslides and Avalanches*, Volume 2. Elsevier, New York, pp.1–92.

Patton, F.D. (1966) *Multiple Modes of Shear Failure in Rock and Related Materials*. PhD Dissertation, University of Illinois, p.282.

Peng, S.S. (1973) Time-dependent aspects of rock behavior as measured by a Servocontrolled Hydraulic testing machine. *International Journal Rock Mechanics and Mining Science & Geomechanics Abstracts*, 10(3), 235–246.

Peng, S. & Podnieks, E.R. (1972) Relaxation and the behavior of failed Rock *International Journal Rock Mechanics and Mining Science & Geomechanics Abstracts*, 9(6), 699–712.

Perzyna, P. (1963) The constitutive equations for rate sensitive plastic materials. *Quarterly Applied Mathematics*, 20(4), 321–332.

Perzyna, P. (1966) Fundamental problems in viscoplasticity. In: *Advances in Applied Mechanics*, Volume 9, Academic Press, Cambridge, MA, pp.243–277.

Prager, W. (1961) *Introduction to Mechanics of Continua*. Dover, Inc., New York.

Price, N.J. (1966) *Fault and Joint Development in Brittle and Semi-Brittle Rock*. Pergamon, London, p.176.

Rice, J.R. (1975) Continuum mechanics and thermodynamics of plasticity in relation to microscale deformation mechanisms. In: *Constitutive Equations in Plasticity*. MIT Press, Cambridge, pp.23–75.

Rummel, F. & Fairhurst, C. (1970) Determination of the post-failure behavior of brittle Rock using a servo-controlled testing machine. *Rock Mechanics*, 2, 189–204.

Serdengecti, S. & Boozer, G.D. (1961) The effects of strain rate and temperature on the behavior of rocks subjected to triaxial compression. In: *Proceedings 4th U.S. Symposium Rock Mechanics*. Mineral Industries Experiment Station Bulletin 76, Pennsylvania State University, University Park, PA pp 83–103.

Scholz, C.H. (1968) Microfracturing and the inelastic deformation of Rock in compression. *Journal of Geophysical Research*, 73(4), 1417–1432.

Singh, M.M. (1989) Strength of Rock. In: *Physical Properties of Rocks and Minerals*, Volume. II-2, Hemisphere Pub. Corp., New York, pp.83–121.

Singh, U.K., Digby, P.J. & Stephansson, O.J. (1987) Constitutive equation for progressive failure of brittle Rock. In: *Constitutive Laws for Engineering Materials: Theory and Applications*. Elsevier, New York, pp.923–930.

Terzaghi, K. (1962) Stability of steep slopes on hard unweathered Rock. *Geotechnique*, 12(4), 251–270.

Wawersik, W.R. & Brown, W.S. (1971) *Creep Fracture in Rock in Uniaxial Compression. Department of Mechanical Engineering*. University of Utah, Salt Lake City, UT.

Zavodni, Z.M. & Broadbent, C.D. (1980) Slope failure kinematics. *CIM Bulletin*, 73(816), 60–65.

Chapter 8

Poroplasticity

Poroplasticity begins with poroelasticity, which deals with reversible deformation of a fluid infiltrated porous, fractured material that allows for fluid flow through connected voids. Poroplasticity implies deformation beyond the elastic limit as in conventional plasticity theory.

There is an extensive literature concerning poroelasticity and poroplasticity, especially in the realm of soil mechanics where the concept of *effective stress* originated. The many books (Bear and Verruijt, 1987; Istok, 1989; Barenblatt *et al*, 1990; Nikolaevskij, 1990; Charles, 1991, 1997; Coussy, 1995; Lewis and Schrefler, 1998) and symposia proceedings (e.g., Rock Fractures and Fluid Flow, 1996; Poro-mechanics: A Tribute to Maurice A. Biot, 1998) attest to the maturity of the subject and to the continuing development of numerical methods for solving practical engineering problems. Coupled fluid flow and solid deformation in rock mechanics came after much study in soil mechanics, but even so, the literature is still large and mainly follows soil mechanics formulations where Hooke's law governs solid deformation and Darcy's law governs fluid flow. However, there is a major difference between soil and rock, and that is in the nature of connected void space. In soil mechanics, the connected void space consists mainly of pores between solid particles. In rock mechanics, the connected voids may also be pores but are most likely to be cracks, fractures, and joints. Connectivity of pores, cracks, fractures, and joints necessary for fluid flow at various scales poses a major challenge to characterization of rock masses with respect to fluid flow and gives rise to *equivalent* properties. Problem solving is also challenging and, in all but the simplest cases, requires numerical methods. The popular finite element method is especially suited for solving engineering design problems involving poromechanics. Examples aid in understanding the fundamentals outlined here.

Several basic weight and volume relationships are essential to the description of a porous solid, which is illustrated schematically in Figure 8.1. Two volume relationships of importance are porosity n and void ratio e. By definition, $n = Vv/V$ and $e = Vv/Vs$ where $V = Vs + Vv$. Saturation S is defined by $S = Vw/Vv$ where Vw is volume of water or other liquid in the porous solid and Va is volume of air or other gas. Saturation is often described as a percent. Moisture content mc is given in terms of weight: $mc = Ww/Ws$. Specific weights dry and with water present, γ_d and γ, respectively are γ_d = Ws/V and $\gamma = W/V$. Specific gravity of solids is usually designated as G and is the ratio $G = \gamma_s/\gamma_w$ where γ_w is the specific weight of water. Specific weight is sometimes referred to as weight density in distinction from mass density ρ. The two are related by $\gamma = \rho g$ where g is the acceleration of gravity. Many other relationships follow from definitions, for example, $n = e/(1+e)$ and $e = n/(1-n)$.

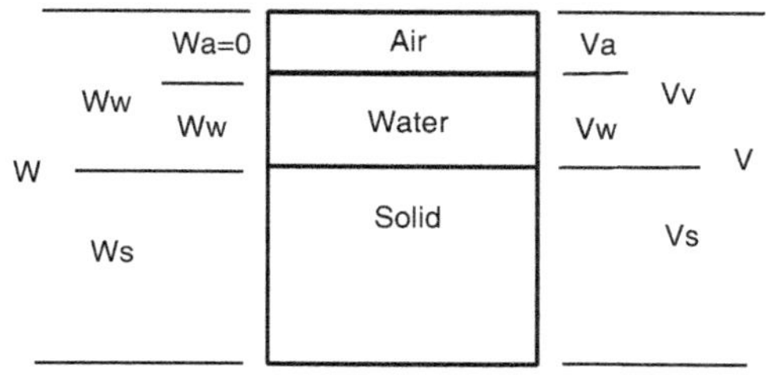

Figure 8.1 Schematic illustration of a porous, fractured solid with connected voids filled with liquid and gas.

8.1 Effective stress

The concept of *effective stress* is central to poroelastic/plastic theory. Coupling between fluid and solid may be two-way. If so, the fluid flow and pressure in the connected void spaces of the material affect solid stress and deformation that in turn affect the fluid response. Effective stress is the force per unit area transmitted through the solid skeleton of a material containing connected pores and cracks that allow for fluid flow. The concept is vital to understanding the mechanics of soils and is largely attributed to Terzaghi in texts on soil mechanics (Terzaghi and Peck, 1948; Taylor, 1948; Hough, 1957). The concept is illustrated in Figure 8.2 where the total force N acts normal to a plane of interest with area A. The total force N is composed of a force N' transmitted through the solid contacts and a force P transmitted through the fluid. The contacts between solid grains are point contacts and thus have zero area; the fluid occupies area A. Forces per unit area are:

$$\begin{aligned} N/A &= (N'+P)/A \\ \sigma &= \sigma' + p \end{aligned} \tag{8.1}$$

Thus, the total stress σ is the sum of the effective stress σ' and the pore fluid pressure p (compression is positive). Both are absolute measures, although pressure p is often relative to atmospheric pressure and referred to as *excess pore pressure*. In a water-saturated column of sand h (ft, m) high where γ_w is the specific weight of water, fluid pressure at the column bottom is given by $p = \gamma_w h$. Total stress at the column bottom is given by $\sigma = \gamma h$ where γ is specific weight of the solid and water combined.

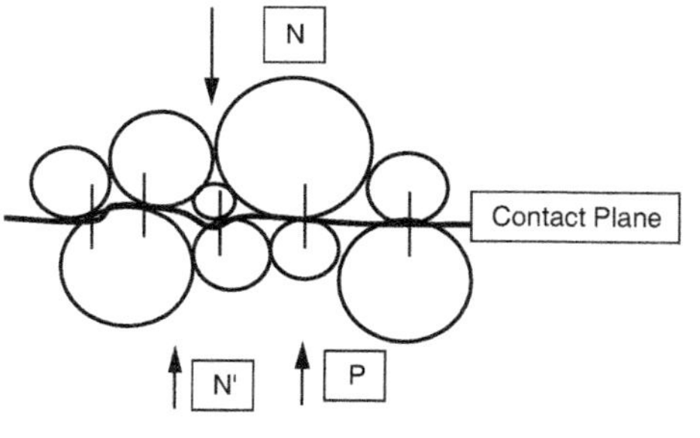

Figure 8.2 Schematic illustration of forces acting across a plane in a porous material.

Effective stress allows for mobilization of frictional resistance to shear along a contact plane; the pore fluid offers no shear resistance. Total stress enters consideration of equilibrium; effective stress enters consideration of strength. If the solid contacts are cemented, then cohesion c between grains is present and strength has a cohesive component as well as a frictional component. The frictional contribution to shear strength is not sliding friction but rather is *internal friction*, which is characterized by an angle of internal friction ϕ. Strength in shear may then be expressed as a sum $\tau = \sigma' \tan(\phi) + c$ which is easily recognized as the famous Mohr-Coulomb criterion. Of course, other criteria are also possible. While the concept of effective stress in Figure 8.2 is certainly imperfect, the key point is the recognition of the importance of effective stress in determining strength of porous, fractured rock and soil.

With reference to Figure 8.3, which shows a plane through a section of a porous media that cuts across grains and fluid-filled voids unlike the pseudo-plane in Figure 8.2, which threads through contacts between grains. The total area of the triangle is composed of an area of solid grains with area A_S and a void space with area A_V. A force N acting normal to the area A is the sum of forces N_S and N_V acting normal to the solid and fluid-filled void areas. Equilibrium requires:

$$
\begin{aligned}
N &= N_S + N_F, \text{ that is,} \\
\sigma A &= \sigma_S A_S + \sigma_F A_V, \text{ or}
\end{aligned}
\tag{8.2a}
$$

$$
\begin{aligned}
\sigma &= \sigma_S (A_S / A) + \sigma_F (A_V / A) \\
&= (\sigma_S A_S) / A + (\sigma_F A_V) / A \\
\sigma &= \sigma'_S + \sigma'_F
\end{aligned}
\tag{8.2b}
$$

There are other forms of Equation (8.2a,b) that are of interest:

$$
\begin{aligned}
\sigma &= \sigma_S f_S + \sigma_F f_V \\
\sigma &= \sigma_S (1 - n_A) + \sigma_F n_A \\
\sigma &= \sigma'_S + \sigma'_F
\end{aligned}
\tag{8.2c}
$$

where f_S and f_V are area fractions of the total area and n_A is an area porosity ($n_A = A_V / A$). These parameters are often taken to be volume fractions and volume porosity. The last of Equation (8.2) shows all stresses computed as forces per unit of *total area* and that the total stress is the sum of the effective stresses σ'_S and σ'_F transmitted through the solid grains and fluid-filled voids.

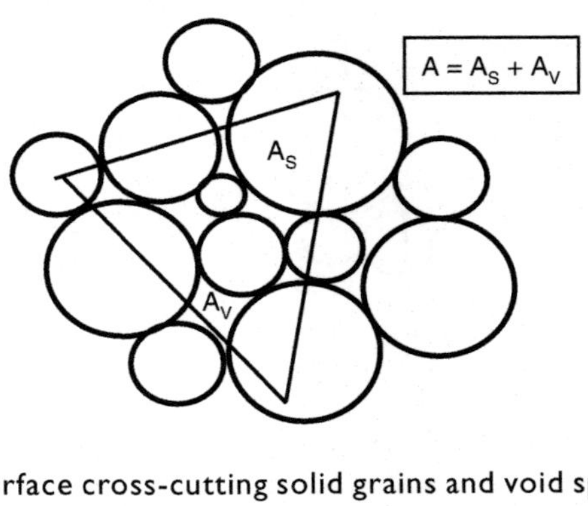

Figure 8.3 A triangular surface cross-cutting solid grains and void space in a porous material.

A similar analysis of a shearing force could also be done, but in consideration of the lack of shearing resistance by the fluid in the void space, the total shearing resistance is simply the resistance provided by the solid. Thus, the effective (solid) shear stress is the total shear stress, and the effective fluid shear stress is zero. A tacit assumption in this view is the fluid lacks viscosity.

If the triangle in Figure 8.3 were a coordinate plane and a force acted at an angle to the plane, then components of that force per unit area would be (total) stress components, say, $\sigma_{xx}\ \tau_{xy}\ \tau_{xz}$ in the usual notation where the yz-plane is the coordinate plane with a normal direction parallel to the x-axis. The total stress is a second-order tensor that follows the usual axis rotation formulas. Interestingly, the effective stresses also are second-order tensors when they are computed per unit total area. In some analyses, stresses transmitted through solid and fluid are computed per unit of solid and per unit of fluid area, respectively. These *partial* stresses are not tensors.

Biot (1941) is credited with placing the concept of effective stress on a firm continuum mechanics foundation. In this formulation, the concept was expressed as $\sigma = \sigma' + \alpha p$ where σ, σ' and p are total stress, effective stress transmitted through the solid skeleton, and fluid pressure, respectively. The coefficient α is an elastic property of the porous material (Biot's coefficient) about which much has been written. Biot's 1941 formulation was subsequently improved by computing stresses per unit total area (Biot, 1955, 1956), so $\sigma = \sigma' + \sigma''$ where σ, σ' and σ'' are total stress, effective (solid) stress, and "fluid" stress, all computed per unit total area. The fluid stress is related to fluid pressure by $\sigma'' = n_A p$ where $n_A = A_V / A$ is area porosity. Thus, $\sigma'' A = A_V p$. This usage eliminates the need for a "coefficient" and the inconsistency which occurs when the total and effective solid stresses are computed per unit total area while the fluid pressure is not. Use of an *effective* fluid stress allows for consistency and treatment as tensors. The system of equations describing consolidation also describes other important poromechanics problems such as slope stability in civil and mining engineering, shaft and tunnel stability in wet ground, *stope* stability in underground hardrock mine excavations using hydraulic fills, and gas outbursts in softrock underground mines.

8.2 Poroelastic/plastic models

The material model involves deformation of the composite of solid and fluid together and movement of the fluid relative to the solid. In the "dry" case, the composite model is simply Hooke's law because total and effective stresses are the same. Thus:

$$\{\sigma\} = [C]\{\varepsilon\} \tag{8.3}$$

where $\{\sigma\}$, $[C]$, and $\{\varepsilon\}$ are stress, elastic stiffness, and strain matrices and tension is positive.

In the "wet" case with the porous solid at saturation at fluid pressure π, the total stress is:

$$\{\sigma\} = [C]\{\varepsilon\} + \{c\}(\pi) \tag{8.4}$$

where $\{\sigma\}$ is now total stress and $\{c\}$ is a column matrix of pore pressure coefficients in the notation of Sandhu and Wilson (1969) and Ghaboussi and Wilson (1973). The first term on the right-hand side of Equation (8.4) is the effective solid stress; the second term is the effective fluid stress. Both are referred to the dimensions of the continuum. Because of pore fluid flow, often water, Darcy's law usually applies. Darcy's law was advanced in his study of fountains in a city in France (Darcy, 1856). Thus:

$$\begin{Bmatrix} \dot{w}_x \\ \dot{w}_y \\ \dot{w}_z \end{Bmatrix} = -[k](\begin{Bmatrix} \partial\pi/\partial x \\ \partial\pi/\partial y \\ \partial\pi/\partial z \end{Bmatrix} - \begin{Bmatrix} \gamma_x \\ \gamma_y \\ \gamma_z \end{Bmatrix}) \tag{8.5a}$$

or in more compact form:

$$\{\dot{w}\} = -[k]\{h\} \tag{8.5b}$$

where the negative sign indicates flow against the hydraulic gradient, $\{w\} = \{U\} - \{u\}$ is displacement of the fluid relative to the solid, $\{U\}$ is the displacement of the fluid, $\{u\}$ is the displacement of the solid, the superior dot indicates time differentiation $\partial(\,)/\partial t$, $[k]$ is a matrix of hydraulic conductivities, π is effective fluid stress (pressure), γ is specific weight of the fluid flowing, and $\{h\}$ is the gradient of the hydraulic potential.

The velocities are "nominal" velocities such that components of flow volume of fluid relative to the solid are $\{q\} = \{w\}A$ where $\{q\}$ is a vector of flow rate components. The discharge of fluid is the component of flow normal to the area of interest is the scalar product, $q = \{q\} \bullet \{n\}$ where $\{n\}$ is a unit vector normal to the area considered. Actual flow velocity is given by $v = w/n_A$ where n_A is area porosity and would be used when calculating fluid transport times. As a reminder, the actual fluid pressure p is related to the effective fluid stress (pressure) reckoned per total area in a similar manner: $p = \pi / n_A$.

A change in fluid volume per unit of volume relative to the solid is $\zeta = (\partial w_x/\partial x + \partial w_y/\partial y + \partial w_z/\partial z)$ and which in consideration of $\{w\} = \{U\} - \{u\}$ is given by:

$$-\zeta = (\{c\}^t\{\varepsilon\} - (c)\pi) \tag{8.6}$$

where (c) is a compressibility coefficient of the fluid, the parentheses indicate a scalar, the negative sign indicates contraction, and a superior t indicates transpose.

The overall poroelastic material model may be expressed as:

$$\begin{pmatrix} \{\sigma\} \\ -\zeta \end{pmatrix} = \begin{bmatrix} [C] & \{c\} \\ \{c\}^t & (c) \end{bmatrix} \begin{pmatrix} \{\varepsilon\} \\ \pi \end{pmatrix} \tag{8.7}$$

where the 7×7 matrix in square brackets [ ] is symmetric. Other forms are possible and when considered lead to distinctions between "drained" and "undrained" properties. When $\pi = 0$, the process is "drained"; when $\zeta = 0$, the process is "undrained." Both may be used in laboratory determination of material properties. Details concerning poroelastic properties are discussed by Detournay and Cheng (1993) and concisely by Rice (1998).

8.3 Finite element model

As always, the governing equations follow from physical laws, kinematics, and material laws. Application of the divergence theorem in the forms of the principle of virtual work and power lead to a finite element model of poromechanics. Thus:

$$\begin{aligned} \{F_u\} &= [K_{uu}]\{U\} + [K_{u\pi}]\{\pi\} \\ \{F_\pi\} &= [K_{\pi u}]\{\dot{U}\} - [K'_{\pi\pi}]\{\dot{\pi}\} - [K_{\pi\pi}]\{\pi\} \end{aligned} \tag{8.8a,b}$$

where the subscripts u and π refer to porous solid and fluid, respectively, a superior dot indicates time derivative, and a prime is used to distinguish the two fluid stiffnesses in Equation (8.8b). Curly brackets are column matrices. Matrices $\{U\}$ and $\{\pi\}$ are $3n \times 1$ and $n \times 1$ column matrices of node displacements and node pressure (effective fluid stress) respectively, with dimensions $3n$ and n (n is equal to the number of nodes in the finite element mesh). The variables $\{U\}$ and $\{\pi\}$ are the unknowns that are sought.

Time integration is required during the solution process. In case of plasticity, the system is nonlinear so incremental loading is required during the solution process. Although separate load and time increments may be used, using the same increment for both simplifies the solution procedure. Large systems of equations require an iterative solution technique; small systems may be solved by elimination. However, beyond the poroelastic domain and into the poroelastic/plastic domain, some form of an iterative solution scheme is necessary.

Using conventional notation one has at the element level:

$$\begin{aligned}\{u\} &= [N]\{u_n\}, \quad \{\pi\} = [N']\{\pi_n\},\\ \{\varepsilon\} &= [B]\{u_n\}, \quad \{\partial\pi\} = [B']\{\pi_n\},\end{aligned} \tag{8.9}$$

where a subscript n indicates a nodal value, a prime indicates of pore fluid value, and $\{\partial\pi\}$ is a 3×1 column matrix of partial derivatives.

Additional symbolic details of the overall or global finite element model include forces:

$$\begin{aligned}\{F_u\} &= \int_S [N_u]^t \{T\} dS + \int_V [N_u]^t \{\gamma\} dV\\ \{F_\pi\} &= -\int_S [N_\pi]^t (q) dS + \int_V [B_\pi]^t [k] \{\gamma_w\} dV\end{aligned} \tag{8.10}$$

Here, discharge through the surface S of the considered region is a scalar q given by $q = \{w\}^t \{n\}$ with $\{n\}$ being a unit outward normal vector. Relevant stiffness matrices are:

$$\begin{aligned}&[K_{uu}] = \int_V [B]^t [C][B] dV, \quad [K_{u\pi}] = \int_V [B]^t \{c\}][N_\pi] dV\\ &[K_{\pi u}] = [K_{u\pi}]^t, \qquad\qquad [K'_{\pi\pi}] = \int_V [N_\upsilon]^t (c)][N_\pi] dV\\ &[K_{\pi\pi=}] = \int_V [B_\pi]^t [k]][B_\pi] dV\end{aligned} \tag{8.11}$$

Boundary conditions suggested by the formulas for forces in Equation (8.10) include surface forces or displacements and fluid velocities or fluid pressure. Sometimes $\{\partial\pi\} - \gamma$ in Darcy's law is set to zero to insure a no-flow boundary condition. Mixed conditions are possible, for example, a normal force and shear displacements. Example problems illustrate application including necessary input properties: elastic moduli, strengths, hydraulic conductivities, fluid compressibility, and specific weights. In case of jointed rock, the usual case in rock engineering, joint geometry and joint properties are needed.

8.4 Equivalent properties

Large rock masses generally contain too many joints (cracks, fractures, small faults, or offsets) to take into account individually. Consequently, the properties of joints and intact rock between must be considered in combination to form properties of an *equivalent* material. Equivalent properties are computed through a *homogenization* process that leads to the same average stress, strain, fluid pressure, and velocity that occur in the original, heterogeneous jointed rock sample. The concept applied to elastic compliances is illustrated in Figure 8.4, where the same relationship between overall average stress and strain exists in the heterogeneous and homogenized volumes. The angle brackets indicate volume averages and $[C^*]$ is a 6×6 symmetric matrix of equivalent moduli. In a finite element analysis, each element is homogenized according to the joints and intact rock in the element. Joints are treated as thin layers of material with very different properties than the adjacent intact rock and are represented as ordinary finite elements during homogenization.

Element size is not a concern and need not be a "representative volume element" or RVE, one where addition of one more joint is of negligible consequence to equivalent moduli values. The assumption of an RVE is common to almost all equivalent properties material models but is not practical in rock mechanics. A nonrepresentative volume element (NRVE) approach is needed in rock mechanics. To be sure, joint elements are not represented explicitly in a finite element mesh but rather are virtual elements, so to speak. Thus, storage requirements are only increased by the number of elements in a mesh to allow each element to have unique moduli and other properties.

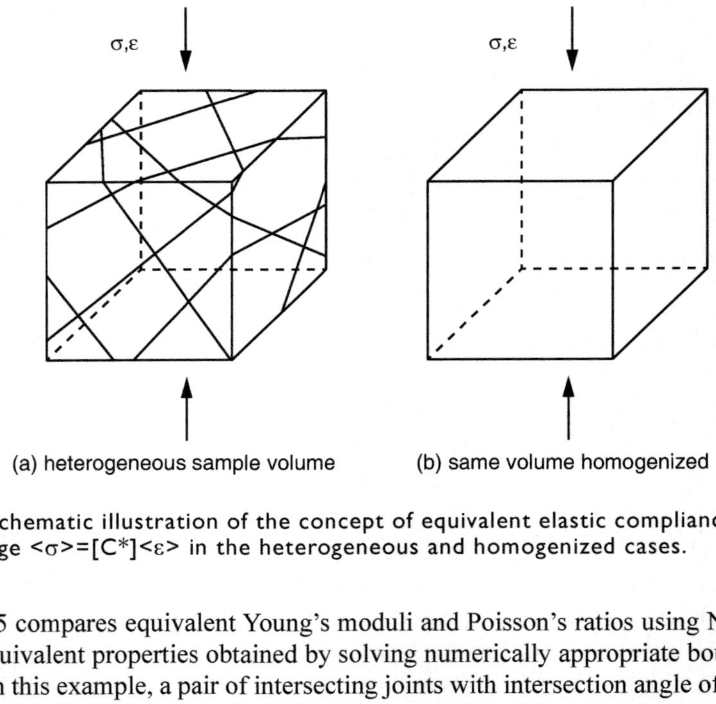

Figure 8.4 Schematic illustration of the concept of equivalent elastic compliances. On average $<\sigma>=[C^*]<\varepsilon>$ in the heterogeneous and homogenized cases.

Figure 8.5 compares equivalent Young's moduli and Poisson's ratios using NRVE theory with true equivalent properties obtained by solving numerically appropriate boundary value problems. In this example, a pair of intersecting joints with intersection angle of 15° transect the sample cube. The bisector of the 15° joint intersection angle varies from 0° to 90°. Joint Young's modulus and shear modulus are 1/100th of corresponding rock moduli. Poisson's ratios are the same. Jointing induces anisotropy as the results in the figure show, for example, moduli in the coordinate directions are very different.

Figure 8.6 shows a sample cube containing four joints and the equivalent hydraulic conductivities computed from NRVE theory. The four joints result in an almost isotropic hydraulic conductivity matrix, which is why the error boxes have such a small range on the vertical scale. Again, NRVE theory agrees well with true equivalent properties, hydraulic conductivities.

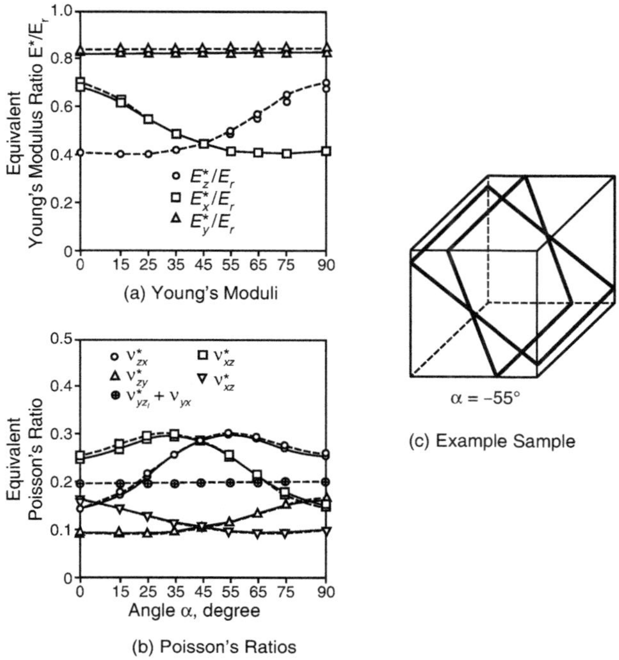

Figure 8.5 Equivalent Young's moduli and Poisson's ratio for a sample containing two joints intersecting at 15° as functions of the dip of the two joint set. Solid=true values, dotted=NRVE theory values, symbols denote coordinate directions (Pariseau, 1993).

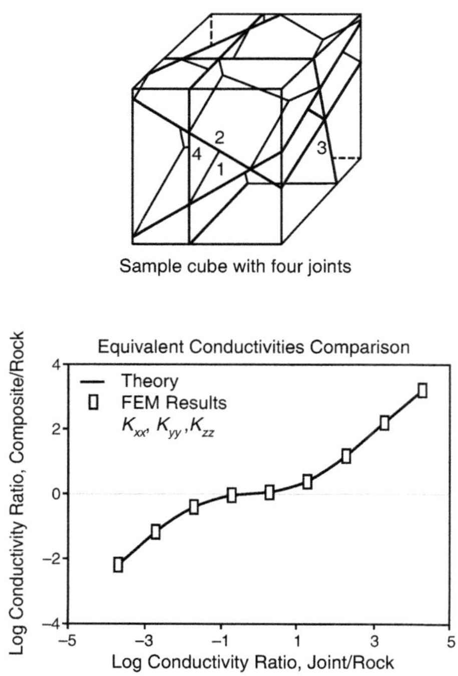

Figure 8.6 Comparison of equivalent hydraulic conductivities in a sample cube containing four joints over eight orders of magnitude of the ratio of joint to rock hydraulic conductivity (Pariseau, 1993).

Strength of jointed rock is not amenable to homogenization. No single set of strengths can accommodate intact and joint failure. The two have separate strengths, so stress in intact rock and joints must be computed separately to determine yielding within an element. This determination is accomplished with the aid of influence functions. These functions relate local (sub-element) averages to global (element) averages. Element averages are then obtained directly from a finite element solution when a linear displacement element is used. Details are given by Pariseau (1993) in case of jointed poroelastic rock where equivalent hydraulic conductivities are computed. In case of conventional (dry) elastic-plastic analyses, details may be found in Pariseau (1999a).

Although no equivalent strength is possible for jointed rock, when very large finite element models (millions of elements) are considered, an empirical approach of some kind is necessary for realism. Jointed rock is simply weaker (and more compliant) than intact rock and more often anisotropic as well. A conventional approach to the problem is "back analysis" where engineering-scale observations, for example, displacement measurements using multi-anchored borehole extensometers, are used to "calibrate" a model. Anchor loss indicates extent of yielding in the vicinity of an excavation. Other extensometer readings may be used to guide reduction in elastic moduli, which largely control displacements in finite element models. Such guidance leads to a scale factor for elastic moduli by which the laboratory moduli used in a first analysis are multiplied to obtain field-scale moduli for use in a second more realistic analysis that gives a close match to extensometer measurements.

An aid to empirical scaling is a quasi-dimensionless approach that supposes the strain to failure is the same in the field and in the finite element model. In consideration of a uniaxial test, this rule is $\varepsilon_f = (C_o / E)_{lab} = (C_o / E)_{field}$ where ε_f is strain to failure, E is Young's modulus, and C_o is unconfined compressive strength. According to this rule, a strength scale factor has the same magnitude as the modulus scale factor. A second empirical rule is to suppose that the energy to failure is the same. Thus, $2U = (C_o)(\varepsilon_f)_{lab} = (C_o)(C_o / E)_{field}$. By this rule, strength scales as the square root of the modulus scale factor. For example, if $(E_{field} / E)_{lab} = 0.36$, then

$$(C_o)_{field} / (C_o)_{lab} = \sqrt{E_{field} / (E)_{lab}} = 0.6.$$

8.5 Example problems

Example problems serve two important purposes; they serve to validate finite element models and to provide guidance to engineering designs. Validation consists mainly of quantitative comparisons of finite element results with known solutions. Important categories of engineering problems that often involve saturated ground include safety and stability of foundations, slopes, shafts, and tunnels. All of the example problems are solved numerically using the finite element method embodied in the UT4 program. This program has evolved from the pioneering efforts and original programs by Dahl (1969) in rock mechanics and Wilson (1963) in structures. The program is not in the public domain but has been used by students and professional colleagues in pursuit of their academic and research goals. Periodic benchmarking has insured continuing reliability of the program. Of course, there is commercially available software from several companies that offer technical support for investigation of poroelastic/plastic design analysis.

Foundation problems are of great importance in geotechnical engineering and are essentially *consolidation* problems. Consolidation involves settlement of foundations and thus expulsion of pore fluid during construction and subsequent loading. Two consolidation problems of a column of porous material H ft deep, which is loaded under a uniform compressive stress distributed across a porous loading plate, serve to verify UT4 reliability. The loading plate allows fluid to escape upwards even as the plate compacts the column downwards. The first consolidation problem is solved with the assumption of an incompressible pore fluid. Figure 8.7 compares results using UT4 with results from Ghaboussi and Wilson (1973) using the finite element method and with the exact solution (Detournay and Cheng, 1993). The plot is in dimensionless coordinates with a dimensionless time as a parameter. The x-axis is essentially excess pore pressure. Dimensionless time is $\tau = C_v t / H^2$ where t is real time and C_v is a consolidation coefficient such that $C_v = 2Gk / \gamma_w$. In this case, Poisson's ratio is zero.

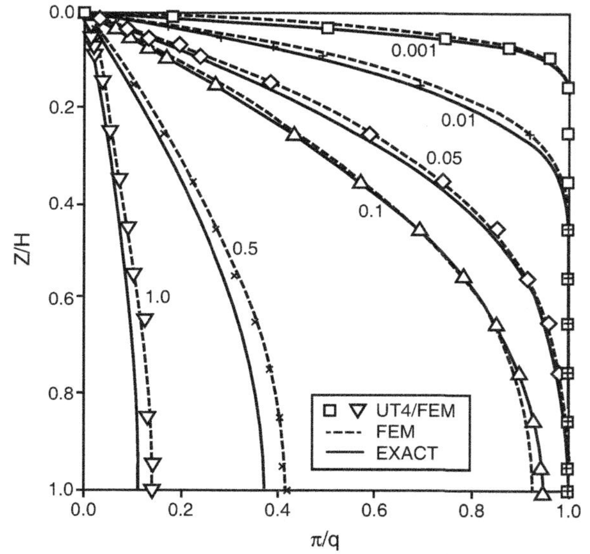

Figure 8.7 Fluid pressure as a function of depth with time as a parameter. Incompressible case (Pariseau, 1994).

The second consolidation problem allows the pore fluid to be compressible. Results are presented in Figure 8.8. A noticeable difference from the incompressible case is in the near instantaneous drop in excess pore pressure. This feature is a consequence of fluid compressibility. Incompressibility is often a reasonable assumption in soil mechanics but is almost never reasonable in rock mechanics when the fluid (water and certainly gas) is much more compressible than the host rock.

Settlement is certainly of interest and is shown in Figure 8.9. The parameter M is a bulk modulus of the pore fluid, the reciprocal of compressibility ($c = 1/M$). In case of an incompressible pore fluid, M and η becomes indefinitely large and c approaches zero. Close agreement of UT4 results with the other two sets of results is evident in the figure.

Consolidation under a strip load reveals another interesting feature of poromechanics and that is a decrease in pore pressure that is not monotonic as in the preceding two example problems. This is the famous Mandel-Cryer effect (Mandel, 1953; Cryer, 1963). Figure 8.10

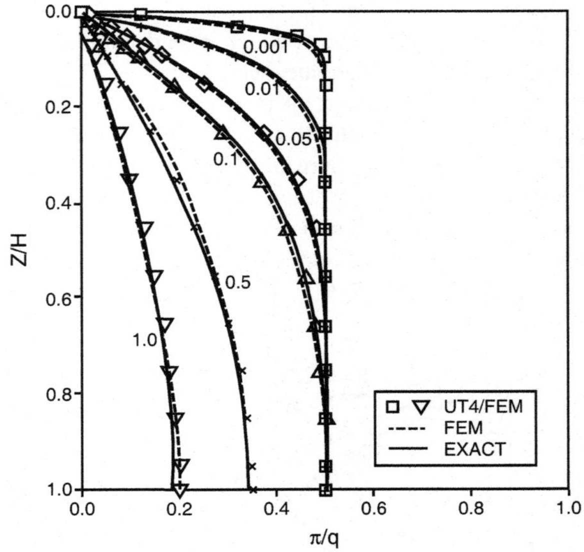

Figure 8.8 Fluid pressure as a function of depth with time as a parameter. Compressible fluid case (Pariseau, 1994).

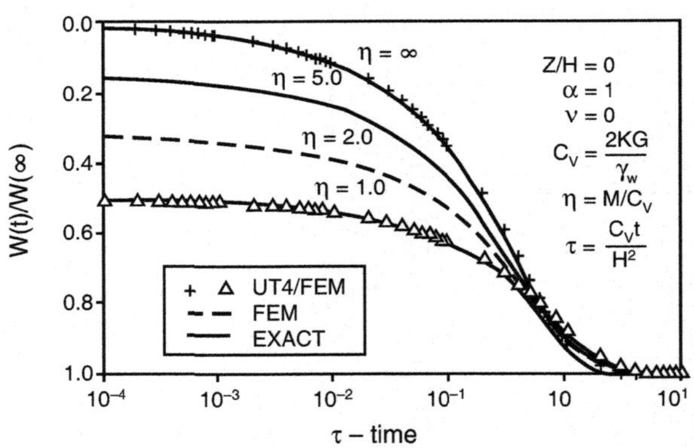

Figure 8.9 Surface displacement as a function of time during one-dimensional consolidation in case of incompressible and compressible pore fluid (Pariseau, 1994).

shows the mesh used for the analysis. Figure 8.11 shows pore pressure history across the mesh near the loaded strip; Figure 8.12 shows the history of excess pore pressure as a function of depth. The problem has an exact solution and has also been addressed by Ghaboussi and Wilson (1973) using the finite element method. Comparison with UT4 results shows close agreement.

Another problem related to consolidation is excavation of a circular tunnel or shaft. A mesh for this two-dimensional problem is shown in Figure 8.13.

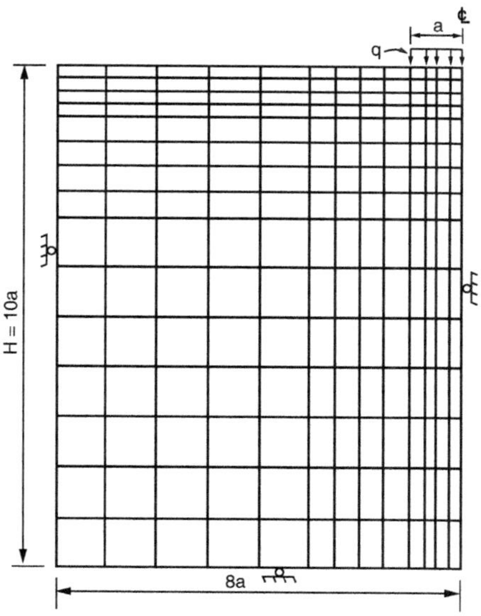

Figure 8.10 Finite element mesh for a strip loaded half space analysis (Pariseau, 1994).

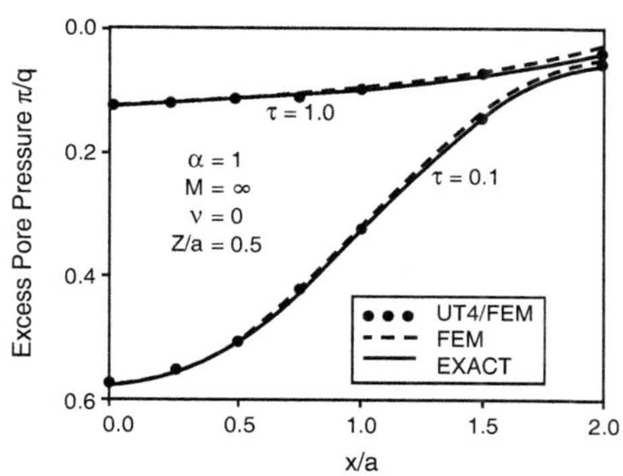

Figure 8.11 Fluid pressure as a function of horizontal position and two times (Pariseau, 1994).

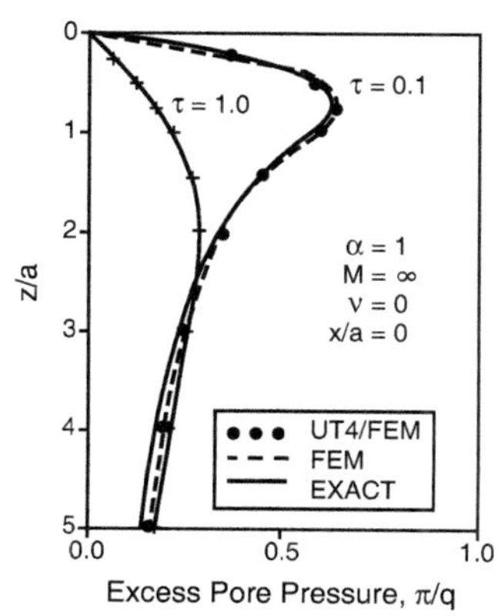

Figure 8.12 Vertical distribution of fluid pressure as a function of depth below a strip loaded half space (Pariseau, 1994).

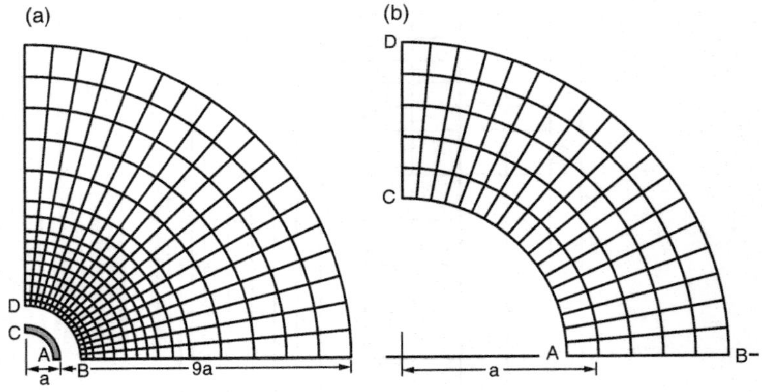

Figure 8.13 Finite element mesh for analysis of a circular excavation. (a) Overall, (b) near wall (Pariseau, 1994).

Fluid pressure as a function of distance along a radial line into the excavation wall at three angular positions is shown in Figure 8.14. The pre-excavation stress is non-hydrostatic with a major principal stress acting in the y-direction and a minor principal stress acting in the x-direction.

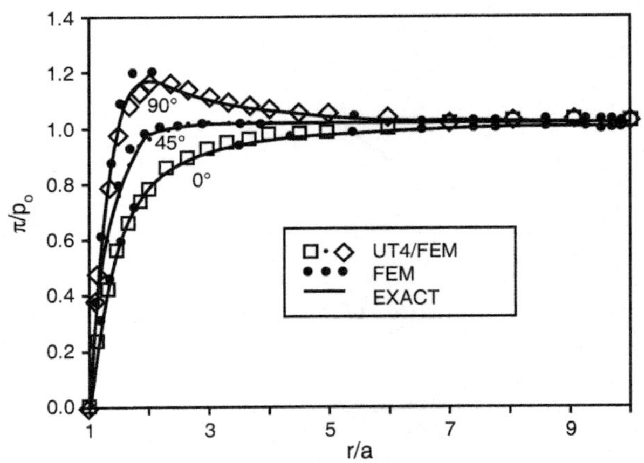

Figure 8.14 Fluid pressure as a function of distance along a line at three angular orientations about a circular excavation (Pariseau, 1994).

Yielding might occur at the outset or at some later time during consolidation. The problem of poroelastic/plastic consolidation has an exact solution, as do the two poroelastic consolidation problems presented previously. There is an interesting difference, however. In the two poroelastic problems, the column was of finite depth, perhaps depth to bedrock as a practical matter. Column depth in the poroelastic/plastic problem is of infinite depth. The difference is of no engineering consequence. Figure 8.15 illustrates two new poroelastic/plastic consolidation problems, one where the material is considered weightless and is loaded at the surface only; the other is loaded only by self-weight. In both cases, the total vertical stress required for equilibrium is independent of time. Consequently, solid deformation and fluid flow are unlinked; fluid flow is then governed by a diffusion equation $\partial\pi / \partial t = c_v^2 \partial^2 \pi / \partial z^2$, which has the same form in the elastic and elastic/plastic domains but with different diffusivities or consolidation coefficients (c_v) where use of the square is convenient.

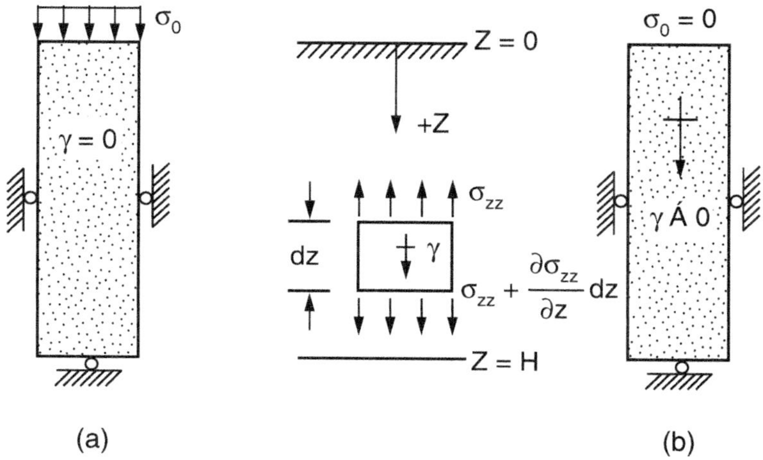

Figure 8.15 Two consolidation problems. (a) Surface load only, (b) weight only (Pariseau 1999).

Solution to the problem in case of surface loading only is:

$$
\begin{aligned}
&Z(t) = 2\beta\sqrt{t} && : \text{depth to the elastic/plastic interface} \\
&\pi_p = P_1 + \frac{(\pi_{EP} - P_1)}{erf(\beta / c_p)} erf[(\beta / c_p)(z / Z)] && : 0 < z < Z(t),\ t > 0 \\
&\pi_e = P_0 + \frac{(\pi_{EP} - P_0)}{erf(\beta / c_e)} erf[(\beta / c_e)(z / Z)] && : z < Z(t) < \infty,\ t > 0 \qquad (8.12) \\
&\pi_p = P_1 \ : z = 0, \qquad \pi_e = P_0 \ : t = 0+, && \\
&\pi_p = \pi_e \text{ and } \partial\pi_p / \partial z = \partial\pi_e / \partial z && : z = Z(t)
\end{aligned}
$$

where *erf* means error function. Details are available in Pariseau (1999b).

As in almost all elastic/plastic problems, location of the elastic-plastic interface in terms of external boundary conditions is key to the problem solution. In case of a surface load only, yielding occurs at the surface first using a Drucker-Prager or a Mohr-Coulomb failure criterion. Depth to the bottom of the plastic zone is $Z(t) = 2\beta\sqrt{t}$ where β is determined by continuity of stress at the elastic-plastic interface. Figure 8.16 shows $Z(t)$ analytically and numerical through a finite element solution using UT4. Material is elastic below the bottom of the plastic domain. Surface displacement as a function of time follows a similar curve.

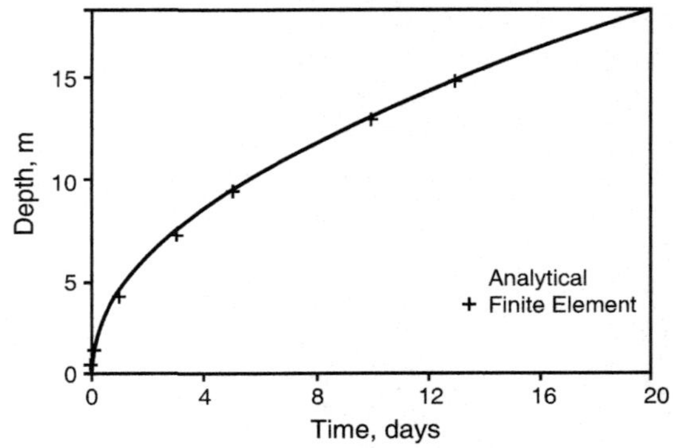

Figure 8.16 Depth to the bottom of the plastic domain as a function of time (Pariseau, 1999).

Effective vertical stress distribution to a depth of 30 m is shown in Figure 8.17. Time is a parameter in the plot. The vertical line serves to locate the elastic/plastic interface. This line defines the depth to the elastic/plastic interface where it crosses a plotted curve. The plastic domain is above an intersection point; below the intersection point, deformation is purely elastic. Solid curves are from the analytical solution Equation (8.12); dotted lines are from the UT4 finite element solution. Agreement is excellent and further indicates reliability of UT4. Demonstration of reliability lends confidence to engineering applications made for design guidance.

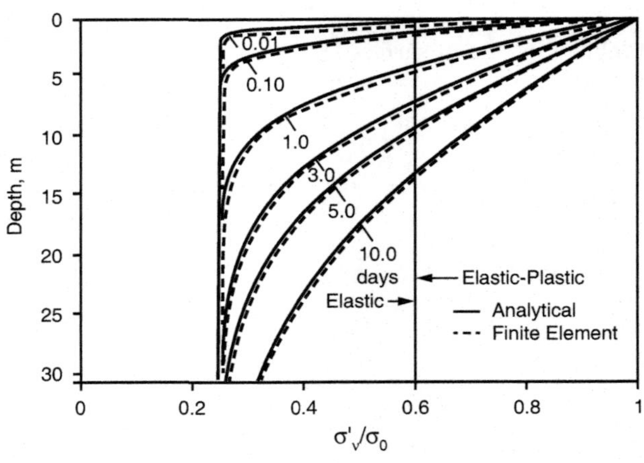

Figure 8.17 Effective vertical stress as a fraction of the surface load versus depth with time as a parameter (Pariseau, 1999).

Slope stability is one of the most important engineering design challenges. Soil slopes have been studied in great detail for many years; rock slopes have also been studied in much detail, especially in deep surface mines where slope angle is a major determinant of profitability. Computer programs abound in soil mechanics that accommodate poromechanics analysis, especially those based on the well-known method of slices. In rock mechanics,

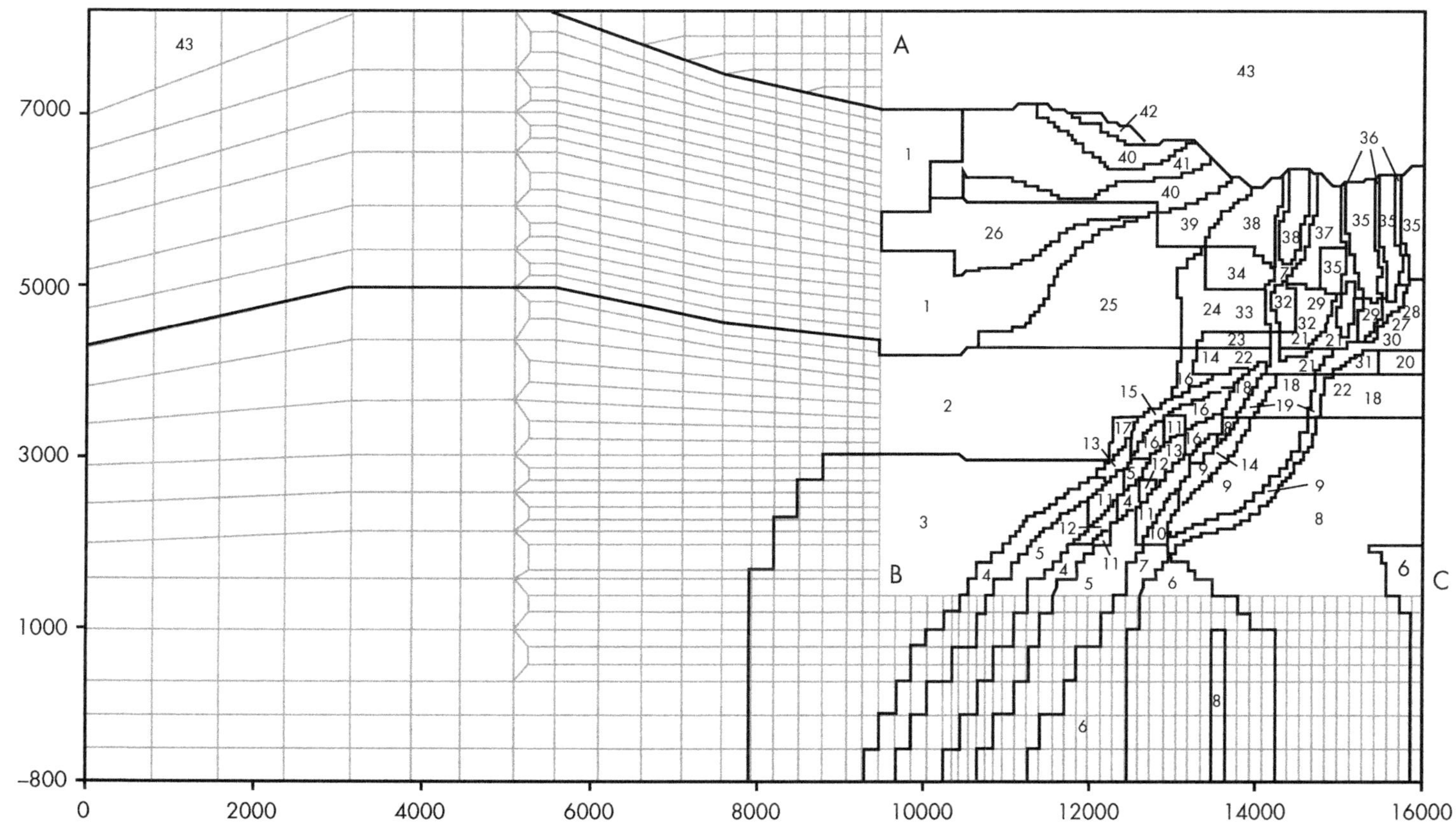

Figure 8.18 Finite element mesh in vertical section of a deep open pit mine before mining (Schmelter and Pariseau, 1997).

there are only a few coupled programs for slope stability analysis based on poromechanics that have capability for simulating sequential excavation. Although numerical models in soil and rock mechanics share fundamentals, an important distinction is found in the assumption of fluid (water) compressibility. Water has a bulk modulus, say, of 1.7 GPa ($2.4x10^5$ psi); a soil bulk modulus may be 0.17 GPa ($2.4x10^4$ psi), and a rock bulk modulus may be 17 GPa ($2.4x10^6$ psi). Thus, water may reasonably be assumed to be incompressible in soil mechanics, but assumption of (water) incompressibility in rock mechanics would not be justified.

An example of poroelastic/plastic analysis of a rock slope is given by Schmelter and Pariseau (1997). Details of the analysis are given in this reference. The finite element mesh and geology of a large open pit copper mine is shown in vertical section in Figure 8.18. Element boundaries are omitted for viewing geological details near the pit slope in the upper right side of the figure. The mesh is 16,000 ft (4.87 km) wide. There are 13 rock types and 11 hydro units that combine to form 43 sets of material properties in the mesh.

Figure 8.19 shows distributions of element safety factors in the dry case (a) and in the wet case (b) for comparison after considerable mining that extends the pit to a depth of approximately 2000 ft (610 m) in the model. In the dry case (a), there are zones where the safety factor is less than 1 behind the crest of the slope and near the toe. In the wet case, the zones of low safety factor (less than 1) near the crest and the toe are noticeably larger than in the dry case and confirm the importance of water and drainage to stability of the slope.

An example of poroelastic/plastic analysis in underground mining is discussed by Sheik and Pariseau (2001). The mine was a small, inactive operation in northern Idaho and was being used as an experimental site for rock mechanics investigations. Material properties were obtained from mine measurements and laboratory tests. In particular, hydraulic conductivities were determined *in situ* from borehole packer measurements, as were elastic moduli using a Goodman jack. Rock strengths were empirically scaled from laboratory values to mine scale values by an order of magnitude reduction. Figure 8.20 shows a vertical cross-section through the mine. There are three stopes (excavations for ore) in the figure, which also shows the excavation (cut) sequence used in finite element analyses.

The poroelastic/plastic finite element code incorporated a special feature that allowed for desaturation. Excavation by blasting in rock is almost instantaneous and allows expansion of excavation walls into the newly formed opening. High suctions (tension!) are produced when the pore fluid is incompressible. Consequently, effective stresses that determine strength become tensile and lead to large zones of failure, an unrealistic outcome. Desaturation in this study involved depressurization, which set pore pressure to zero instead of allowing tensile pore fluid pressure. A second provision was to assume a compressible pore fluid (water). Excavation of a stope was followed by back filling with saturated sand (hydraulic fill). The seven cuts and three fills were each followed by six consolidation runs involving 10 time steps. Equilibrium was approached in approximately 10 days. The finite element mesh for the study is shown in Figure 8.21. The heavy inclined line near the mesh top is the ground surface. Access to the mine was through the lower tunnel portal at ground level.

The shallow depth of the mine workings and the relatively high rock strength combined for a very safe, stable underground environment. However, some yielding of backfill was indicated at the excavation walls that subsequently spread to the center of the fill. Evidently, the increase of fill strength in time with drainage was not sufficient to overcome the load transfer to the fill by stope wall closure (relative displacement between walls). Figure 8.22 shows the distribution of safety factors 10 days after mining and Figure 8.23 shows water pressure 10 days after mining.

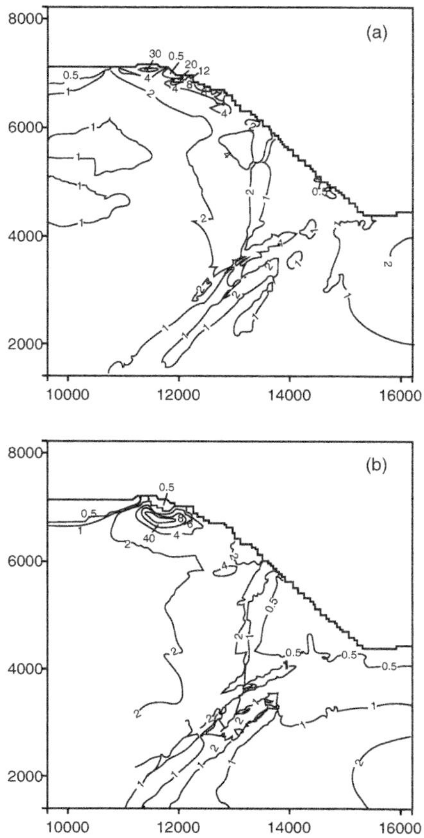

Figure 8.19 Safety factor distributions immediately after excavation: (a) dry, (b) wet (Schmelter and Pariseau, 1997).

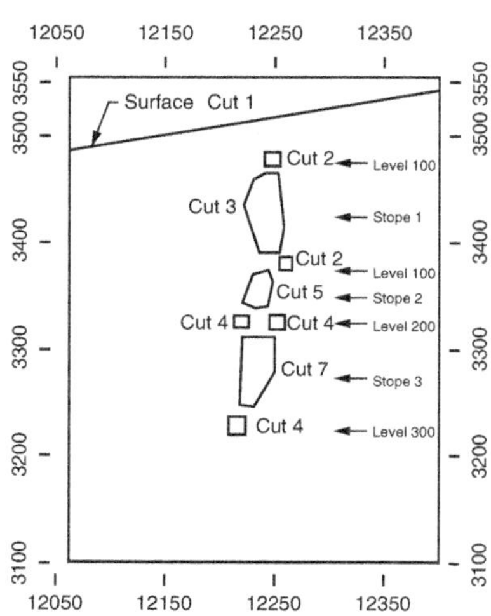

Figure 8.20 Vertical section through the mine showing the excavation sequence. Units=ft (Sheik and Pariseau, 2001).

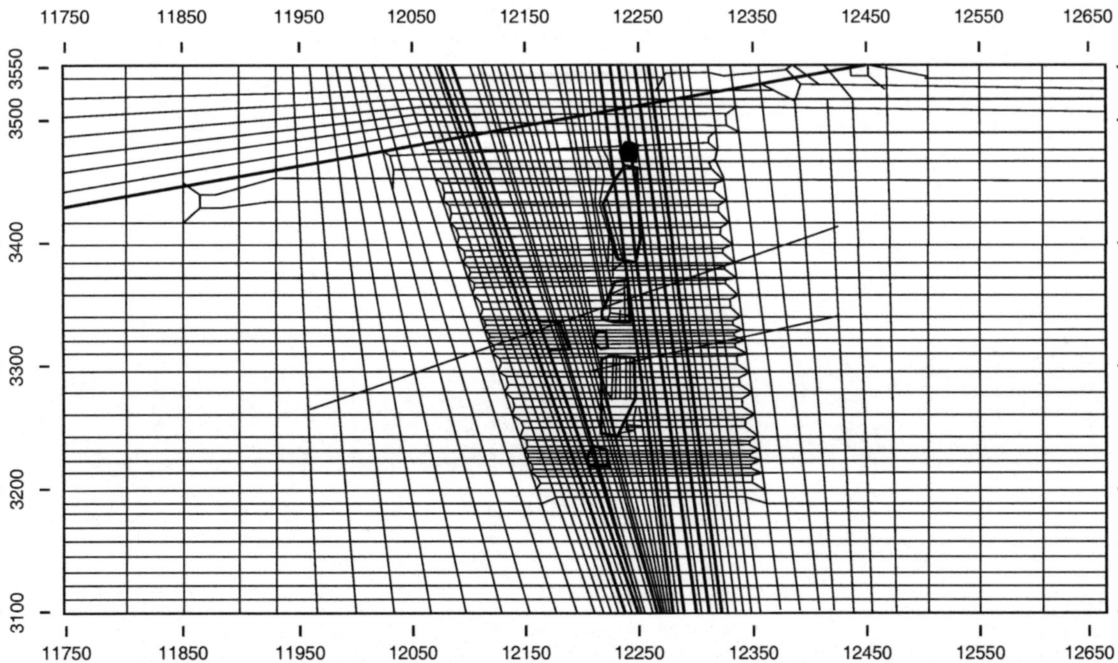

Figure 8.21 Finite element mesh of a small mine cross-section used for underground poromechanics study. Units are ft (Sheik and Pariseau, 2001).

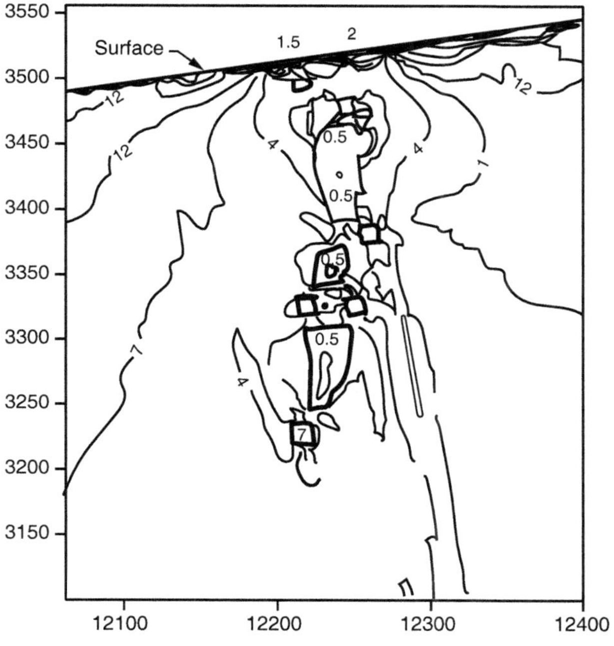

Figure 8.22 Safety factor distribution 10 days after excavation (Sheik and Pariseau, 2001).

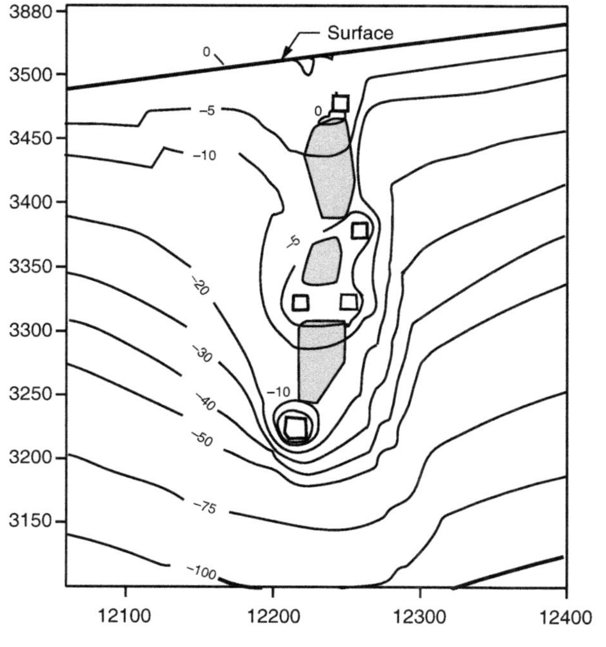

Figure 8.23 Calculated water pressure distribution 10 days after excavation (Sheik and Pariseau, 2001).

A useful feature of poromechanics analysis is the capability for calculating water flows. In this study, a calculation of water discharge from the mine from inflows into the access tunnels and out the portal is shown in Figure 8.24. Discharge is slow in the beginning and then increases rapidly during excavation and then decreases in time. Two sections involving two different meshes were used in the study as seen in the figure. Calculated discharge in the figure is in cubic meters per day. In fact, the mine was dry for the most part. Only a small discharge was evident underground, and that was from a fault that transected the workings. The reason for the great disparity between calculation and reality is in the connectivity of the joints, which are the major conduits of water flow where the intact rock between joints is almost impermeable. Despite borehole packer measurements over lengths of 5 ft (1.6 m), the *in situ* permeability was greatly overestimated. A realistic joint model that reveals water channels over mine scale distances in contrast to meter scale measurements would be needed for realistic discharge estimation. Indeed, what to do about the joints is an ongoing key issue in rock mechanics and in practice distinguishes rock mechanics as a unique field of endeavor in the mechanics.

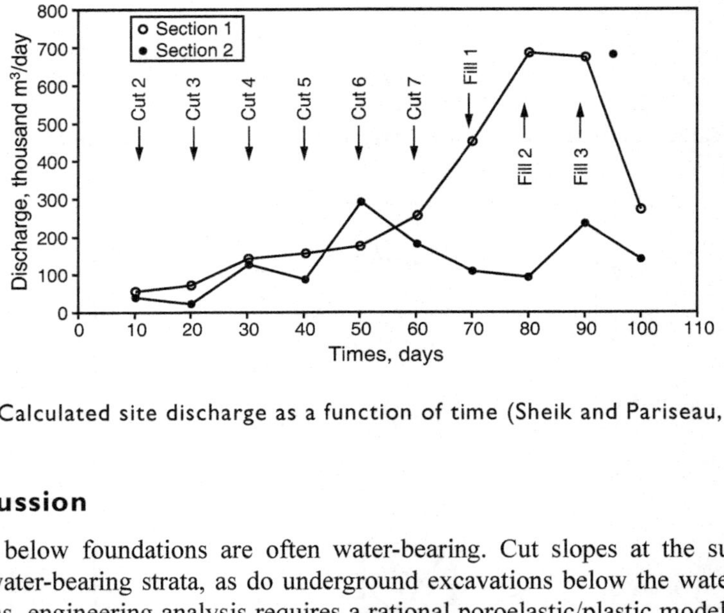

Figure 8.24 Calculated site discharge as a function of time (Sheik and Pariseau, 2001).

8.6 Discussion

Formations below foundations are often water-bearing. Cut slopes at the surface often encounter water-bearing strata, as do underground excavations below the water table. For these reasons, engineering analysis requires a rational poroelastic/plastic model of material behavior and a reliable computational tool for design. The range of purely elastic deformation is surely limited in soil and rock mechanics, so strength must be taken into account. The concept of effective stress is essential to the task. Only a few simple, one-dimensional consolidation problems are amenable to analytical solution, but these solutions provide important tests for reliability of numerical solutions.

A number of computer programs are available from specialty software firms. In this regard, the popular finite-element method has been shown to be especially well suited to poroelastic/plastic analysis, but other techniques are also used.

Example problems of surface and underground excavation in wet rock demonstrate practical application where laboratory measurements and field observations were made by the author and students in cooperative research studies with mining companies and government agencies. In this regard, poromechanics applied to rock encounter a major challenge in accounting for effects of joints. In these studies and finite element models, equivalent properties serve the purpose. Practical results allow for quantifying the role of water on excavation stability under site-specific conditions. Of necessity, much detail is relegated to references. The literature concerning poromechanics is vast, and no attempt was made to do a formal review of the subject. Only the few references of direct relevance were used here.

References

Barenblatt, G.I., Entov, V.M. & Ryzhik, V.M. (1990) *Theory of Fluid Flows Through Natural Rocks*. Kluwer Academic Publishers, Boston.

Bear, J. & Verruijt, A. (1987) *Modeling Groundwater Flow and Pollution*. D. Reidel Publishing Company, Boston.

Biot, M.A. (1941) General theory of three-dimensional consolidation. *Journal of Applied Physics*, 12, 155–164.

Biot, M.A. (1955) Theory of elasticity and consolidation of porous anisotropic solid. *Journal of Applied Physics*, 26, 182–185.

Biot, M.A. (1956) General solutions of the equations of elasticity and consolidation for a porous material. *Journal of Applied Mechanics*, 23, 91–95.

Charles, P.A. (1991) *Rock Mechanics*, Volume 1, *Theoretical Fundamentals*. Editions Technip, Paris.

Charles, P.A. (1997) *Rock Mechanics*, Volume 2. *Petroleum Applications*. Editions Technip, Paris.

Coussy, O. (1995) *Mechanics of Porous Continua*. John Wiley & Sons, New York.

Cryer, C.W. (1963) A comparison of the three-dimensional consolidation theories of Biot and Terzaghi. *The Quarterly Journal of Mechanics and Applied Mathematics*, 16, 401–412.

Dahl, H.D. (1969) *Finite Element Model for Anisotropic Yielding in Gravity Loaded Rock*. PhD Thesis, The Pennsylvania State University, University Park, PA.

Darcy, H. (1856) *Les Fontaines Publiques de la Ville de Dijon*. Libraire des Corps Imperiaux des Ponts et Chausses et des Mines, Paris.

Detournay, E. & Cheng, A.H.-D. (1993) Fundamentals of poroelasticity. *Comprehensive Rock Engineering*, Volume 2. Pergamon Press, New York. pp.113–171.

Ghaboussi, J. & Wilson, E.L. (1973) Flow of incompressible fluid in porous elastic media. *International Journal for Numerical Methods in Engineering*, 5, 419–442.

Hough, B.K. (1957) *Basic Soils Engineering*. Ronald Press, New York.

Istok, J. (1989) *Groundwater Modeling by the Finite Element Method*. American Geophysical, Washington, DC.

Lewis, R.W. & Schrefler, B.A. (1998) *The Finite Element Method in the Static and Dynamic Deformation and Consolidation of Porous Media* (2nd ed.). John Wiley & Sons, New York.

Mandel, J. (1953) Consolidation des Sols (Etude Mathematique). *Geotechnique*, 3, 287–299.

National Research Council (1996) *Rock Fractures and Fluid Flow*. National Academy Press, Washington, DC.

Nikolaevskij, V.N. (1990) *Mechanics of Porous and Fractured Media*. World Scientific Publishing Co., Singapore.

Pariseau, W.G. (1993) Equivalent properties of a jointed Biot material. *International Journal of Rock Mechanics and Mining Sciences & Geomechanics Abstracts*, 30(7), 1151–1157.

Pariseau, W.G. (1994) Design considerations for stopes in wet mines. *Proceedings of the 12th Annual GMTC Workshop on Mines Systems Design and Ground Control, 1994*. pp.37–48.

Pariseau, W.G. (1999a) An equivalent plasticity theory for jointed rock masses. *International Journal of Rock Mechanics and Mining Sciences*, 36(7), 907–918.

Pariseau, W.G. (1999b) Poroelastic-plastic consolidation – analytical solution. *International Journal for Numerical and Analytical Methods in Geomechanics*, 23(7), 577–594.

Rice, J.R. (1998) *Elasticity of Fluid-infiltrated Porous Solids (Poroelasticity)*. Available from: Esag.harvad.edu/rice/e2_Poroelasticity.pdf

Sandhu, R.S. & Wilson, E.L. (1969) Finite element analysis of seepage in elastic media. *Journal of the Engineering Mechanics Division*. EM3, Proc. American Society of Civil Engineers, pp.641–652.

Schmelter, S.C. & Pariseau, W.G. (1997) Coupled finite element modeling of slope stability. *SME Preprint 97–3, SME Annual Meeting*. Denver, CO, February 24–27.

Sheik, A.K. & Pariseau, W.G. (2001) Role of water in the stability of a shallow, underground mine. *Mining Engineering*, 53(1), 37–42.

Taylor, D.W. (1948) *Fundamentals of Soil Mechanics*. John Wiley & Sons, New York.

Terzaghi, K. & Peck, R.B. (1948) *Soil Mechanics in Engineering Practice*. John Wiley & Sons, New York.

Thimus, J.F., Abousleiman, Y.U., Cheng, A.H.-D., Coussy, O. & Detornay, E. (Eds). (1998) *Poro-Mechanics: A Tribute to Maurice A. Biot*. Balkema, Rotterdam.

Wilson, E.L. (1963) *Finite Element Analysis of Two-dimensional Structures*. PhD Thesis, University of California – Berkeley.

Epilogue

Developments in theoretical soil plasticity in the early 1960s led to study of flows of particulate materials, powders, sands, crushed rock, and so forth. Model studies (Johanson and Jenike, 1961) and pilot-plant scale studies (Jenike, 1961) led to important applications in industry and eventually to the formation of the renowned firm of Jenike and Johanson, now just Jenike. While traditional applications of plasticity theory to retaining walls and foundations had stability as a primary objective, the opposite result – flow –is the desirable outcome in application to chutes, bins, silos, and ore passes. Similarities in stress fields are present in both applications. However, time is real in applications of geoplasticity theory to particulate flows; the velocity field is real.

Of particular interest are the evolutions of stress and velocity fields from the instant a hopper or chute gate is opened to the time flow fields are established, if established, as the material stored in the bin above is drawn down and eventually emptied. Specific weight of material is obviously important to the quantification of such gravity flows. Acceleration should also be taken into account. Both require numerical procedures for solution of the characteristic system of stress and velocity even in the plane problem. A simple iterative technique suffices in pursuit of steady state fields, which are justified with continued filling of material to replenish the material drawn.

Many early sand models used horizontal layers that while qualitative clearly demonstrated the development of velocity discontinuities. In study of sand models of ore passes, a luminous point method was used that allowed for quantitative analysis of deformation in the vicinity of a slot outlet in a plane strain mode (Pariseau, 1966). Figure E1 is a photograph of point positions after an increment of outlet discharge. The model was laid flat, a side removed, a template with holes set down and then sprinkled with a luminous powder before replacing the side and standing upright. Polaroid photographs were then taken at intervals of discharge with the lights out. After enlargement by a factor of 20 and connecting the dots, literally, on profile paper, the deformation of an initially square grid was recorded, which by definition is a "motion."

Data reduction was guided by the simple relationship $\theta = (1/2)\,\mathrm{Arctan}[\dot{\gamma}_{xy} / (\dot{\varepsilon}_{xx} - \dot{\varepsilon}_{yy})]$ where θ defines the direction of the major principal stress (assuming isotropy). Principal stress directions are shown in Figure E2. Directions of sliplines, stress characteristics, follow $(\theta \pm \mu)$ where $2\mu = \pi/2 - \phi$. Figure E3 shows experimentally determined sliplines in the vicinity of the ore pass model outlet.

With the advent of the digital age, a series of theoretical calculations of stress and velocity fields were carried out on a CDC 1604 computer (Control Data Corporation) with input via a punched tape. Regions of computation are shown in Figure E4. Results clearly indicated

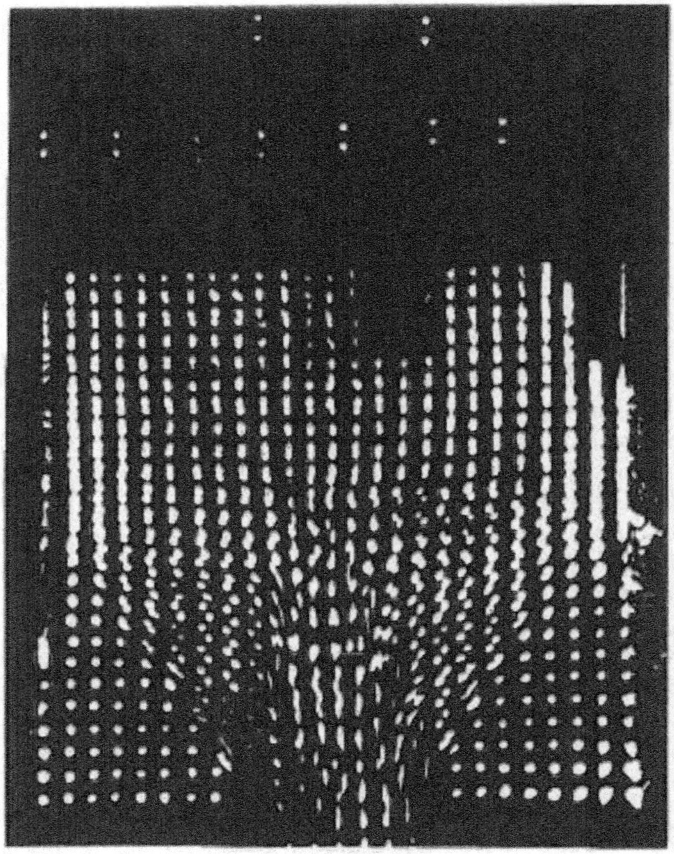

Figure E1 An example of a sand model experiment showing motion of points near a slot outlet in a model ore pass. The outlet is one third the width of the ore pass.

Source: Photograph is doubly exposed.

the need for nonassociated rules of flow. Not too surprisingly, expansion required by associated flow rules was far in excess of experimental observations. Indeed, an incompressible assumption ($\phi_v = 0$) was in much better agreement with experiment.

Many industrial facilities for handling particulate materials are terminated with hoppers that have sloping walls intended to aid discharge. This observation and the success of sand model experiments in testing theory motivated study of sand models with sloping walls (Pariseau, 1976). A new procedure for observing the motion during drawdown was also developed. The new procedure used a photosensitive dye instead of luminous powder for marking an initially undeformed grid. Data were recorded by camera. Stills from the camera film were then used to reduce data from tracings as before.

Figures E5 and E6 show some results in the case of plane strain sand models in hoppers with sloping walls. An isotropic point occurs in Figure E5 on the centerline above the outlet

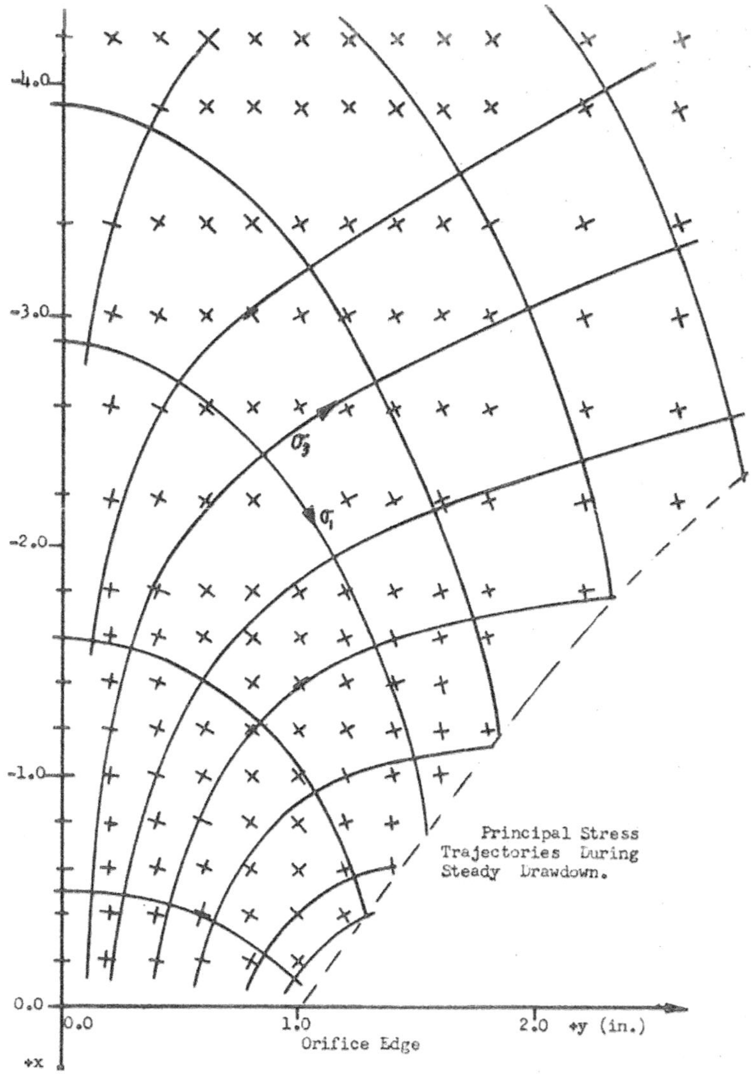

Figure E2 Experimentally determined principal stress directions near the ore pass model outlet.

where the major principal stress direction begins to curve upward from downwards. The regions in Figure E6 are separated by discontinuities in velocity. The flow direction changes to sharply downwards, near vertical, across a steeply downward trending division while flow changes towards parallelism with hopper walls across a gently downward trending division.

In case of axial symmetry, there is almost no experimental data for comparison with theory. However, Jenike and Johanson used a split conical hopper with layers to observe flow in axial symmetry as shown in Figure E7. The photograph shows formation of a pipe over the

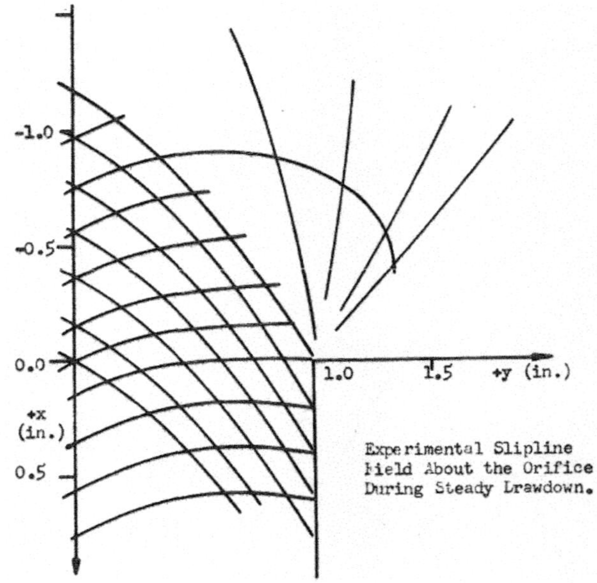

Figure E3 Sliplines in the vicinity of the model ore pass outlet during steady drawdown.

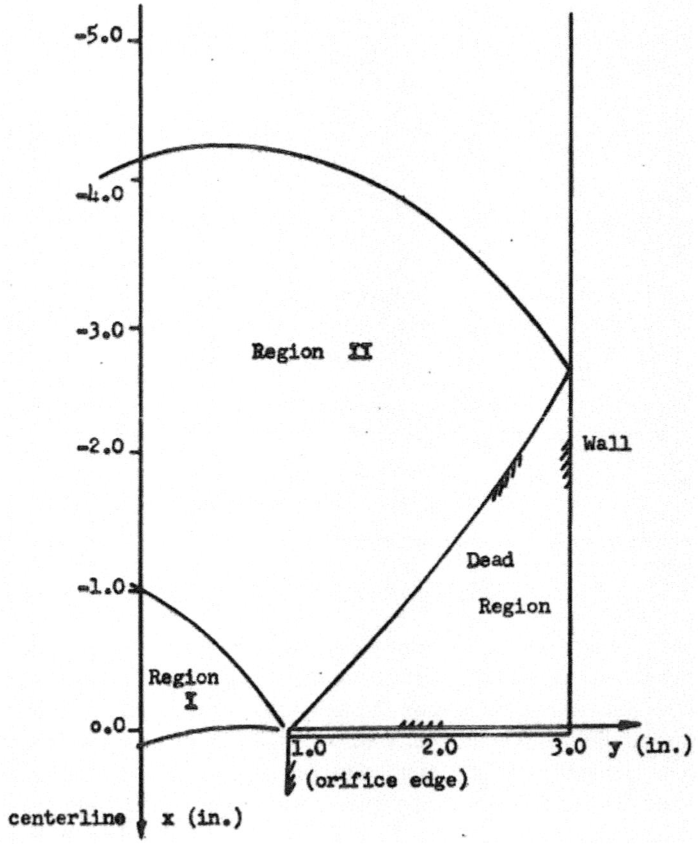

Figure E4 Computational regions for theoretical calculations of model stress and velocity fields.

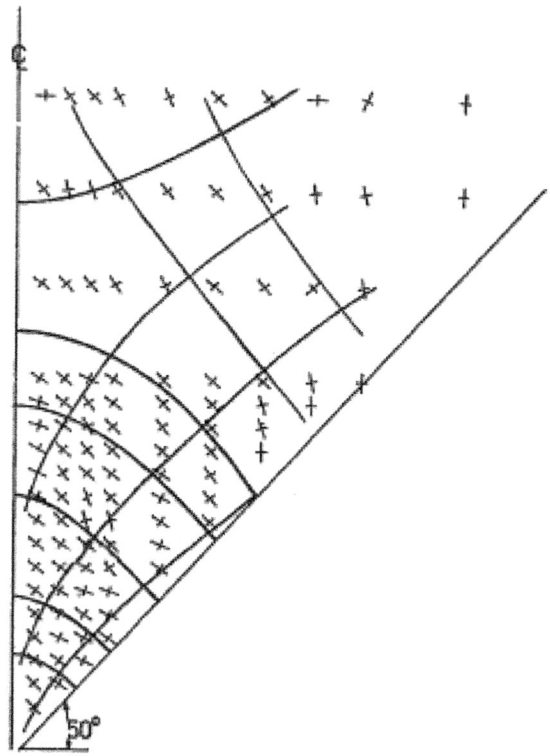

Figure E5 Experimental principal direction field (Pariseau, 1976).

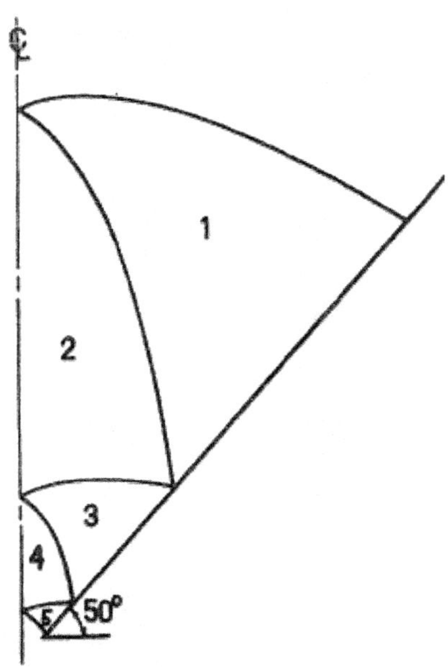

Figure E6 Characteristic regions in a sand hopper model (Pariseau, 1976).

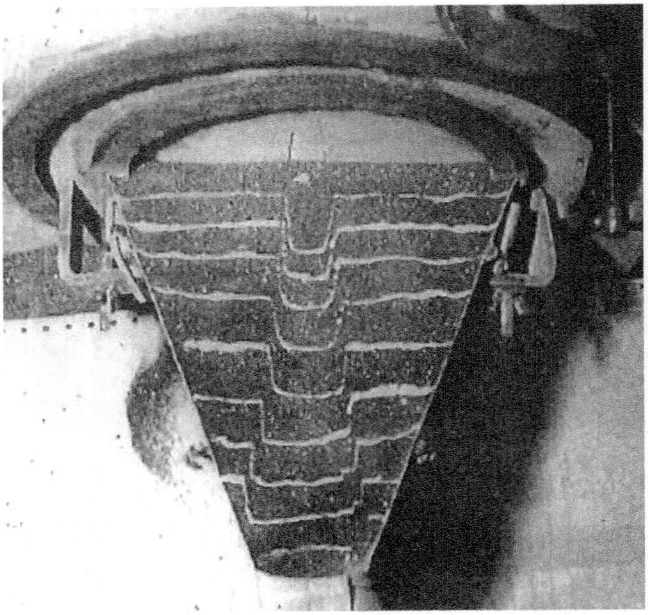

Figure E7 Photograph of an axially symmetric sand model after drawdown with replacement. Wall slope is 50° (Jenike and Johanson, 1961).

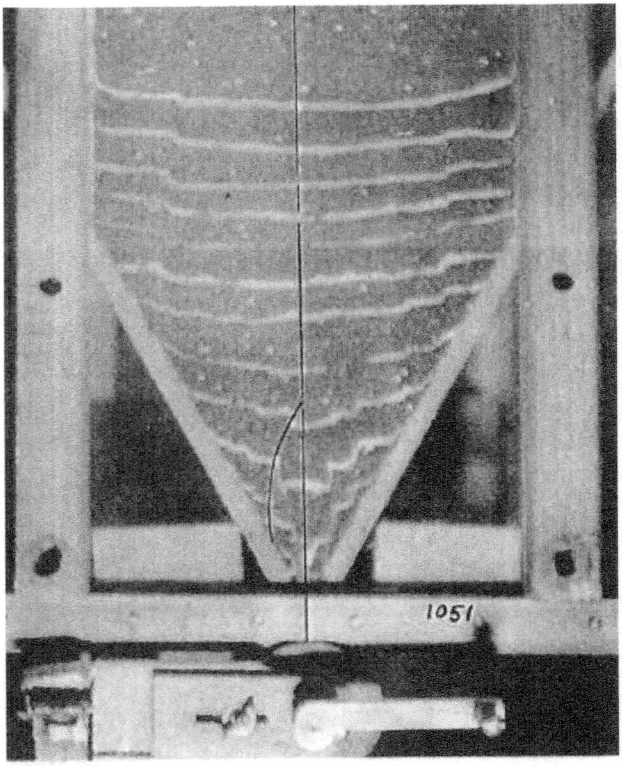

Figure E8 A plane strain sand model hopper with inclined walls (50°) and with a vertical bin above (Jenike and Johanson, 1961).

outlet with a diminishing diameter with distance above the outlet. Pipe walls suggest envelopes of sliplines. For comparison, a photograph of a plane strain sand model result is shown in Figure E8 where several velocity discontinuities are evident.

These and other early studies of flows of particulate materials provided insight and experimental justification for application of plane strain plasticity theory to many important industrial processes.

Although physical data from experiments as done on sand models remains essential to progress and, indeed, has given rise to a sub-discipline of the physics of granular flows, modern computer methods provide far superior experimental platforms for study of particulate materials. Stability analyses are readily done using reliable numerical methods in three dimensions, such as the finite element method (FEM) and the finite difference (FDM) method. Flow models are also readily done using numerical methods based on discrete and distinct element (DEM) methods, and others such as discontinuous deformation analysis (DDA) and the numerical manifold method (NMM). These methods allow elements of various shapes, but mainly balls of different sizes, to translate and rotate in accordance with equations of motion subject to contact with adjacent balls. Contact is usually frictional but may include adhesion, and balls may be deformable. Considerable application of such models is found in materials handling from flow in bins and hoppers to chutes feeding conveyor belts and so on, including ore passes, loading chutes, and finger raises in block caving mines. Computer methods are certainly the way forward.

References

Jenike, A.W. (1961) Gravity Flow of Bulk Solids. Bulletin No. 108 of the Utah Engineering Experiment Station, Salt Lake City, UT.

Johanson, J.R. & Jenike, A.W. (1961) Stress and Velocity Fields in Gravity Flow of Bulk Solids. Bulletin No. 116 of the Utah Engineering Experiment Station, Salt Lake City, UT.

Pariseau, W.G. (1966) *The Gravity-Induced Movement of Materials in ore Passes Analyzed as a Problem in Coulomb Plasticity*. PhD. Thesis, University of Minnesota.

Pariseau, W.G. (1976) Experimental observations of velocity discontinuities in flowing sand. *The Effects of Voids on Material Deformation*. Volume 16, American Society of Mechanical Engineers, New York.

Appendix A

Axial symmetry discussion

Axial symmetry implies in an $(r\omega z)$ system of cylindrical coordinate all derivatives with respect to ω vanish. All rz planes are identical. Moreover, the circumferential stress σ_ω is a principal stress; no shear stress acts on rz planes. The situation is illustrated in Figure A1, where all stresses are positive; specifically, compression is positive.

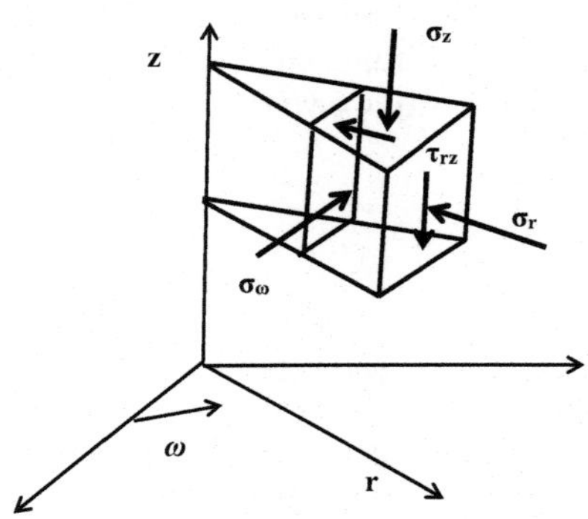

Figure A1 Axial symmetric stress.

Because the rz planes are identical, one need only to use coordinates in a single plane of reference as shown in Figure A2. The principal stresses in the rz plane are shown in the figure with the major principal stress σ_1 inclined at a counterclockwise angle θ to the r-axis. The Cartesian stresses are given by the usual formulas:

$$\begin{aligned}\sigma_{rr} &= (1/2)(\sigma_1+\sigma_3)+(1/2)(\sigma_1-\sigma_3)\cos(2\theta)\\ \sigma_{zz} &= (1/2)(\sigma_1+\sigma_3)-(1/2)(\sigma_1-\sigma_3)\cos(2\theta)\\ \tau_{rz} &= (1/2)(\sigma_1-\sigma_3)\sin(2\theta)\end{aligned} \tag{A1}$$

The axially symmetric form of the equations of motion is:

$$\begin{aligned}\sigma_{rr,r}+\tau_{rz,z}+(1/r)(\sigma_{rr}-\sigma_{\omega\omega}) &= \gamma_r-\rho\dot{u}\\ \tau_{rz,r}+\sigma_{zz,z}+(1/r)\tau_{rz} &= \gamma_z-\rho\dot{v}\end{aligned} \tag{A2}$$

where the comma denotes differentiation, γ_r and γ_z are body forces per unit volume in the r and z directions, respectively, ρ is mass density, and $\dot{u}$ and $\dot{v}$ are velocities in the r and z

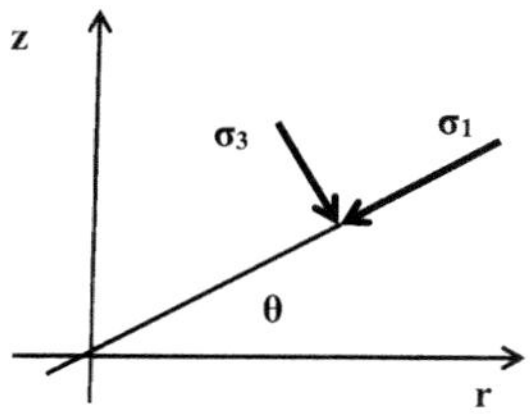

Figure A2 Principal stresses in an *rz* plane.

directions, respectively. Evidently, as the radius of curvature r tends to infinitely, the system Equations (A2) tends to the plane strain equations of motion.

In the isotropic case, the yield condition $Y = 0$ is a function of the stress invariants I_1, I_2 I_3 given by:

$$I_1 = \sigma_1 + \sigma_2 + \sigma_3,\ I_2 = \sigma_1\sigma_2 + \sigma_2\sigma_3 + \sigma_3\sigma_1,\ I_3 = \sigma_1\sigma_2\sigma_3 \tag{A3}$$

Hence:

$$Y(I_1, I_2, I_3) = Y(\sigma_1, \sigma_2, \sigma_3) = 0 \tag{A4}$$

After a change of variables:

$$\begin{aligned} &\sigma = \sigma_{\omega\omega}, \\ &\sigma_m = (1/2)(\sigma_1 + \sigma_3),\ \tau_m = (1/2)(\sigma_1 - \sigma_3) \\ &\sigma_1 = (1/2)(\sigma_m + \tau_m),\ \sigma_3 = (1/2)(\sigma_m - \tau_m) \end{aligned} \tag{A4}$$

the yield condition $Y = Y(\sigma, \sigma_m, \tau_m) = 0$ or explicitly as:

$$\sigma_m = g(\sigma, \tau_m) \tag{A5}$$

Using the yield condition (Equation A5) in the equations of transformation (Equation A1), one obtains with the addition of the circumferential stress:

$$\begin{aligned} &\sigma_{rr} = g + \tau_m \cos(2\theta) \\ &\sigma_{zz} = g - \tau_m \cos(2\theta) \\ &\tau_{rz} = \tau_m \sin(2\theta) \\ &\sigma_{\omega\omega} = \sigma \end{aligned} \tag{A6}$$

To be sure, Equations (A6) imply the yield condition is satisfied.

Derivatives of stress needed in the equations of motion are:

$$\begin{aligned} &\sigma_{rr,r} = \tau_{m,r}[\partial g / \partial \tau_m + \cos(2\theta)] - [2\tau_m \sin(2\theta)]\theta_{,r} + (\partial g / \partial \sigma)\sigma_{,r} \\ &\tau_{rz,z} = \tau_{m,z} \sin(2\theta) + [2\tau_m \cos(2\theta)]\theta_{,z} \\ &\tau_{rz,r} = \tau_{m,r} \sin(2\theta) + [2\tau_m \cos(2\theta)]\theta_{,r} \\ &\sigma_{zz,z} = \tau_{m,z}[\partial g / \partial \tau_m - \cos(2\theta)] + [2\tau_m \sin(2\theta)]\theta_{,z} + (\partial g / \partial \sigma)\sigma_{,z} \end{aligned} \tag{A7}$$

Also:

$$(1/r)(\sigma_{rr} - \sigma_{\omega\omega}) = (1/r)[g + \tau_m \cos(2\theta) - \sigma]$$
$$(1/r)(\tau_{rz}) = (1/2)[\tau_m \sin(2\theta)] \qquad \text{(A8)}$$

The equations of motion with yield satisfied now have the form:

$$\begin{aligned} A_1\tau_{m,r} + B_1\tau_{m,z} + C_1\theta_{,r} + D_1\theta_{,z} &= E_1^* \\ A_2\tau_{m,r} + B_2\tau_{m,z} + C_2\theta_{,r} + D_2\theta_{,z} &= E_2^* \\ \tau_{m,r}dr + \tau_{m,z}dz &= d\tau_m \\ \theta_{,r}dr + \theta_{,z}dz &= d\theta \end{aligned} \qquad \text{(A9)}$$

The letters in Equation (A9) are:

$$\begin{aligned} &A_1 = [\cot(\psi) + \cos(2\theta)], \quad A_2 = \sin(2\theta)] \\ &B_1 = \sin(2\theta), \qquad B_2 = [\cot(\psi) - \cos(2\theta)] \\ &C_1 = -[2\tau_m \sin(2\theta)] \qquad C_2 = [2\tau_m \cos(2\theta)] \\ &D_1 = [2\tau_m \cos(2\theta)] \qquad D_2 = [2\tau_m \sin(2\theta)] \\ &E_1^* = E_1 - (\partial g/\partial\sigma)(\sigma_{,r}) - (1/r)[g + \tau_m \cos(2\theta) - \sigma] \\ &E_2^* = E_2 - (\partial g/\partial\sigma)(\sigma_{,z}) - (1/r)[\tau_m \sin(2\theta)] \end{aligned} \qquad \text{(A10)}$$

where $\cot(\psi) = \partial g / \partial \psi_m$ and $E_1 = \rho\dot{u} - \gamma_r$ and $E_2 = \rho\dot{v} - \gamma_z$.

A.1 Stress subsystem

The stress subsystem in axial symmetry is derived in much the same way as in the plane strain case that began with a determinant solution for derivatives of τ_m and θ. Formally: $\tau_{m,r} = N_1/\Delta$, $\tau_{m,z} = N_2/\Delta$, $\theta_{,r} = N_3/\Delta$, $\theta_{,z} = N_4/\Delta$ where Δ is a determinant of the coefficients in Equation (A10) and the *N*s are determinants formed by replacing columns by non-homogeneous terms in Equation (A9). Thus:

$$\Delta = \begin{vmatrix} A_1 & B_1 & C_1 & D_1 \\ A_2 & B_2 & C_2 & D_2 \\ dr & dz & 0 & 0 \\ 0 & 0 & dr & dz \end{vmatrix}, \quad N_1 = \begin{vmatrix} E_1^* & B_1 & C_1 & D_1 \\ E_2^* & B_2 & C_2 & D_2 \\ d\tau_m & dz & 0 & 0 \\ d\theta & 0 & dr & dz \end{vmatrix}, \text{ etc.} \qquad \text{(A11)}$$

In the event $\Delta = 0$, one must also have $N_1 = N_2 = N_3 = N_4 = 0$ for a solution. Expansion of $\Delta = 0$ gives the condition:

$$(dz/dr)^2[AC] - ([AD] + [BC])(dz/dr) + [BD] = 0 \qquad \text{(A12)}$$

where the notation $[XY] = X_1Y_2 - X_2Y_1$. The coefficients in Equation (A12) are:

$$\begin{aligned} &[AC] = 2\tau_m(1 + \cot(\psi)\cos(2\theta)) \quad [AD] = 2\tau_m(\cot(\psi)\sin(2\theta)) \\ &[BC] = 2\tau_m(\cot(\psi)\sin(2\theta)) \qquad [BD] = 2\tau_m(1 - \cot(\psi)\cos(2\theta)) \end{aligned} \qquad \text{(A13)}$$

Hence:

$$dz / dr = \frac{\cot(\psi)\sin(2\theta)}{1+\cot(\psi)\cos(2\theta)} \pm [(\frac{\cot(\psi)\sin(2\theta)}{1+\cot(\psi)\cos(2\theta)})^2 - \frac{(1-\cot(\psi)\cos(2\theta)}{1+\cot(\psi)\cos(2\theta)}]^{1/2} \tag{A14}$$

Thus:

$$dz / dr = \frac{\sin(2\theta)}{\tan(\psi)+\cos(2\theta)} \pm [1-\tan^2(\psi)]^{1/2} \tag{A15}$$

If $\tan(\psi) < 1$, then the characteristic directions dz / dr are real and the system is hyperbolic. If $\tan(\psi) = 1$, then the system is parabolic with one real characteristic direction. Otherwise, the characteristics are imaginary and the system is elliptic. These results are similar to the plane strain case. With the assumption $\tan(\psi) \le 1$ and $\tan(\psi) = \sin(\phi)$ one obtains after the substitution $2\mu = \pi / 2 - \phi$ into Equation (A15):

$$dz / dr = \tan(\theta \pm \mu) \tag{A16a,b}$$

where Equation (A16a) defines the first C_1 family of characteristic curves and (A16b) defines the C_2 second family of characteristic curves.

When Equation (A16) holds, a solution requires $N_1 = 0$. Hence:

$$d\tau_m \{[BC]-[bd](dz / dr)\} - d\theta[CD] - [E*C]dz - [E*D]dr = 0 \tag{A17}$$

where:

$$\begin{aligned} [CD] &= -4\tau_m^2 \\ [E*C] &= E_1^* 2\tau_m \cos(2\theta) - E_2^* 2\tau_m \sin(2\theta) \\ [E*D] &= E_1^* 2\tau_m \sin(2\theta) - E_2^* 2\tau_m \cos(2\theta) \end{aligned} \tag{A18}$$

After some algebra, one obtains:

$$\pm d\tau_m \cot(\phi) + d\theta(2\tau_m) + [E_1^* \sin(\theta \pm \mu) - E_2^* \cos(\theta \pm \mu)]ds = 0 \tag{A19a,b}$$

where ds is an element of arc length along a characteristic curve (Equation A16), that is, $dz = ds\sin(\theta \pm \mu)$, $dr = ds\cos(\theta \pm \mu)$. Equations (A16a,b) and (A19a,b) are the axially symmetric stress subsystem of equations.

A.2 Discussion of stress

The stress subsystem of equations appears similar to the plane strain system. However, the similarity is superficial because of the non-homogeneous terms in Equation (A19a,b). In the plane strain case, these terms are $E_1 = \rho\dot{u} - \gamma_x$ and $E_2 = \rho\dot{v} - \gamma_y$. In axial symmetry:

$$\begin{aligned} E_1^* &= E_1 - (1/r)(g + \tau_m \cos(2\theta) - \sigma) - (\partial g / \partial\sigma)(\partial\sigma / \partial r) \\ E_2^* &= E_2 - (1/r)(\tau_m \sin(2\theta)) - (\partial g / \partial\sigma)(\partial\sigma / \partial z) \end{aligned} \tag{A20}$$

The first terms on the right-hand side are body force and inertia terms; the second terms are curvature terms, and the third terms arise whenever all three principal stresses affect yield.

In this regard, no stipulation of magnitude of σ , the circumferential stress, has been made, although the notation $\sigma_1 \geq \sigma_3$ is imposed on principal stresses in the rz plane, but that is all.

Several interesting possibilities arise:

$$\text{If} \begin{array}{ll} \sigma = \sigma_1 & \text{then } g - \sigma = -\tau_m \\ \sigma = \sigma_3 & \text{then } g - \sigma = +\tau_m \\ \sigma = \sigma_m & \text{then } g - \sigma = 0 \\ \sigma = \sigma_m \pm \tau_m \sin(\phi) & \text{then } g - \sigma = \pm\tau_m \sin(\phi) \end{array} \tag{A21}$$

The first two of Equation (A21) are often assumed. The second is reminiscent of an assumption in plane strain metal plasticity $(\phi = 0)$, where the stress acting normal to the plane of deformation tends towards the mean of the principal stresses in the plane of interest. In addition to the hypothesis that σ is equal to one of the principal stresses in the rz plane, independence of yielding on the intermediate principal stress, σ, is often assumed. These assumptions are mathematically expedient but may have some physical utility.

The stress subsystem of equations in axial symmetry is composed of two equations of equilibrium and one of yield. There are thus three equations involving four non-zero stresses $\sigma_{rr}, \tau_{rz}, \sigma_{zz}$ and $\sigma_{\omega\omega}$. Consequently, commitment to a material law is unavoidable. An elastic-plastic stress-strain law is required, unless additional assumptions as indicated previously are invoked to make up for the deficit of information. Even in plane strain where the stress subsystem appears statically determinant and gives rise to the concept of limiting equilibrium, velocity boundary conditions should be taken into account for an acceptable solution. To be sure, velocities arise in consideration of a stress-strain law and the geometry of incremental strain or strain rates. A complete system is then composed of a velocity subsystem and a stress subsystem.

Axial symmetry also brings out the fact that the Mohr-Coulomb theory may be considered to be a plane strain form of a three-dimensional yield function in stress space, or it may be assumed outright at the beginning. In the latter case, as applied to axially symmetric problems, various plastic regimes should be distinguished according to the ordering of the magnitudes of principal stresses (Cox *et al.*, 1961). Worth noting is the fact that there are many yield functions in principal stress space that can be reduced to equality with the Mohr-Coulomb yield function under plane strain conditions.

A.3 Velocity subsystem

From the plastic stress-strain relationship with Y as the plastic potential or the normality principle:

$$\begin{aligned} -\partial\dot{u} / \partial r &= \dot{e}_r + \lambda\partial Y / \partial\sigma_r \\ -\dot{u} / r &= \dot{e}_\omega + \lambda\partial Y / \partial\sigma_\omega \\ -\partial\dot{v} / \partial z &= \dot{e}_z + \lambda\partial Y / \partial\sigma_z \\ -(1/2)(\partial\dot{u} / \partial z + \partial\dot{v} / \partial r) &= \dot{e}_{rz} + \lambda\partial Y / \partial\tau_{rz} \end{aligned} \tag{A22}$$

where $\dot{u}, \dot{v}, \dot{e}_r, \dot{e}_\omega, \dot{e}_z, \lambda, Y$ are the r and z velocities, elastic strain parts, scalar function of proportionality, and plastic potential, respectively. The negative sign is used because compression is reckoned positive. The yield function is assumed to be continuously differentiable almost everywhere. When using the Mohr-Coulomb criterion, edge regimes are quite possible and even likely. The right-hand side of Equation (A22) will then have additional parts in the form of yield function derivatives.

After elimination of λ from Equation (A22):

$$
\begin{aligned}
&\dot{u}_{,r}Y_{,rz} - (1/2)(\dot{u}_{,z} + \dot{v}_{,r})Y_{,r} = \dot{e}_{rz}Y_{,r} - \dot{e}_{r}Y_{,rz} \\
&(\dot{u}/r)Y_{,rz} - (1/2)(\dot{u}_{,z} + \dot{v}_{,r})Y_{,\omega} = \dot{e}_{rz}Y_{,\omega} - \dot{e}_{\omega}Y_{\omega} \\
&\dot{v}_{,z}Y_{,rz} - (1/2)(\dot{u}_{,z} + \dot{v}_{,r})Y_{,z} = \dot{e}_{rz}Y_{,r} - \dot{e}_{r}Y_{,rz}
\end{aligned}
\tag{A23}
$$

where $\partial Y/\partial \sigma_r = Y_{,r}$ and so on, and the comma denotes partial differentiation as before. Subtracting the second from the first of Equation (A23) and with addition of the differentials of velocities, one obtains:

$$
\begin{array}{llllll}
\dot{u}_{,r}Y_{,rz} - (1/2)(\dot{u}_{,z})(Y_{,r} - Y_{,\omega}) & - (1/2)(\dot{v}_{,r})(Y_{,r} - Y_{,\omega}) & 0 & & = \dot{e}_{rz}(Y_{,r} - Y_{,\omega}) - (\dot{e}_{r} - \dot{e}_{\omega})Y_{,rz} \\
0 \quad (1/2)(\dot{u}_{,z})Y_{,z} & - \quad (1/2)(\dot{v}_{,r})(Y_{,z}) & + & \dot{v}_{,z}(Y_{,rz}) & = \dot{e}_{rz}Y_{,r} - \dot{e}_{r}Y_{,rz} \\
\dot{u}_{,r}dr + \dot{u}_{,z}dz & + & 0 & 0 & = d\dot{u} \\
0 + 0 & + & \dot{v}_{,r}dr & + \dot{v}_{z}dz & = d\dot{v}
\end{array}
\tag{A24}
$$

which has the familiar form:

$$
\begin{aligned}
&A_1\dot{u}_{,r} + B_1\dot{u}_{,z} + C_1\dot{v}_{,r} + D_1\dot{v}_{,\omega} = E_1 \\
&A_2\dot{u}_{,r} + B_2\dot{u}_{,z} + C_2\dot{v}_{,r} + D_2\dot{v}_{,\omega} = E_2 \\
&\dot{u}_{,r}dr + \dot{u}_{,z}dz + 0 \quad 0 \quad = d\dot{u} \\
&0 + 0 \quad + \dot{v}_{,r}dr + \dot{v}_{z}dz \quad = d\dot{v}
\end{aligned}
\tag{A25}
$$

with the solution:

$$
(dz/dr)^2[AC] - ([AD] + [BC])(dz/dr) + [BD] = 0 \tag{A26}
$$

where the notation $[XY] = X_1Y_2 - X_2Y_1$. The coefficients in Equation (A26) are:

$$
\begin{aligned}
&[AC] = -Y_{,rz}Y_{,z} \quad [AD] = +Y_{,rz}Y_{,rz} \\
&[BC] = 0 \qquad\quad\; [BD] = -(1/2)(Y_{,r} - Y_{,\omega})Y_{,rz}
\end{aligned}
\tag{A27}
$$

Also:

$$
\begin{aligned}
&[CE] = (1/2)(Y_{,rz})[(\dot{u}/r)Y_{,z} + \dot{e}_{z}(Y_{,r} - Y_{,\omega}) - (\dot{e}_{r} - \dot{e}_{\omega})Y_{,z}] \\
&[DE] = -(Y_{,rz})[\dot{e}_{rz}(Y_{,r} - Y_{,\omega}) - (\dot{e}_{r} - \dot{e}_{\omega})Y_{,rz} + (\dot{u}/r)Y_{rz}]
\end{aligned}
\tag{A28}
$$

The requirement that the characteristic curves of the velocity subsystem be real can be inferred from Equation (A26). Thus:

$$\begin{aligned}&(Y_{,rz})^2-(Y_{,z})(Y_{,r}-Y_{,\omega})\geq 0 \text{ or}\\&(\dot{e}_{,rz})^2-(\dot{e}_{,z})(\dot{e}_{,r}-\dot{e}_{,\omega})\geq 0 \text{ or}\\&\dot{e}_1\dot{e}_3\leq \dot{e}_z\dot{e}_\omega\end{aligned} \tag{A29}$$

where the strain rates are the plastic parts of the total strain rate. Recall that if there are no real directions of the characteristics, then the determinant of the coefficients is not zero and the velocity subsystem is elliptic. Otherwise, the subsystem is hyperbolic or parabolic, depending on whether there are two or one real characteristics, respectively.

There are numerous special cases that one may consider. Perhaps the most common case is one where an assumption is made such that the intermediate principal stress does not affect yield (e.g., Mohr-Coulomb yield) *and* the circumferential stress is the intermediate principal stress, so $Y_{,\omega}=0$. With these assumptions, the relationship between velocities along the characteristic curves is given by:

$$du+dv\tan(\phi)+(u/r)\cos^2(\theta\pm\mu)=0 \tag{A30}$$

where the assumption made in the plane strain case that the elastic and plastic parts of the strain rate tensor bear a constant ratio to each other so the right-hand side of Equation (A30) vanishes. Also, the velocities have been replaced by increments of displacement through division by dt.

A.4 Discussion of velocity

With the yield function Y as the plastic potential, one has for principal plastic strain rates:

$$\dot{e}_i^p=\partial Y/\sigma_i \qquad (i=1,2,3) \tag{A31}$$

where Y is regular, that is, smooth or has a continuously turning tangent plane almost everywhere. In the event that the yield function is independent of the intermediate principal stress, a physically plausible event, then $\partial Y/\partial\sigma_{II}=0$ with ordering of the principal stresses $\sigma_I\geq\sigma_{II}\geq\sigma_{III}$. However, this result cannot be taken to imply $\dot{e}_{II}^p=0$, which would imply plane strain and not axial symmetry. In fact, the yield function may be a composite of several intersecting planes, for example, the famous Tresca yield criterion $Y=\sigma_I-\sigma_{III}-2k=0$, which is a hexagon in principal stress space centered on the hydrostatic line. There are several possibilities:

$$\begin{aligned}&\sigma_1\geq\sigma_2\geq\sigma_3\ \ \sigma_2\geq\sigma_3\geq\sigma_1\ \ \sigma_3\geq\sigma_1\geq\sigma_2\\&\sigma_1\geq\sigma_3\geq\sigma_2\ \ \sigma_2\geq\sigma_1\geq\sigma_3\ \ \sigma_3\geq\sigma_2\geq\sigma_1\end{aligned} \tag{A32}$$

which correspond to:

$$\begin{aligned}&\sigma_1-\sigma_3=2k\ \ \sigma_2-\sigma_1=2k\ \ \sigma_3-\sigma_2=2k\\&\sigma_1-\sigma_2=2k\ \ \sigma_2-\sigma_3=2k\ \ \sigma_3-\sigma_1=2k\end{aligned} \tag{A33}$$

At the intersection of yield planes ("loading surfaces"), the possibilities are $\sigma_1 = \sigma_2$, $\sigma_1 = \sigma_3$, $\sigma_3 = \sigma_2$. The flow rule, normality, requires the strain rate vector to point in the interval between normals to the intersecting planes, so that it is not necessary that the intermediate strain rate be zero.

The similarity between axial symmetry and plane strain is not nearly as close as one would assume at the start of analysis. Indeed, assumptions that may lack physical justification are necessary to move forward. In fact, the situation points to a need for a comprehensive numerical approach beyond limiting equilibrium. The finite element method is one such approach to elastic-plastic analysis in geomechanics.

Reference

Cox, A.D., Eason, G. & Hopkins, H.G. (1961) Axially symmetric plastic Deformations in soils. *Philosophical Transactions of the Royal Society of London A*, 254(1036), 1–45.

Appendix B

Finite element review in brief

The finite element method (FEM) is a numerical technique for solving differential equations. Development of the method occurred simultaneously with the development of digital computers beginning in the mid-1950s. Applications to ground control problems in mine engineering appeared in the mid-1960s. In this regard, origins of the finite element program UT3/PC can be traced to the dissertations by Dahl (1969) in application to mining and to Wilson (1963) in civil engineering. FEM is an enormously popular numerical technique for solving differential equations described in many books (Desai and Abel, 1972; Oden, 1972; Cook, 1974; Segerlind, 1976; Zienkiewicz and Cheung, 1967; Zienkiewicz, 1977; Bathe, 1982; Logan, 1986). While the finite element method is a well-understood numerical technique and has been in undergraduate engineering curricula for many years, a brief outline of the method is helpful in understanding FEM application to rock mechanics and rock engineering.

B.1 Finite element concept

A finite element is simply a subdivision of the region of interest. Almost always, an element is homogeneous. A triangle is a simple two-dimensional element that illustrates the concept. Consider the triangle in Figure B1. Vertices, corners or nodes, are numbered 1, 2, 3 and have the coordinates indicated.

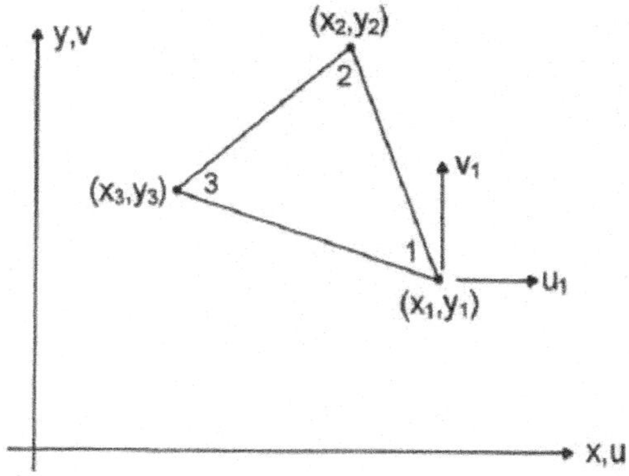

Figure B1 A plane triangular finite element.

Displacements in the x and y directions are u and v, respectively, and are considered known. Displacements in the interior are unknown but may be estimated by interpolation from the known displacements. A linear interpolation is the simplest. Thus:

$$u = a_o + a_1 x + a_2 y, \quad v = b_o + b_1 x + b_2 y \tag{B1}$$

where x and y are coordinates of an interior point where the displacements are sought. The constants a_o, a_1, etc., must be known, of course. Also, the displacements given by Equation (B1) must agree with the known displacements at the nodes. Thus:

$$\begin{aligned} u_1 &= a_o + a_1 x_1 + a_2 y_1 \\ u_2 &= a_o + a_1 x_2 + a_2 y_2 \text{ or in matrix notation} \\ u_3 &= a_o + a_1 x_3 + a_2 y_3 \end{aligned} \tag{B1a}$$

$$\begin{Bmatrix} u_1 \\ u_2 \\ u_3 \end{Bmatrix} = \begin{bmatrix} 1 & x_1 & y_1 \\ 1 & x_2 & y_2 \\ 1 & x_3 & y_3 \end{bmatrix} \begin{Bmatrix} a_o \\ a_1 \\ a_2 \end{Bmatrix} \tag{B1b}$$

which can be solved for the desired constants and similarly for the bs. Inspection of Equation (B1) shows that the constants are given in terms of the node coordinates and displacements. When the solution for the constant terms is substituted back into Equation (B1), the result has the matrix form:

$$\begin{Bmatrix} u \\ v \end{Bmatrix} = [N]\{\delta\} \tag{B2}$$

where $\{\delta\}$ is a column matrix (vector) of the known node displacements. The matrix $[N]$, a common notation, is thus an interpolation matrix for displacements. Elements of $[N]$ are linear functions $N(x, y)$ in this example.

Strains in the finite element method follow from the definition of strains in terms of displacement derivatives and the finite element approximation to displacements (Equation B2). The result has the form:

$$\begin{Bmatrix} \varepsilon_{xx} \\ \varepsilon_{yy} \\ \gamma_{xy} \end{Bmatrix} = \{\varepsilon\} = [B]\{\delta\} \tag{B3}$$

where the elements of $[B]$ are derivatives of $[N]$ and the known displacements $\{\delta\}$ are constant. The elements of $[B]$ are themselves constant, so *this triangle is a constant strain triangle.*

Stresses follow from strains via Hooke's law in the purely elastic case and via the elastic-plastic relationship beyond the yield point. In compact incremental form:

$$\{\Delta\sigma\} = [E]\{\Delta\varepsilon\} \tag{B4}$$

The increments in Equation (B4) are associated with increments of load from gravity and excavation in a problem of interest. These increments approximate differentials.

There are other types of triangles and other types of elements in two dimensions. If nodes are added to a triangle at midpoints along the sides of the triangle in Figure B1, then interpolation must be of degree two. Strains then vary linearly over the triangle. Quadrilateral elements are also possible with corner nodes and mid-side nodes. One very useful quadrilateral for analysis of plane (two-dimensional) problems is one composed of four constant strain triangles. This element is a 4CST quadrilateral.

Three-dimensional analogs of triangular and quadrilateral elements are tetrahedrons and rectangular parallelepipeds (brick shape). One could consider a tetrahedral analog of the 4CST element by adding a center node to a tetrahedron and thus making it a composite of the internal tetrahedrons. However, the result is simply more tetrahedrons, unlike the 4CST result. One could also consider subdividing a brick-shaped element into tetrahedrons as a quadrilateral is divided into four triangles to form the 4CST element. Interestingly, there are just two ways to divide a brick-shaped element into five tetrahedrons; there are many ways of dividing a brick into more, say, six tetrahedrons.

Another interesting and important element type is an *isoparametric* element. This element involves a transformation from generic space to finite element space, as illustrated in Figure B2.

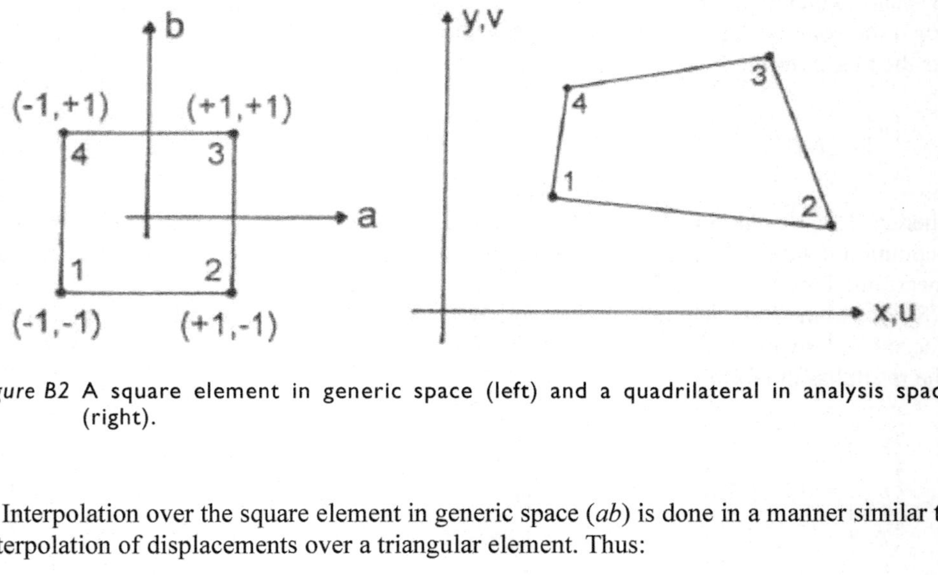

Figure B2 A square element in generic space (left) and a quadrilateral in analysis space (right).

Interpolation over the square element in generic space (ab) is done in a manner similar to interpolation of displacements over a triangular element. Thus:

$$u = \alpha_o + \alpha_1 a + \alpha_2 b + \alpha_3 ab \quad \text{(B5a)}$$

where u is the displacement component in the a direction and an ab term is added for the square (quadrilateral) element. After solving for the coefficients, the αs, displacements in the generic square element have the matrix form:

$$\begin{Bmatrix} u \\ v \end{Bmatrix} = [N(a,b)]\{\delta\} \quad \text{(B5b)}$$

where $\{\delta\}$ is an 8×1 column matrix of displacements at the nodes of the square. A coordinate transformation between (ab) and (xy) is needed to connect the two elements in Figure B2. Symbolically, the transformation of coordinates is:

$$x = F(a,b),\ \ y = G(a,b),\ a = f(x,y),\ b = g(x,y) \tag{B5c}$$

Coordinates in (xy) may be computed similarly to the way displacements are computed in (ab). Thus:

$$x = S_1x_1 + S_2x_2 + S_3x_3 + S_4x_4,\ y = S_1y_1 + S_2y_2 + S_3y_3 + S_4y_4 \tag{B5d}$$

where x_1, y_1, etc., are coordinates of the element nodes in (xy). The functions S_1, S_2, S_3, S_4 are *shape* functions. When the interpolation and shape functions are the same, that is, when $N_1(a,b) = S_1(a,b)$, etc., the element is an isoparametric element.

B.2 Element equilibrium

Element equilibrium is at the core of the finite element method. The most direct way of obtaining element equilibrium in finite element form is through the application of the principle of virtual work or more generally the divergence theorem. This theorem states that the integral of the normal component of a vector over the surface of a body is equal to the integral of the divergence of the vector throughout the volume of the body. Thus:

$$\int_S (U \bullet n)dS = \int_V (\nabla \bullet U)dV \tag{B6a}$$

where S and V are surface and volume, respectively; $(U \bullet n)$ and $(\nabla \bullet U)$ are the inner product (dot product) of U on S and divergence of U in V. In two dimensions:

$$U \bullet n = U_x n_x + U_y n_y \text{ and } \nabla \bullet U = \partial U / \partial x + \partial U / \partial y \tag{B6b}$$

In mechanical terms, Equation (B6a) is:

$$\begin{aligned} &\int_S (T \bullet \delta)dS + \int_V (\gamma \bullet \delta)dV = \int_V (\varepsilon \bullet \sigma)dV, \text{ that is,} \\ &\int_S (T_x u + T_y v)dS + \int_V (\gamma_x u + \gamma_y v)dV = \int_V (\varepsilon_{xx}\sigma_{xx} + \varepsilon_{yy}\sigma_{yy} + \gamma_{xy}\tau_{xy})dV \end{aligned} \tag{B6c}$$

In matrix form, Equation (B6) is:

$$\int_S \begin{Bmatrix} u \\ v \end{Bmatrix}^t \begin{Bmatrix} T_x \\ T_y \end{Bmatrix} dS + \int_V \begin{Bmatrix} u \\ v \end{Bmatrix}^t \begin{Bmatrix} \gamma_x \\ \gamma_y \end{Bmatrix} dV = \int_V \{\varepsilon\}^t \{\sigma\} dV \tag{B6d}$$

The superscript t means transpose. In finite element form, one has:

$$\{\delta\}^t \int_S [N]^t \{T\} dS + \{\delta\}^t \int_V [N]^t \{\gamma\} dV = \{\delta\}^t \int_V [B]^t \{\sigma\} dV \tag{B7}$$

where the constant node displacements are moved outside the integral signs. The result (Equation B7) implies:

$$\int_S [N]^t \{T\} dS + \int_V [N]^t \{\gamma\} dV = \int_V [B]^t \{\sigma\} dV = \int_V [B]^t [E][B] dV \{\delta\} \tag{B8}$$

or:

$$\{f\} = [k]\{\delta\} \tag{B9}$$

where $\{f\}$ = left-hand side of Equation (B8), a force vector, and $[k]\{\delta\}$ = right-hand side of Equation (B8), a product of element stiffness $[k]$ and element node displacement vector $\{\delta\}$. Solution of Equation (B9) under prescribed forces allows for the determination of strains and then stresses in the element.

The element stiffness is an integral, that is:

$$k = \int_V [B]^t [E][B] dV \tag{B10}$$

which is readily integrated in the case of a linear displacement, constant strain element because all the terms under the integral sign are constants. Thus, $k = [B]^t [E][B] V$ and the integration can be explicit. Stiffness of higher order elements generally requires numerical integration, as does stiffness of isoparametric elements.

B.3 Global equilibrium

Global equilibrium refers to an assemblage of elements, a finite element *mesh* as illustrated in Figure B3. The assemblage of elements in the figure covers a rectangular region in the *xy* plane. As an aside, data at any interior point on the rectangular grid could be determined by interpolation from known data at the nodes of the triangles. Indeed, the nodes were determined by the data point locations. The data could be temperatures, elevations, ore grade, or some other quantity of interest such as water or gas pressure. However, interest here is in displacements that allow for determination of strains and stresses.

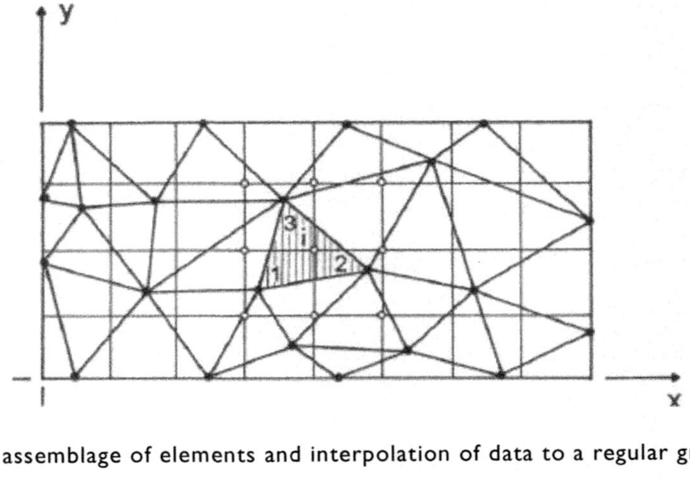

Figure B3 An assemblage of elements and interpolation of data to a regular grid.

Assembly of elements into a mesh requires a global numbering system in addition to the local numbering system of nodes. Figure B4 illustrates the concept. The figure also indicates a possibility of mixing element types, for example, quadrilaterals and triangles.

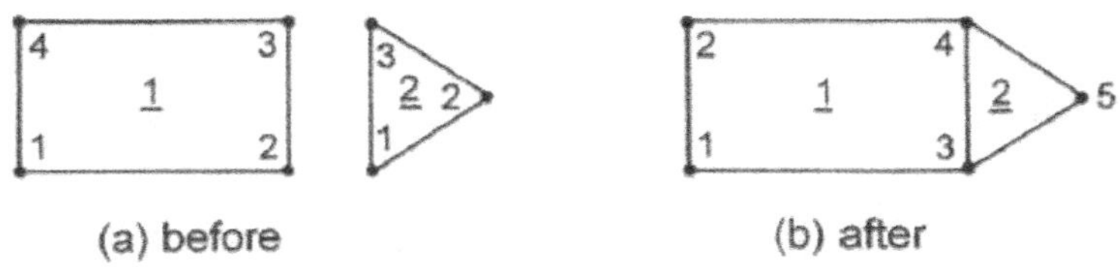

Figure B4 Local (a) and global (b) numbering of nodes.

The assembly process has a physical basis and that is the force at any node shared by adjacent elements is simply the sum of forces contributed by each element. Symbolically, the total force at a node in a mesh is:

$$F_i = \sum_e f_i \tag{B11a}$$

where F_i is the total force at node i that is shared by elements e with element node forces f_i. In view of the element equilibrium equation:

$$F_i = \sum_e \sum_{j=1}^{j=n} k_{ij}^e \delta_j \tag{B11b}$$

where the inner summation is over the n nodes of element e (4 for quadrilaterals, 3 for triangles). After arrangement of the nodes in the mesh in order from 1 to N, the result is global equilibrium. Thus:

$$\{F\} = [K]\{\Delta\} \tag{B12}$$

where $\{F\}$ and $\{\Delta\}$ are vectors of global node forces and displacements; $[K]$ is a *master stiffness matrix*. Solution of Equation (B12) gives the node displacements and thus allows for determination of element strains and subsequently element stresses throughout a mesh.

In two dimensions, there are two forces and two displacements at each node, so the force and displacement vectors have dimensions of $2N\times1$. The master stiffness matrix is $2N\times2N$. If there are 500,000 nodes in a mesh, then the master stiffness matrix dimension is $10^6 \times 10^6$. Computer storage of such a large array is obviously a serious challenge even in two dimensions and so is inversion of such a large matrix in any numerical analysis. Fortunately, the master stiffness matrix is sparsely populated, meaning that there are many zero entries. This feature gives rise to compact storage schemes that allow FEM to be implemented in a practical way that includes clever solution schemes. Very large problems are usually solved with iterative schemes rather than elimination schemes. Efficient and accurate solution strategies are much discussed in the literature concerning numerical methods and are often closely linked to computer architecture, topics beyond the scope of this discussion. Suffice to say that the traditional Gauss-Seidel (GS) and conjugate gradient (CG) iterative schemes work quite well in solving FEM problems involving coal mine strata mechanics. The latter is particularly appealing because of the possibility of parallelization and thus speedup in solution time that is often 95% of a problem runtime.

B.4 Boundary conditions

Specification of displacements that prevent rigid body motion at mesh boundaries is essential. This specification is easily done by fixing two nodes. Forces (tractions) may be prescribed over a portion of a mesh boundary and displacements elsewhere on a boundary as well. Mixed forces and displacement conditions are possible, but the same components of force and displacement cannot be prescribed. If a force and displacement normal to a boundary are prescribed, an inconsistency occurs that may be resolved according to computer programming, or the run may simply be aborted, again, according to programming. Of course, such an inconsistency should be avoided at the outset. In UTH3/PC, displacements prevail. Often displacements normal to an external boundary to a mesh are prescribed (fixed), while forces are prescribed at excavation (internal) boundaries. When displacements normal to an external boundary are set to zero, the boundary is "rollered" and indicated by a roller symbol in a diagram of the mesh for the problem at hand. Excavation forces are usually computed automatically, as are gravity loads. When the pre-excavation stresses (initial stresses) are known, then forces associated with excavation can be automatically computed. However, the region to be excavated must be specified, usually as an input file of elements to be mined. Mined elements are effectively "cut" from the mesh and are "cut" elements. A file of cut elements is generated automatically during mesh generation.

B.5 Practical considerations

Several practical considerations arise in connection with finite element analysis. One is in element size; another is mesh size. Yet another is what to do with the voluminous output that is generated during a program run involving a mesh, say, of a million elements. An element file containing element stresses could easily be one million lines long and similarly for element strain and displacement output files. In a multi-step excavation sequence, output data may be truly enormous. Filtering output data for important design guidance is not at all a simple or trivial activity.

Several rules of thumb are helpful in deciding on element size and mesh size. *One rule* states that elements should have an aspect ratio no greater than four or five for numerical accuracy, that is, the ratio of greatest to least element edge length should be no greater than five. Although higher aspect ratios are permissible, solution time may be lengthened considerably at high aspect ratios. An aspect ratio of three or less is desirable. *Another rule* of thumb is that no less than five elements should be used across the least dimension of an excavation to obtain a reasonable approximation of stress distribution along the excavation walls. A *third rule* of thumb is that the external boundaries of mesh enclosing an excavation should be about five times "excavation size" away from an opening. In case of a circle, the excavation size is simply the circle diameter. In case of rectangular openings, excavation size is the long dimension of the opening. In case of three-dimensional cavern-like or large brick-shaped openings, excavation size is the intermediate edge length of the opening. In all cases, one may examine the change in stress in elements near the external boundaries of a mesh to see whether the changes in stress induced by excavation are small or negligible, in which case the mesh is large enough. These rules of thumb are based on computational experience and the fact that stress concentration in the elastic domain, while greatest at excavation boundaries, decreases rapidly with distance from the excavation boundaries. Often a "1-D" (one dee) rule is invoked where D is the dimension used to quantify "excavation size" that in case of a circular excavation is the circle diameter, D.

References

Bathe, K.-L. (1982) *Finite Element Procedures in Engineering Analysis*. Prentice-Hall, Englewood Cliffs, NJ.

Cook, R.D. (1974) *Concepts and Applications of Finite Element Analysis*. John Wiley & Sons, New York.

Dahl, H.D. (1969) *A Finite Element Model for Anisotropic Yielding in Gravity Loaded Rock*. PhD Dissertation, The Pennsylvania State University, The College of Mineral Sciences Experiment Station, University Park, PA.

Desai, C.S. & Abel, J.F. (1972) *Introduction to the Finite Element Method, A Numerical Method for Engineering Analysis*. Van Nostrand Reinhold Company, New York.

Logan, D.L. (1986) *A First Course in the Finite Element Method*. PWS-KENT Publishing Company, Boston.

Oden, J.T. (1972) *Finite Elements of Nonlinear Continua*. McGraw-Hill, New York.

Segerlind, L.J. (1976) *Applied Finite Element Analysis*. John Wiley & Sons, New York.

Wilson, E.L. (1963) *Finite Element Analysis of Two-Dimensional Structures*. PhD Dissertation, University of California – Berkeley, University Microfilms, Ann Arbor, MI, p. 72.

Zienkiewicz, O.C. (1977) *The Finite Element Method* (3rd ed.). McGraw-Hill, London.

Zienkiewicz, O.C. & Cheung, Y.K. (1967) *The Finite Element Method in Structural and Continuum Mechanics*. McGraw-Hill, London.

Index

Note: page numbers in *italics* indicate figures; page numbers in **bold** indicate tables